'*Entangled Life* is gorgeous! En[joying it very much.]'

Margaret Atwood (on Twitter)

'Enthralling and completely mind-blowing... Dazzlingly good'

Sunday Times

'A thrilling and fascinating insight into the living world... I hope and trust that it will become an instant classic'

George Monbiot

'Few books blew my mind like Merlin Sheldrake's *Entangled Life*... Go and get swept up in a new world'

Andrea Wulf

'Nearly every page of this book contained either an observation so interesting or a turn of phrase so lovely that I was moved to slow down, stop, and reread'

Science

'Sheldrake engages us in a miraculous web of connections, interactions and communication that changes the way we need to look at life, the planet and ourselves'

Isabella Tree

'A magical journey deep into the roots of Nature by an expert storyteller, weaving the tale of our co-evolution with fungi into a scientific adventure. A must-read'

Paul Stamets

'A joy... A captivating trip into the weird and wonderful world around us – and inside us'

Daily Mail

'Mind-boggling and fascinating. Sheldrake asks us to consider a life-form that is radically alien to ours, yet vibrant and lively underfoot'

Hans Ulrich Obrist

'This book is like one surprise after another'

David Byrne

'Expands our conception of the living world... It is laced with intriguing details'

Financial Times

'Fungi are everywhere, and Merlin Sheldrake is an ideal guide to their mysteries. He's passionate, deeply knowledgeable and a wonderful writer'

Elizabeth Kolbert

'I fell in love with this book. Merlin is a scientist with the imagination of a poet, and a beautiful writer'

Michael Pollan

'Reading this book, I felt surrounded by a web of wonder. The natural world is more fantastic than any fantasy, so long as you have the means to perceive it. This book provides the means'

Jaron Lanier

'Deeply engaging and constantly surprising... The magic of mushrooms is not merely mind-expanding...it might expand the very concept of mind'

Prospect

MERLIN SHELDRAKE

Merlin Sheldrake is a biologist and a writer. He received a Ph.D. in Tropical Ecology from Cambridge University for his work on underground fungal networks in tropical forests in Panama, where he was a predoctoral research fellow of the Smithsonian Tropical Research Institute. He is a musician and keen fermenter.

Entangled Life, his first book, is an international bestseller and was nominated for a host of prizes, including the British Book Awards Book of the Year 2021 for Narrative Non-Fiction, the Rathbones Folio Prize 2021 and the Wainwright Prize 2021. *Entangled Life* was also featured as Radio 4's *Book of the Week* and selected as a Book of the Year in *The Times*, *Daily Telegraph*, *Sunday Times*, *New Statesman* and *Time*, among others. It has been translated into twenty languages.

merlinsheldrake.com
Find Merlin on Twitter @MerlinSheldrake
And on Instagram @merlin.sheldrake

MERLIN SHELDRAKE

Entangled Life

How Fungi Make Our Worlds, Change
Our Minds, and Shape Our Futures

VINTAGE

17 19 20 18 16

Vintage is part of the Penguin Random House group of companies
whose addresses can be found at global.penguinrandomhouse.com

Penguin
Random House
UK

First published in Vintage in 2021
First published in hardback by The Bodley Head in 2020

penguin.co.uk/vintage

A CIP catalogue record for this book is available
from the British Library

ISBN 9781784708276

Printed and bound in Great Britain by Clays Ltd, Elcograf S.p.A.

The authorised representative in the EEA is Penguin Random House
Ireland, Morrison Chambers, 32 Nassau Street, Dublin D02 YH68

Penguin Random House is committed to a sustainable future
for our business, our readers and our planet. This book is
made from Forest Stewardship Council® certified paper.

*With gratitude to the fungi
from which I have learned*

Contents

Prologue

I looked up towards the top of the tree. Ferns and orchids sprouted from its trunk, which vanished into a tangle of lianas in the canopy. High above me, a toucan flapped off its perch with a croak, and a troupe of howler monkeys worked themselves into a slow roar. The rain had only just stopped, and the leaves above me shed heavy drops of water in sudden showers. A low mist hung over the ground.

The tree's roots wound outwards from the base of its trunk, soon vanishing into the thick drifts of fallen leaves that covered the floor of the jungle. I used a stick to tap the ground for snakes. A tarantula scuttled off, and I knelt, feeling my way down the tree's trunk and along one of its roots into a mass of spongy debris where the finer roots matted into a thick red and brown tangle. A rich smell drifted upwards. Termites clambered through the labyrinth, and a millipede coiled up, playing dead. My root vanished into the ground, and with a trowel I cleared the area around the spot. I used my hands and a spoon to loosen the top layer of earth, and dug as gently as I could, slowly uncovering it as it ranged out from the tree and twisted along just below the surface of the soil.

After an hour, I had travelled about a metre. My root was now thinner than string and had started to proliferate wildly.

It was hard to keep track of as it knotted with its neighbours, so I lay down on my stomach and lowered my face into the shallow trench I had made. Some roots smell sharp and nutty and others woody and bitter, but the roots of my tree had a spicy resinous kick when I scratched them with a fingernail. For several hours I inched along the ground, scratching and sniffing every few centimetres to make sure I hadn't lost the thread.

As the day went on, more filaments sprang out from the root I'd uncovered and I chose a few of them to follow all the way to the tips, where they burrowed into fragments of rotting leaf or twig. I dipped the ends in a vial of water to wash off the mud and looked at them through a loupe. The rootlets branched like a small tree and their surface was covered with a filmy layer which appeared fresh and sticky. It was these delicate structures I wanted to examine. From these roots, a fungal network laced out into the soil and around the roots of nearby trees. Without this fungal web my tree would not exist. Without similar fungal webs no plant would exist anywhere. All life on land, including my own, depended on these networks. I tugged lightly on my root and felt the ground move.

Introduction

What Is It Like To Be A Fungus?

There are moments in moist love when heaven
is jealous of what we on Earth can do.

Hafiz

Fungi are everywhere but they are easy to miss. They are inside you and around you. They sustain you and all that you depend on. As you read these words, fungi are changing the way that life happens, as they have done for more than a billion years. They are eating rock, making soil, digesting pollutants, nourishing and killing plants, surviving in space, inducing visions, producing food, making medicines, manipulating animal behaviour and influencing the composition of the Earth's atmosphere. Fungi provide a key to understanding the planet on which we live, and the ways we think, feel and behave. Yet they live their lives largely hidden from view, and more than 90 per cent of their species remain undocumented. The more we learn about fungi, the less makes sense without them.

Fungi make up one of life's kingdoms – as broad and busy a category as 'animals' or 'plants'. Microscopic yeasts are fungi, as are the sprawling networks of honey fungi, or *Armillaria*, which are among the largest organisms in the world. The current record holder, in Oregon, weighs hundreds of tonnes, spills across 10 square kilometres, and is somewhere between 2,000 and 8,000 years old. There are probably many larger, older specimens that remain undiscovered.[1]

Many of the most dramatic events on Earth have been – and continue to be – a result of fungal activity. Plants only made it out of the water around 500 million years ago because of their collaboration with fungi, which served as their root systems for tens of million years until plants could evolve their own. Today, over 90 per cent of plants depend on mycorrhizal fungi – from the Greek words for fungus (*mykes*) and root (*rhiza*) – which can link trees in shared networks sometimes referred to as the 'Wood Wide Web'. This ancient association gave rise to all recognisable life on land, the future of which depends on the continued ability of plants and fungi to form healthy relationships.

Plants may have greened the planet, but if we could cast our eyes back to the Devonian period, 400 million years ago, we'd be struck by another life form: *Prototaxites*. These living spires were scattered across the landscape. Many were taller than a two-storey building. Nothing else got anywhere close to this size: plants existed but were no more than a metre tall, and no animal with a backbone had yet moved out of the water. Small insects made their homes in the giant trunks, chewing out rooms and corridors. This enigmatic group of organisms – thought to have been enormous fungi – were the largest living structures on dry land for at least 40 million years, twenty times longer than the genus *Homo* has existed.[2]

To this day, new ecosystems on land are founded by fungi. When volcanic islands are made or glaciers retreat to reveal bare rock, lichens (pronounced *LY-kens*) – a union of fungi and algae or bacteria – are the first organisms to establish themselves, and to make the soil in which plants subsequently take root. In well-developed ecosystems soil would be rapidly sluiced off by rain were it not for the dense mesh of fungal tissue that holds it together. From deep sediments on the sea floor, to the surface of deserts, to frozen valleys in Antarctica, to our guts and orifices, there are few pockets of the globe where fungi can't be found. Tens to hundreds of species can exist in the leaves and stems of a single plant. These fungi weave themselves through the gaps between plant cells in an intimate brocade and help to defend plants against disease. No plant grown under natural conditions has been found without these fungi; they are as much a part of planthood as leaves or roots.[3]

The ability of fungi to prosper in such a variety of habitats depends on their diverse metabolic abilities. Metabolism is the art of chemical transformation. Fungi are metabolic wizards and can explore, scavenge and salvage ingeniously, their abilities rivalled only by bacteria. Using cocktails of potent enzymes and acids, fungi can break down some of the most stubborn substances on the planet, from lignin, wood's toughest component, to rock, crude oil, polyurethane plastics and the explosive TNT. Few environments are too extreme. A species isolated from mining waste is one of the most radiation-resistant organisms ever discovered, and may help to clean up nuclear waste sites. The blasted nuclear reactor at Chernobyl is home to a large population of such fungi. A number of these radio-tolerant species even grow towards radioactive 'hot' particles, and appear to be able to harness radiation as a source of energy, as plants use the energy in sunlight.[4]

Mushrooms dominate the popular fungal imagination, but just as the fruits of plants are one part of a much larger structure that includes branches and roots, so mushrooms are only the fruiting bodies of fungi, the place where spores are produced. Fungi use spores like plants use seeds: to disperse themselves.

Mushrooms are a fungus's way to entreat the more-than-fungal world, from wind to squirrel, to assist with the dispersal of spores, or prevent it from interfering with this process. They are the parts of fungi made visible, pungent, covetable, delicious, poisonous. However, mushrooms are only one approach among many: the overwhelming majority of fungal species release spores without producing mushrooms at all.

We all live and breathe fungi, thanks to the prolific abilities of fungal fruiting bodies to disperse spores. Some species discharge spores explosively, which accelerate 10,000 times faster than a Space Shuttle directly after launch, reaching speeds of up to a hundred kilometres per hour – some of the quickest movements achieved by any living organism. Other species of fungi create their own microclimates: spores are carried upwards by a current of wind generated by mushrooms as water evaporates from their gills. Fungi produce around fifty megatonnes of spores each year – equivalent to the weight of 500,000 blue whales – making them the largest source of living particles in the air. Spores are found in clouds and influence the weather by triggering the formation of the water droplets that form rain, and ice crystals that form snow, sleet and hail.[5]

Some fungi, like the yeasts that ferment sugar into alcohol and cause bread to rise, consist of single cells that multiply by budding into two. However, most fungi form networks of

Spores

many cells known as hyphae (pronounced *HY-fee*): fine tubular structures that branch, fuse and tangle into the anarchic
filigree of mycelium. Mycelium describes the most common of
fungal habits, better thought of not as a thing, but as a process
– an exploratory, irregular tendency. Water and nutrients flow
through ecosystems within mycelial networks. The mycelium of
some fungal species is electrically excitable and conducts waves
of electrical activity along hyphae, analogous to the electrical
impulses in animal nerve cells.[6]

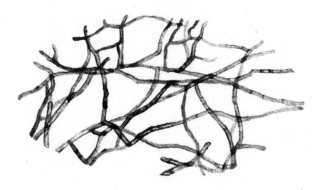

Mycelium

Hyphae make mycelium, but they also make more specialised
structures. Fruiting bodies, such as mushrooms, arise from the
felting together of hyphal strands. These organs can perform
many feats besides expelling spores. Some, like truffles, produce aromas that have made them among the most expensive
foods in the world. Others, like shaggy ink cap mushrooms
(*Coprinus comatus*), can push their way through asphalt and
lift heavy paving stones, although they are not themselves a
tough material. Pick an ink cap and you can fry it up and eat
it. Leave it in a jar, and its bright white flesh will deliquesce
into a pitch-black ink over the course of a few days (the illustrations in this book were drawn with *Coprinus* ink).[7]

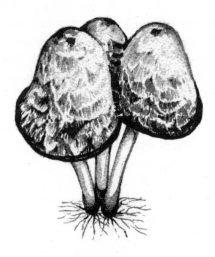

Shaggy ink cap mushrooms, *Coprinus comatus*, drawn with ink
made from shaggy ink cap mushrooms

Their metabolic ingenuity allows fungi to forge a wide variety
of relationships. Whether in their roots or shoots, plants have
relied on fungi for nutrition and defence for as long as there
have been plants. Animals, too, depend on fungi. After humans,
the animals that form some of the largest and most complex
societies on Earth are leafcutter ants. Colonies can reach sizes
of over eight million individuals, with underground nests that
grow larger than 30 metres across. The lives of leafcutter ants
revolve around a fungus which they cultivate in cavernous
chambers and feed with fragments of leaf.[8]

Human societies are no less entwined with fungi. Diseases
caused by fungi cause billions of dollars of losses – the rice
blast fungus ruins a quantity of rice large enough to feed
more than sixty million people every year. Fungal diseases of
trees, from Dutch elm disease to chestnut blight, transform
forests and landscapes. Romans prayed to the god of mildew,
Robigus, to avert fungal diseases but weren't able to stop the
famines that contributed to the decline of the Roman Empire.
The impact of fungal diseases is increasing across the world:

unsustainable agricultural practices reduce the ability of plants to form relationships with the beneficial fungi on which they depend. The widespread use of antifungal chemicals has led to an unprecedented rise in new fungal superbugs that threaten both human and plant health. As humans disperse disease-causing fungi, we create new opportunities for their evolution. Over the last fifty years, the most deadly disease ever recorded – a fungus that infects amphibians – has been spread around the world by human trade. It has driven ninety species of amphibian to extinction and threatens to wipe out over a hundred more. The variety of banana that accounts for 99 per cent of global banana shipments, the Cavendish, is being decimated by a fungal disease and faces extinction in the coming decades.[9]

Like leafcutter ants, however, humans have worked out how to use fungi to solve a range of pressing problems. In fact, we have probably deployed fungal solutions for longer than we have been *Homo sapiens*. In 2017, researchers reconstructed the diets of Neanderthals, cousins of modern humans who went extinct approximately 50,000 years ago. They found that an individual with a dental abscess had been eating a type of fungus, a penicillin-producing mould, implying knowledge of its antibiotic properties. There are other less ancient examples, including 'the Iceman', an exquisitely well-preserved Neolithic corpse found in glacial ice, dating from around 5,000 years ago. On the day he died, the Iceman was carrying a pouch stuffed with wads of the tinder fungus (*Fomes fomentarius*) that he almost certainly used to make fire, and carefully prepared fragments of the birch polypore mushroom (*Fomitopsis betulina*) most probably used as a medicine.[10]

The indigenous peoples of Australia treated wounds with moulds harvested from the shaded side of eucalyptus trees. The Jewish Talmud features a mould cure known as 'chamka', consisting of mouldy corn soaked in date wine. Ancient Egyptian papyruses from 1500 BCE refer to the curative properties of mould, and in 1640 the King's herbalist in London, John Parkinson, described the use of moulds to treat wounds. But it was only in 1928 that Alexander Fleming discovered that a mould produced a bacteria-killing chemical called penicillin. Penicillin became the first modern antibiotic and has since saved countless lives. Fleming's discovery is widely credited as one of the defining moments of modern medicine, and arguably helped to shift the balance of power in the Second World War.[11]

Penicillin, a compound that could defend fungi from bacterial infection, turned out to defend humans as well. This is not unusual: although fungi have long been lumped together with plants, they are actually more closely related to animals – an example of the kind of category mistake researchers regularly make in their struggle to understand fungal lives. At a molecular level, fungi and humans are similar enough to benefit from many of the same biochemical innovations. When we use drugs produced by fungi we are often borrowing a fungal solution and rehousing it within our own bodies. Fungi are pharmaceutically prolific, and today we depend on them for many other chemicals besides penicillin: cyclosporine (an immunosuppressant drug that makes organ transplants possible), cholesterol-lowering statins, a host of powerful antiviral and anti-cancer compounds (including the multi-billion-dollar drug Taxol, originally extracted from the fungi that live within yew trees), not to mention alcohol (fermented by a yeast) and psilocybin (the active component in psychedelic mushrooms recently shown in clinical trials to be capable of lifting severe depression and anxiety). Sixty per cent of the enzymes used in industry are generated by fungi, and 15 per cent of all vaccines are produced by engineered strains of yeast. Citric acid, produced by fungi, is used in all fizzy drinks. The global market for edible fungi is booming, and projected to increase from $42 billion in 2018 to $69 billion by 2024. Sales of medicinal mushrooms are increasing yearly.[12]

Fungal solutions don't stop at human health. Radical fungal technologies can help us respond to some of the many problems that arise from ongoing environmental devastation. Antiviral compounds produced by fungal mycelium reduce colony collapse disorder in honeybees. Voracious fungal appetites can be deployed to break down pollutants such as crude oil from oil spills, in a process known as 'mycoremediation'. In 'mycofiltration', contaminated water is passed through mats of mycelium which filter out heavy metals and break down toxins. In 'mycofabrication', building materials and textiles are grown out of mycelium and replace plastics and leather in many applications. Fungal melanins, the pigments produced by radio-tolerant fungi, are a promising new source of radiation-resistant biomaterials.[13]

Human societies have always pivoted around prodigious fungal metabolisms. A full litany of the chemical accomplishments of fungi would take months to recite. Yet despite their promise, and central role in many ancient human fascinations, fungi have received a tiny fraction of the attention given to animals and plants. The best estimate suggests that there are between 2.2 and 3.8 million species of fungi in the world – six to ten times the estimated number of plant species – meaning that a mere 6 per cent of all fungal species have been described. We are only just beginning to understand the intricacies and sophistications of fungal lives.[14]

For as long as I can remember I've been fascinated by fungi and the transformations they provoke. A solid log becomes soil, a lump of dough rises into bread, a mushroom erupts overnight – but how? As a teenager I dealt with my bafflement by finding ways to involve myself with fungi. I picked mushrooms, and grew mushrooms in my bedroom. Later, I brewed alcohol in the hope that I might learn more about yeast and its influence on me. I marvelled at the transformation of honey into mead, and fruit juice into wine – and at how the product of these transformations could transform my own senses and those of my friends.

By the time my formal study of fungi began, when I became an undergraduate at Cambridge in the Department of Plant Sciences – there is no Department of Fungal Sciences – I had become fascinated by symbiosis – the close relationships that form between unrelated organisms. The history of life turned out to be full of intimate collaborations. Most plants, I learned, depend on fungi to provide them with nutrients from the soil, such as phosphorus or nitrogen, in exchange for energy-giving sugars and lipids produced in photosynthesis – the process by which plants eat light and carbon dioxide from the air. The relationship between plants and fungi gave rise to the biosphere as we know it and supports life on land to this day, but we seemed to understand so little. How did these relationships arise? How do plants and fungi communicate with one another? How could I learn more about the lives of these organisms?

I accepted the offer of a PhD to study mycorrhizal relationships in tropical forests in Panama. Soon afterwards, I moved to a field station on an island run by the Smithsonian Tropical Research Institute. The island and surrounding peninsulas were part of a nature reserve entirely covered by forest, apart from a clearing for dormitories, a canteen and lab buildings. There were greenhouses for growing plants, drying cupboards filled with bags of leaf litter, a room lined with microscopes, and a walk-in freezer packed with samples: bottles of tree sap, dead bats, tubes containing ticks pulled from the backs of spiny rats and boa constrictors. Posters on the noticeboard offered cash rewards to anyone who could source fresh ocelot droppings from the forest.

The jungle bristled with life. There were sloths, pumas, snakes, crocodiles; there were basilisk lizards that could run across the surface of water without sinking. In just a few hectares there lived as many woody plant species as in the whole of Europe. The diversity of the forest was reflected in the rich variety of field biologists who came there to study it. Some climbed trees and observed ants. Some set out at dawn every day to follow the monkeys. Some tracked the lightning that struck trees during tropical storms. Some spent their days suspended from a crane measuring ozone concentrations in the forest canopy. Some warmed up the soil using electrical elements to see how bacteria might respond to global heating. Some studied the way beetles navigate using the stars. Bumblebees, orchids, butterflies – there seemed to be no aspect of life in the forest that someone wasn't observing.

I was struck by the creativity and humour of this community of researchers. Lab biologists spend most of their time in charge of the pieces of life they study. Their own human lives are lived outside the flasks that contain their subject matter. Field biologists rarely have so much control. The world is the flask and they're inside it. The balance of power is different. Storms wash away the flags that mark their experiments. Trees fall on their plots. Sloths die where they planned to measure the nutrients in the soil. Bullet ants sting them as they crash past. The forest and its inhabitants dispel any illusions that scientists are in charge. Humility quickly sets in.

The relationships between plants and mycorrhizal fungi are key to understanding how ecosystems work. I wanted to learn more about the way nutrients passed through fungal networks, but I became dizzy when I thought about what was going on underground. Plants and mycorrhizal fungi are promiscuous: many fungi can live within the roots of a single plant, and many plants can connect with a single fungal network. In this way a variety of substances, from nutrients to signalling compounds, can pass between plants via fungal connections. In simple terms, plants are socially networked by fungi. This is what is meant by the 'Wood Wide Web'. The tropical forests I worked in contained hundreds of plant and fungal species. These networks are inconceivably complicated, their implications huge and still poorly understood. Imagine the puzzlement of an extraterrestrial anthropologist who discovered, after decades of studying modern humanity, that we had something called the Internet. It's a bit like that for contemporary ecologists.

In my efforts to investigate the networks of mycorrhizal fungi that strung their way through the soil, I collected thousands of soil samples and tree root trimmings and mashed them into pastes to extract their fats, or DNA. I grew hundreds of plants in pots with different communities of mycorrhizal fungus and measured how big their leaves grew. I sprinkled thick rings of black pepper around the greenhouses to deter cats from creeping in and bringing with them rogue fungal communities from outside. I dosed plants with chemical labels and traced these chemicals through roots and into the soil so that I might measure how much must have passed to their fungal associates – more mashing and more pastes. I spluttered around the forested peninsulas in a small motorboat that often broke down, climbed up waterfalls looking for rare plants, trudged for miles down muddy paths carrying a backpack full of waterlogged soil, and drove trucks into drifts of thick red jungle mud.

Of the many organisms that lived in the rainforest, I was most enthralled by a species of small flower that sprouted from the ground. These plants were the height of a coffee cup, their stalks spindly and pale white with a single bright blue flower balanced on top. They were a species of jungle gentian called *Voyria*, and had long ago lost the ability to photosynthesise. In doing so they had lost their chlorophyll, the pigment that makes photosynthesis possible and gives plants their green colour. I was perplexed by *Voyria*. Photosynthesis is one of the things that makes plants plants. How could these plants survive without it?

I suspected *Voyria's* relationships with their fungal partners were unusual, and I wondered whether these flowers might tell me something about what was going on below the surface of the soil. I spent many weeks searching for *Voyria* in the jungle. Some flowers grew in open stretches of the forest and were easy to spot. Others hid, tucked behind buttressed tree roots. Within plots a quarter of the size of a football pitch there could be hundreds of flowers, and I had to count them all. The forest was rarely open or flat, so this meant scrambling and stooping. In fact, it meant almost anything but walking. Each evening I returned to the field station filthy and exhausted. Over supper my Dutch ecologist friends cracked jokes about my cute blossoms with their frail stems. They studied the ways that tropical forests stored carbon. While I scuffed along squinting at the ground in search of tiny flowers, they measured the girth of trees. In a carbon budget of the forest, *Voyria* were inconsequential. My Dutch friends teased me about my small ecology and my dainty fascinations. I teased them about their brute ecology and their machismo. At dawn the next day, I would set off once again, peering at the floor in the hope that these curious plants could help me find my way underground, into this hidden, teeming world.

Whether in forests, labs or kitchens, fungi have changed my understanding of how life happens. These organisms make questions of our categories and thinking about them makes the world look different. It was my growing delight in their power to do so that led me to write this book. I have tried to find ways to enjoy the ambiguities that fungi present, but it's not always easy to be comfortable in the space created by open questions. Agoraphobia can set in. It's tempting to hide in small rooms built from quick answers. I have done my best to hold back.

A friend of mine, the philosopher and magician David Abram, used to be the house magician at Alice's Restaurant, in Massachusetts (made famous by the Arlo Guthrie song). Every night he passed around the tables; coins walked through his fingers, reappeared exactly where they shouldn't, disappeared again, divided in two, vanished into nothing. One evening, two customers returned to the restaurant shortly after leaving and pulled David aside, looking troubled. When they left the restaurant, they said, the sky had appeared shockingly blue and the clouds large and vivid. Had he put something in their drinks? As the weeks went by, it continued to happen – customers returned to say the traffic had seemed louder than it was before, the streetlights brighter, the patterns on the sidewalk more fascinating, the rain more refreshing. The magic tricks were changing the way people experienced the world.

David explained to me why he thought this happened. Our perceptions work in large part by expectation. It takes less cognitive effort to make sense of the world using preconceived images updated with a small amount of new sensory information than to constantly form entirely new perceptions from scratch. It is our preconceptions that create the blind spots in which magicians do their work. By attrition, coin tricks loosen the grip of our expectations about the way hands and coins work. Eventually, they loosen the grip of our expectations on our perceptions more generally. On leaving the restaurant, the sky looked different because the diners saw the sky as it was there and then, rather than as they expected it to be. Tricked out of our expectations, we fall back on our senses. What's astonishing is the gulf between what we expect to find, and what we find when we actually look.[15]

Fungi, too, trick us out of our preconceptions. Their lives and behaviours are startling. The more I've studied fungi, the more my expectations have loosened, and the more familiar concepts have started to appear unfamiliar. Two fast-growing fields of biological enquiry have helped me both navigate these states of surprise and provide frameworks that have guided my exploration of the fungal world.

The first is a growing awareness of the many sophisticated, problem-solving behaviours that have evolved in brainless organisms outside the animal kingdom. The best-known examples are slime moulds, such as *Physarum polycephalum* (though they are amoeba, not fungi, as true moulds are). As we'll see, slime moulds have no monopoly on brainless problem-solving, but they are easy to study and have become poster organisms that have opened up new avenues of research. *Physarum* form exploratory networks made of tentacle-like veins, and have no central nervous system – or anything that resembles one. Yet they can 'make decisions' by comparing a range of possible courses of action, and can find the shortest path between two points in a labyrinth. Japanese researchers released slime moulds into petri dishes modelled on the Greater Tokyo area. Oat flakes marked major urban hubs and bright lights represented obstacles such as mountains – slime moulds don't like light. After a day, the slime mould had found the most efficient route between the oats, emanating into a network almost identical to Tokyo's existing rail network. In similar experiments, slime moulds have re-created the motorway network of the United States and the network of Roman roads in central Europe. A slime mould enthusiast told me about a test he had performed. He frequently got lost in IKEA stores, and would spend many minutes trying to find the exit. He decided to challenge his slime moulds with the same problem, and built a maze based on the floor plan of his local IKEA. Sure enough, without any signs or staff to direct them, the slime moulds soon found the shortest path to the exit. 'You see,' he said with a laugh, 'they're cleverer than me.'[16]

Whether one calls slime moulds, fungi and plants 'intelligent' depends on one's point of view. Classical scientific definitions of intelligence use humans as a yardstick by which all other species are measured. According to these anthropocentric definitions, humans are always at the top of the intelligence rankings, followed by animals that look like us (chimpanzees, bonobos, etc.), followed again by other 'higher' animals, and onwards and downwards in a league table – a great chain of intelligence drawn up by the ancient Greeks, which persists one way or another to this day. Because these organisms don't look like us or outwardly behave like us – or have brains – they have traditionally been allocated a position somewhere at the bottom of the scale. Too often they are thought of as the inert backdrop to animal life. Yet many are capable of sophisticated behaviours that prompt us to think in new ways about what it means for organisms to 'solve problems', 'communicate', 'make decisions', 'learn' and 'remember'. As we do so, some of the vexed hierarchies that underpin modern thought start to soften. As they soften, our ruinous attitudes towards the more-than-human world may start to change.[17]

The second field of research that has guided me in this enquiry concerns the way we think about the microscopic organisms

– or microbes – that cover every inch of the planet. In the last four decades, new technologies have granted unprecedented access to microbial lives. The outcome? For your community of microbes – your 'microbiome' – your body is a planet. Some prefer the temperate forest of your scalp, some the arid plains of your forearm, some the tropical forest of your crotch or armpit. Your gut (which if unfolded would occupy an area of 32 square metres), ears, toes, mouth, eyes, skin and every surface, passage and cavity you possess teem with bacteria and fungi. You carry around more microbes than your 'own' cells. There are more bacteria in your gut than stars in our galaxy.[18]

For humans, identifying where one individual stops and another starts is not generally something we think about. It is usually taken for granted – within modern industrial societies, at least – that we start where our bodies begin and stop where our bodies end. Developments in modern medicine, such as organ transplants, worry these distinctions; developments in the microbial sciences shake them at their foundations. We are ecosystems, composed of – and decomposed by – an ecology of microbes, the significance of which is only now coming to light. The 40 trillion-odd microbes that live in and on our bodies allow us to digest food and produce key minerals that nourish us. Like the fungi that live within plants, they protect us from disease. They guide the development of our bodies and immune systems and influence our behaviour. If not kept in check, they can cause illnesses, and even kill us. We are not a special case. Even bacteria have viruses within them (a nanobiome?). Even viruses can contain smaller viruses (a picobiome?). Symbiosis is a ubiquitous feature of life.[19]

I attended a conference in Panama on tropical microbes, and along with many other researchers spent three days becoming increasingly bewildered by the implications of our studies. Someone got up to talk about a group of plants that produced a certain group of chemicals in their leaves. Until then, the chemicals had been thought of as a defining characteristic of that group of plants. However, it transpired that the chemicals were actually made by fungi that lived in the leaves of the plant. Our idea of the plant had to be redrawn. Another researcher interjected, suggesting that it may not be the fungi living inside the leaf that produced these chemicals, but the bacteria living inside the fungus. Things continued along these lines. After two days, the notion of the individual had deepened and expanded beyond recognition. To talk about individuals made no sense any more. Biology – the study of living organisms – had transformed into ecology – the study of the relationships between living organisms. To compound matters, we understood very little. Graphs of microbial populations projected on a screen had large sections labelled 'unknown'. I was reminded of the way that modern physicists portray the universe, more than ninety-five percent of which is described as 'dark matter' and 'dark energy'. Dark matter and energy are dark because we don't know anything about them. This was biological dark matter, or dark life.[20]

Many scientific concepts – from 'time' to 'chemical bonds' to 'genes' to 'species' – lack stable definitions but remain helpful categories to think with. From one perspective, 'individual' is no different: just another category to guide human thought and behaviour. Nonetheless, so much of daily life and experience – not to mention our philosophical, political and economic systems – depends on individuals that it can be hard to stand by and watch the concept dissolve. Where does this leave 'us'? What about 'them'? 'Me'? 'Mine'? 'Everyone'? 'Anyone'? My response to the discussions at the conference was not just intellectual. Like a diner at Alice's Restaurant, I felt different: the familiar had become unfamiliar. The 'loss of a sense of self-identity, delusions of self-identity and experiences of "alien control"', observed an elder statesman in the field of microbiome research, are all potential symptoms of mental illness. It made my head spin to think of how many ideas had to be revisited, not least our culturally treasured notions of identity, autonomy and independence. It is in part this disconcerting feeling that makes the advances in the microbial sciences so exciting. Our microbial relationships are about as intimate as any can be. Learning more about these associations changes our experience of our own bodies and the places we inhabit. 'We' are ecosystems that span boundaries and transgress categories. Our selves emerge from a complex tangle of relationships only now becoming known.[21]

The study of relationships can be confusing. Almost all are ambiguous. Have leafcutter ants domesticated the fungus they depend on, or has the fungus domesticated the ants? Do plants farm the mycorrhizal fungi they live with, or do the fungi farm the plants? Which way does the arrow point? This uncertainty is healthy.

I had a professor called Oliver Rackham, an ecologist and historian, who studied the ways in which ecosystems have shaped – and been shaped by – human cultures for thousands of years. He took us to nearby forests, and told us about the history of these places and their human inhabitants by reading the twists and splits in the branches of old oak trees, by observing where nettles thrived, by noting which plants did or didn't grow in a hedgerow. Under Rackham's influence, the clean line I had imagined dividing 'nature' and 'culture' started to blur.

Later, doing fieldwork in Panama, I came across many complicated relationships between field biologists and the organisms they studied. I joked with the bat scientists that in staying up all night and sleeping all day they were learning bat habits. They asked how the fungi were imprinting themselves on me. I'm still not sure. But I continue to wonder how, in our total dependence on fungi – as regenerators, recyclers, and networkers that stitch worlds together – we might dance to their tune more often than we realise.

If we do, it's easy to forget. Too often, I become detached and see the soil as an abstract place, a vague arena for schematic interactions. My colleagues and I say things like, '(So-and-so) reported an approximate 25 per cent increase in soil carbon from one dry season to the next wet season.' How can we not? We have no way to experience the wilds of the soil and the countless lives that froth away within it.

With the available tools, I tried. Thousands of my samples passed through expensive machines that whisked, irradiated and blasted the contents of the tubes into strings of numbers. I spent whole months staring into a microscope, immersed in rootscapes filled with winding hyphae frozen in ambiguous acts of intercourse with plant cells. Still, the fungi I could see were dead, embalmed and rendered in false colours. I felt like a clumsy sleuth. While I crouched for weeks scraping mud into small tubes, toucans croaked, howler monkeys roared, lianas tangled and anteaters licked. Microbial lives, especially those buried in soil, were not accessible like the bristling, charismatic, above-ground world of the large. Really, to make my findings vivid, to allow them to build and contribute to a general understanding, imagination was required. There was no way around it.

In scientific circles imagination usually goes by the name of speculation and is treated with some suspicion – in publications it is usually served up with a mandatory health warning. Part of writing up research is scrubbing it clean of the flights of fancy, idle play and thousand trials and errors that give rise to even the smallest of findings. Not everyone who reads a study wants to push their way through the fuss. Besides, scientists have to appear credible. Sneak backstage and one might not find people at their most presentable. Even backstage, in the most nocturnal musings I shared with colleagues, it was unusual to get into the details of how we had imagined – accidentally or deliberately – the organisms we studied, whether fish, bromeliad, liana, fungus or bacterium. There was something embarrassing about admitting that the tangle of our unfounded conjectures, fantasies and metaphors might have helped shape our research. Regardless, imagination forms part of the everyday business of enquiring. Science isn't an exercise in cold-blooded rationality. Scientists are – and have always been – emotional, creative, intuitive, whole human beings, asking questions about a world that was never made to be catalogued and systematised. Whenever I asked what these fungi were doing and designed studies to try and understand their behaviours, I necessarily imagined them.

An experiment forced me to peer into the deeper recesses of my scientific imagination. I signed up to take part in a clinical study into the effects of LSD on the problem-solving abilities of scientists, engineers and mathematicians. The study was part of the wide revival of scientific and medical interest in the untapped potential of psychedelic drugs. The researchers wanted to know if LSD could grant scientists access to their professional unconscious and help them approach familiar problems from new angles. Our imaginations, usually brushed aside, were to be the stars of the show, the phenomena being observed and potentially even measured. An eclectic group of young researchers had been enlisted through posters in science departments around the country ('Do you have a meaningful problem that needs solving?'). It was a brave study. Creative breakthroughs are notoriously hard to facilitate anywhere, let alone in the clinical drug trial unit of a hospital.

The researchers running the experiment had arranged psyche-delic hangings on the walls, set up a sound system for music to be piped in and lit the room with coloured 'mood lights'. Their attempts to de-clinicalise the setting made it seem more artificial: an admission of the impact that they – the scientists – might have on their subject matter. It was an arrangement that made visible many of the healthy insecurities faced by researchers on a day-to-day basis. If only the subjects of all biological experi-ments were provided with their equivalent of mood lighting and relaxing music, how differently they might behave.

The nurses made sure I drank the LSD at exactly 9 a.m. They watched me closely until I had swallowed all the liquid, which had been mixed in a small wine glass's worth of water. I lay down on the bed in my hospital room and the nurses sucked a sample of blood through the cannula in my forearm. Three hours later, when I had reached 'cruising altitude', I was gently encouraged by my assistant to start thinking about my 'work-related problem'. Amid the battery of psychometric tests and personality assessments we had completed before the trip, we had been asked to describe our problems in as much detail as possible – those knots in our enquiries we might be fumbling with. Soaking the knots in LSD might help them loosen. All my research questions were fungal ones, and I was comforted by the knowledge that LSD was originally derived from a fungus that lives within crop plants; a fungal solution to my fungal problems. What would happen?

I wanted to use the LSD trial to think more broadly about the lives of the blue flowers, *Voyria,* and their fungal relationships. How did they live without photosynthesis? Almost all plants sustain themselves by drawing minerals from mycorrhizal fungal networks in the soil; so did *Voyria*, judging by the tousled mass of fungi that crowded into their roots. But without photosynthesis, *Voyria* had no way to make the energy-rich sugars and lipids they needed to grow. Where did *Voyria* get their energy from? Could these flowers draw substances from other green plants via the fungal networks? If so, did *Voyria* have anything to give back to their fungal partners in exchange, or were they just parasites – hackers of the Wood Wide Web?

I lay on the hospital bed with my eyes closed and wondered what it was like to be a fungus. I found myself underground, surrounded by growing tips surging across one another. Schools of globular animals grazing – plant roots and their hustle – the Wild West of the soil – all those bandits, brigands, loners, crap shooters. The soil was a horizonless external gut – digestion and salvage everywhere – flocks of bacteria surfing on waves of electrical charge – chemical weather systems – subterranean highways – slimy infective embrace – seething intimate contact on all sides. As I followed a fungal hypha into a cavernous root, I was struck by the sanctuary it offered. Very few other types of fungi were present; certainly no worms or insects. There was less bustle and hassle. It was a haven I could imagine paying for. Perhaps that was what the blue flowers offered the fungi in return for their nutritional support? Shelter from the storm.

I make no claims about the factual validity of these visions. They are at best plausible, and at worst delirious nonsense. Not even wrong. Nonetheless, I learned a valuable lesson. The way I had grown accustomed to thinking about fungi involved abstract 'interactions' between organisms that actually looked like the diagrams schoolteachers drew on the board: semi-automatic entities that behaved according to an early-Nineties Game Boy logic. However, the LSD had forced me to admit that I had an imagination, and I now saw fungi differently. I wanted to understand fungi, not by reducing them to ticking, spinning, bleeping mechanisms, as we so often do. Rather, I wanted to let these organisms lure me out of my well-worn patterns of thought, to imagine the possibilities they face, to let them press against the limits of my understanding, to give myself permission to be amazed – and confused – by their entangled lives.

Fungi inhabit enmeshed worlds; countless threads lead through these labyrinths. I have followed as many as I can, but there are crevices I haven't been able to squeeze through no matter how hard I've tried. Despite their nearness, fungi are so mystifying, their possibilities so *other*. Should this scare us off? Is it possible for humans, with our animal brains and bodies and language, to learn to understand such different organisms? How might we find ourselves changed in the process? In optimistic moods, I've imagined this book to be a portrait of this neglected branch of the tree of life, but it's more tangled than that. It is an account both of my journey towards understanding fungal lives, and of the imprint fungal lives have left on me and the many others I've met along the road, human or otherwise. 'What shall I do with the night and the day, with this life and this death?' writes the poet Robert Bringhurst. 'Every step, every breath rolls like an egg towards the edge of this question.' Fungi roll us towards the edge of many questions. This book comes from my experience of peering over some of these edges. My exploration of the fungal world has made me re-examine much of what I knew. Evolution, ecosystems, individuality, intelligence, life – none are quite what I thought they were. My hope is that this book loosens some of your certainties, as fungi have loosened mine.

Mushroom spore print

1

A Lure

Who's pimping who?

Prince

A heap of Piedmont white truffles, *Tuber magnatum*, sat on the scales on a check-patterned rag. They were scruffy, like unwashed stones; irregular, like potatoes; socketed, like skulls. Two kilograms: €12,000. Their sweet funk filled the room, and in this aroma was their value. It was unabashed and quite unlike anything else: a lure, thick and confusing enough to get lost in.

It was early November, the height of truffle season, and I had travelled to Italy to join two truffle hunters working out of the hills around Bologna. I was lucky. A friend of a friend knew a man who dealt truffles. The dealer had agreed to set me up with his two best hunters, who in turn had consented to let me go out with them. White-truffle hunters are famously secretive. These fungi have never been domesticated and can only be found in the wild.

Truffles are the underground fruiting bodies of several types of mycorrhizal fungi. For most of the year, truffle fungi exist as mycelial networks, sustained in part by the nutrients they obtain from the soil and also by the sugars they draw from plant roots. However, their subterranean habitat confronts them with a basic problem. Truffles are spore-producing organs, analogous to the seed-producing fruit of a plant. Spores evolved to allow fungi to disperse themselves, but underground their spores can't be caught by air currents, and are invisible to the eyes of animals.[1]

Their solution is to smell. But to smell above the olfactory racket of a forest is no small task. Forests are criss-crossed with smells, each a potential fascination or distraction to an animal nose. Truffles must be pungent enough for their scent to penetrate the layers of soil and enter the air, distinctive enough for an animal to take note amid the ambient smell-scape, and delicious enough for that animal to seek it out, dig it up and eat it. Every visual disadvantage that truffles face – being entombed in the soil, difficult to spot once unearthed, and visually unappealing once spotted – they make up for with smell.

Once eaten, a truffle's job is done: an animal has been lured into exploring the soil and recruited to carry the fungus's spores off to a new place and deposit them in its faeces. A truffle's allure is thus the outcome of hundreds of thousands of years of evolutionary entanglement with animal tastes. Natural selection will favour truffle fungi that match the preferences of their finest spore dispersers. Truffles with better 'chemistry' will attract animals more successfully than those with worse. Like the orchids that mimic the appearance of sexually receptive female bees, truffles provide a depiction of animal tastes – an evolutionary portrait-in-scent of animal fascination.

I was in Italy because I wanted to be drawn underground by a fungus into the chemical world in which it lived. We are ill-equipped to participate in the chemical lives of fungi, but ripe truffles speak a language so piercing and simple that even we can understand it. In doing so, these fungi include us for a moment within their chemical ecology. How should we think about the torrents of interaction that occur between organisms underground? How should we understand these spheres of more-than-human communication? Perhaps running after a dog hot on the trail of a truffle and burying my face in the soil was as close as I could get to the chemical tug and promise that fungi use to conduct so many aspects of their lives.

The human sense of smell is extraordinary. Our eyes can distinguish several million colours, our ears can distinguish half a million tones, but our noses can distinguish well over a trillion different odours. Humans can detect virtually all volatile chemicals ever tested. We outperform rodents and dogs in detecting certain odours, and we can follow scent trails. Smells feature in our choice of sexual partners and in our ability to detect fear, anxiety or aggression in others. And smell is woven into the fabric of our memories; it is common for people suffering from post-traumatic stress disorder to have olfactory flashbacks.[2]

Noses are finely tuned instruments. Your olfactory sense can split complex mixtures into their constituent chemicals, just as a prism can split white light into its constituent colours. To do this, it must detect the precise arrangement of atoms within a molecule. Mustard smells mustardy because of bonds between

Piedmont white truffle, *Tuber magnatum*

nitrogen, carbon and sulphur. Fish smell fishy because of bonds between nitrogen and hydrogen. Bonds between carbon and nitrogen smell metallic and oily.[3]

The ability to detect and respond to chemicals is a primordial sensory ability. Most organisms use their chemical senses to explore and make sense of their environment. Plants, fungi and animals all use similar types of receptors to detect chemicals. When molecules bind to these receptors, they trigger a signalling cascade: one molecule triggers a cellular change, which triggers a bigger change, and so on. In this way, small causes can ripple into large effects: human noses can detect some compounds at as low a concentration as 34,000 molecules in one cubic centimetre, the equivalent of a single drop of water in 20,000 Olympic swimming pools.[4]

For an animal to experience a smell, a molecule must land on their olfactory epithelium. In humans, this is a membrane up and behind the nose. The molecule binds to a receptor, and nerves fire. The brain gets involved as chemicals are identified, or trigger thoughts and emotional responses. Fungi are equipped with different kinds of bodies. They don't have noses or brains. Instead, their entire surface behaves like an olfactory epithelium. A mycelial network is one large chemically sensitive membrane: a molecule can bind to a receptor anywhere on its surface and trigger a signalling cascade that alters fungal behaviour.

Fungi live their lives bathed in a rich field of chemical information. Truffle fungi use chemicals to communicate to animals their readiness to be eaten; they also use chemicals to communicate with plants, animals, other fungi – and themselves. It isn't possible to understand fungi without exploring these sensory worlds, but they are hard for us to interpret. Perhaps it doesn't matter. Like fungi, we spend much of our lives being drawn towards things. We know what it is to be attracted or repelled. Through smell, we can participate in the molecular discourse fungi use to organise much of their existence.

*

In human history, truffles have long been associated with sex. The word for truffle in many languages translates to 'testicle', as in the old Castilian *turmas de tierra*, or Earth's testicles. Truffle fungi have evolved to make animals giddy because their lives depend on it. As I spoke with Charles Lefevre, a truffle scientist and cultivator in Oregon, about his work with the Périgord black truffle, he broke off: 'Funny – as I'm saying this I am "bathing" in the virtual aroma of *Tuber melanosporum*. It's as if a cloud of it is filling my office, but there are currently no truffles here. These olfactory flashbacks are common with truffles in my experience. They can even include visual and emotional memories.'[5]

In France, Saint Anthony – the patron saint of lost objects – is regarded as the patron saint of truffles, and truffle masses are celebrated in his honour. Prayers do little to stop the skulduggery. Cheap truffles are stained or flavoured to pass them off as their more valuable cousins. Prized truffle forests are targeted by truffle poachers. Expertly trained dogs worth thousands of euros are stolen. Poisoned meat is strewn around woods to kill the dogs of rival hunters. In 2010, in a crime of passion, a French truffle farmer, Laurent Rambaud, shot dead a truffle thief he encountered while patrolling his truffle

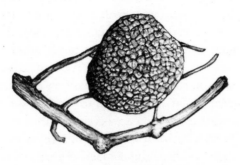

Périgord black truffle, *Tuber melanosporum*

orchards during the night. Following his arrest, 250 supporters marched in support of Rambaud's right to defend his crop, angry at the rise in thefts of both truffles and truffle dogs. The deputy head of the Tricastin truffle growers' union told *La Provence* newspaper that he had advised fellow producers never to patrol their fields with a gun because 'the temptation is too high'. Lefevre put it well. 'Truffles bring out the dark side of people. It's like money lying on the ground, but it's perishable and mercurial.'[6]

Truffles are not the only fungi to attract animal attention. On the west coast of North America, bears upend logs and dig out ditches looking for the prized matsutake mushroom. Oregon mushroom hunters have reported elk with noses bloodied in their hunt for matsutake in sharp pumice soils. Some species of tropical rainforest orchid have evolved to mimic the smell, shape and colour of mushrooms to attract mushroom-loving flies. Mushrooms and other fruit bodies are fungi at their most conspicuous, but mycelium, too, can be a lure. A friend of mine who studies tropical insects showed me a video of orchid bees crowding around a crater in a rotting log. Male orchid bees collect scents from the world and amass them into a cocktail which they use to court females. They are perfume makers. Mating takes seconds, but gathering and blending their scents takes their entire adult lives. Although he hadn't yet tested the hypothesis, my friend had a strong hunch that the bees were harvesting fungal compounds to add to their bouquets. Orchid bees are known to have a taste for complex aromatic chemicals, many of which are produced by fungi that break down wood.[7]

Humans wear perfumes produced by other organisms and it is not uncommon for fungal aromas to be incorporated into our own sexual rituals. Agarwood, or *oudh*, is a fungal infection of *Aquilaria* trees found in India and south-east Asia and one of the most valuable raw materials in the world. It is used to make a scent – dank nuts, dark honey, rich wood – and has been coveted at least since the time of the ancient Greek physician Dioscorides. The best *oudh* is worth more, gram for gram, than gold or platinum – as much as $100,000 per kilogram – and the destructive harvest of *Aquilaria* trees has driven them to near extinction in the wild.[8]

The eighteenth-century French physician Théophile de Bordeu asserted that each organism 'does not fail to spread exhalations, an odour, emanations around itself ... These emanations have taken on its style and its demeanour; they are, in fact, genuine parts of itself.' A truffle's fragrance and an orchid bee's perfume may circulate beyond the flesh of each organism, but these fields of odour make up a part of their chemical bodies which overlap with one another like ghosts at a disco.[9]

I spent several minutes in the truffle weighing room, lost in the aroma. My reverie was interrupted when my host Tony, the truffle dealer, bustled in with one of his clients. He closed the door behind him, sealing in the smell. The client inspected the heap of truffles on the scales and cast an eye over the bowls of unsorted and uncleaned specimens ranged across a grubby work bench. He nodded to Tony, who tied up the corners of the rag. They walked out into the yard, shook hands, and the client drove off in a smart black car.

It had been a dry summer, which had resulted in a poor truffle harvest. Their price reflected their scarcity. Bought directly from Tony, a kilogram would set you back €2,000. The same kilogram purchased at a market or restaurant would cost as much as €6,000. In 2007, a single 1.5-kilogram truffle was sold at auction for £165,000 – like diamonds, the price of truffles increases non-linearly with their size.[10]

Tony had a warm manner and a dealer's bravado. He seemed surprised that I would want to join his hunters, and didn't get my hopes up about our chances of finding any truffles. 'You can go out with my guys, but you probably won't find anything. And it's hard work. Up and down. Through bushes. Through mud. Through streams. Are those the only shoes you have?' I assured him I didn't mind.

Truffle hunters have their turf, sometimes legal, sometimes not. When I arrived, both truffle hunters – Daniele and Paride – were wearing camouflage. I asked whether it helped them to sneak up on the truffles, and they responded in earnest. It allowed them to hunt for truffles without being followed by other truffle hunters. Truffle hunters are in the business of knowing where to look. Their knowledge has value and, like truffles themselves, can be stolen.

Paride was the friendlier of the two, and met me outside with Kika, his favourite truffle dog. He had five dogs, of various ages and states of training, each a specialist in either black or white truffles. Kika was charming, and Paride introduced her proudly. 'My dog is very clever, but I am more clever.' Kika's breed – the Lagotto Romagnolo – is one of the most commonly used for truffle-hunting. She was knee-height and, with hair that fell over her eyes in shaggy ringlets, she resembled a truffle. Indeed, after a morning smelling truffles, meeting a litter of truffle dog puppies, talking truffles, witnessing truffle deals, and eating truffles, even the rounded rocky hills had started to look like truffles. Paride spoke about the subtle cues he and Kika used to communicate with each other. They had learned to read and interpret the tiniest shifts in the other's behaviour and could co-ordinate their movements in near total silence. Truffles had evolved to communicate to animals their readiness to be eaten. Humans and dogs had developed ways to communicate with one another about truffles' chemical propositions.

A truffle's aroma is a complex trait, and seems to emerge out of the relationships the truffle maintains with its community of microbes, and the soil and climate it lives within – its *terroir*. Truffle fruiting bodies house thriving communities of bacteria and yeasts – between a million and a billion bacteria per gram of dry weight. Many members of truffles' microbiomes are able to produce the distinctive volatile compounds that contribute to truffles' aromas, and it is likely that the cocktail of chemicals that reaches your nose is the work of more than a single organism.[11]

The chemical basis of truffles' allure remains uncertain. In 1981, a study published by German researchers found that both Piedmont white truffles (*Tuber magnatum*) and Périgord black truffles (*Tuber melanosporum*) produced androstenol – a steroid with a musky scent – in non-negligible quantities. In pigs, androstenol functions as a sex hormone. It is produced by males and prompts the mating posture in sows. This finding triggered speculation that androstenol might explain the impressive abilities of sows to find truffles buried deep underground. A study published nine years later cast doubt on this possibility. Researchers buried black truffles, a synthetic truffle flavouring and androstenol five centimetres underground, and challenged a pig and five dogs – including the champion of the local county truffle dog contest – to find the samples. All the animals detected the real truffles and the synthetic truffle flavouring. None detected the androstenol.[12]

In a series of further tests, the researchers narrowed truffles' allure down to a single molecule, dimethyl sulphide. It was a neat study, but unlikely to be the whole truth. The smell of a truffle is made up of a flock of different molecules drifting in formation – more than a hundred in white truffles, and around fifty in the other most popular species. These elaborate bouquets are energetically costly and are unlikely to have evolved unless they served some purpose. What's more, animal tastes are diverse. Certainly, not all truffle species are attractive to humans and some are even mildly poisonous. Of the thousand-odd species of truffle in North America, only a handful are of culinary interest. Even these aren't of interest to everyone. As Lefevre explained, a large number of people are offended by the aroma of the otherwise prized species. Some species smell outright repulsive. He told me about *Gautieria*, a genus that produces truffles with a foulsome stench – like 'sewer gas' or 'baby diarrhoea'. His dogs love them, but his wife won't let him bring any into the house, even for taxonomic purposes.[13]

However they do it, truffles create nested layers of attraction around themselves: humans train dogs to find truffles because pigs are so attracted to them that they devour the truffles they find rather than turn them over to their minders. Restaurateurs from New York and Tokyo travel to Italy to build relationships with truffle dealers. Exporters have developed sophisticated chilled packing systems to maintain truffles at optimal conditions as they are washed, packed, hand-delivered to the airport, flown around the world, collected from the airport, carried through customs, repacked and distributed to consumers – all within forty-eight hours. Truffles, like matsutake mushrooms, must arrive fresh on a plate within two to three days of harvest. Truffles' aromas are made in an active process by living, metabolising cells. A truffle's odour increases as its spores develop, and its aroma ceases when its cells die. You can't dry a truffle and expect to taste it later, as you can with some types of mushroom. They are chemically loquacious, vociferous even. Stop the metabolism, and you stop the smell. For this reason, in many restaurants, fresh truffles are grated onto your food before your eyes. Few other organisms are so good at persuading humans to disperse them with such urgency.[14]

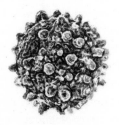

Truffle spore

We piled into Paride's car and drove up a valley on a narrow country road, through the damp yellows and browns of the oak woods that covered the hills. Paride talked about the weather, and cracked jokes about dog-training and the pros and cons of working with a 'bandit' like Daniele. After a few minutes, we turned down a track and pulled over. Kika jumped out of the boot, and we walked along a meadow and into a wood. Daniele had already arrived and was hovering furtively with his dog. There was another truffle hunter nearby, he explained, and we had to be quiet. Daniele's dog was tousled and unkempt and had twigs caught in its curls. It didn't have a name, although Paride said he had heard Daniele call it Diavolo (Devil) earlier that morning. Unlike Kika, who was affectionate and friendly, Diavolo had a tendency to snap and snarl. Paride explained why. Whereas he trained his dogs to hunt for truffles as if it were a game, Daniele trained his by hunger. 'Look' – Paride pointed at Diavolo – 'it's desperate, it's eating acorns.' They bantered for a while. Daniele argued that his dogs were more effective truffle hunters than Paride's well-fed and well-loved 'pets'. Paride stuck up for the reformed school of truffle dog training, summing it up neatly. 'Daniele hunts truffles at night, and I hunt them in the day. He is nervous and I am not. His dog bites, and mine is friendly. His dog is slim, and mine is not slim. He is bad, and I am good.'

All of a sudden, Diavolo darted off. We followed him, Paride providing a commentary as we scrambled along. 'There may be a truffle. Or a mouse. Either way the dog's happy.' We found Diavolo digging and snorting halfway up a muddy bank. Daniele caught up and cleared away the brambles. At this point, Paride explained, the truffle hunter had to read the dog's body language closely. A wagging tail promised truffles, a still tail suggested otherwise. A two-pawed dig indicated white truffles, a one-pawed dig black. The signs looked good, and Daniele began to loosen the soil with a blunt, flat-tipped tool like a giant screwdriver, smelling pinches of soil as he got deeper. He and the dog took it in turns, though he was careful to stop Diavolo digging too vigorously. Paride smiled over at us. 'A hungry dog eats the truffle.'

Finally, about a foot and a half down, Daniele found it lodged in the damp soil. With his fingers and a small metal hook, he pulled away the mud. The truffle's aroma strayed upwards from the hole, brighter and more saturated than in the weighing room. This was its natural habitat, and its scent drifted in easy harmony with the dampness of the ground and the fraying of leaf mould. I imagined being sensitive enough to notice the truffle's aroma at a distance, and compelled enough to drop everything to pursue it. Inhaling its emanations I recalled the passage in Aldous Huxley's *Brave New World* where he describes the performance of a scent organ, an instrument able to give olfactory recitals in the way a musical instrument might. It is a concept easily adapted for truffles – scent organs in a different sense – that perform, in their way, suites of volatile compounds.

How well it had worked. Here we all were, tousled and muddy, standing around a truffle. It had triggered a signalling cascade, tugging a troupe of animals towards it: first a dog, then a truffle-hunting human, then his slower-footed associates. As Daniele picked up the truffle, the ground around it collapsed. 'Look!' Paride cleared the soil aside. 'The house of a mouse.' We had not been the first to arrive.

When we smell a truffle's aroma, we receive a one-way transmission from truffle to world. The process is comparatively nuance-free. To attract an animal the aroma has to be curious, and delicious – yes. But most of all it has to be penetrating and strong. It doesn't really matter whether their spores are scattered by a wild boar or a flying squirrel, so why be picky? Most hungry animals will chase a delicious smell. Moreover, a truffle doesn't change its aroma in response to your immediate attentions. It can excite, but it isn't excitable. Its signal billows out loud and clear and, once begun, it is always on. A ripe truffle broadcasts an unambiguous summons in chemical lingua franca, a pop scent with mass appeal that could cause Daniele, Paride, two dogs, a mouse and me to converge at a single point under a bramble bush on a muddy bank in Italy.

Truffles – like many other highly prized fungal fruiting bodies – are their parent fungus's least sophisticated channels of communication. Much of fungal life, including the growth of mycelium, depends on subtler forms of allure. There are two key moves by which fungal hyphae become a mycelial network. First, they branch. Second, they fuse. (The process by which hyphae merge with each other is known as 'anastomosis', which in Greek means 'to provide with a mouth'.) If hyphae couldn't branch, one hypha could never become many. If hyphae couldn't fuse with one another, they would not be able to grow into complex networks. However, before they fuse, hyphae must find other hyphae, which they do by attracting one another, a phenomenon known as 'homing'. Fusion between hyphae is the linking stitch that makes mycelium mycelium, the most basic networking act. In this sense, the mycelium of any fungus arises from its ability to attract itself to itself.[15]

However, much as a given mycelial network is able to encounter itself, it is able to encounter another. How do fungi maintain a sense of a body subject to continual revision? Hyphae must be able to tell if they are bumping into a branch of themselves, or another fungus entirely. If another, they need to be able to tell whether it is a different – potentially hostile – species, or a sexually compatible member of its own, or neither. Some fungi have tens of thousands of mating types, approximately equivalent to our sexes (the record holder is the split gill fungus, *Schizophyllum commune*, which has over 23,000 mating types, each sexually compatible with nearly every one of the others). The mycelium of many fungi can fuse with other mycelial networks if they are genetically similar enough, even if they aren't sexually compatible. Fungal self-identity matters, but it is not always a binary world. Self can shade off into otherness gradually.[16]

Mycelium growing outwards from a spore. Redrawn
from Buller (1931)

Allure underpins many types of fungal sex, including that of
truffle fungi. Truffles themselves are the outcome of a sexual
encounter: for a truffle fungus like *Tuber melanosporum* to
fruit, the hyphae of one mycelial network must fuse with those
of a separate, sexually compatible network, and pool genetic
material. For most of their lives, as mycelial networks, truffle fungi live as separate mating types, whether '-' or '+' – by
fungal standards, their sexual lives are straightforward. Sex
happens when a '-' hypha attracts and fuses with a '+' hypha.
One partner plays a paternal role, providing genetic material only. The other plays a maternal role, providing genetic
material and growing the flesh that matures into truffles and
spores. Truffles differ from humans in that either '+' or '-' mating types can be maternal or paternal – it is as if all humans
were both male and female and equally able to play the part
of a mother or a father, provided we could have sex with a
partner of the opposite mating type. How the sexual attraction between truffle fungi plays out remains unknown. Closely
related fungi use pheromones to attract mates, and researchers

have a strong suspicion that truffles, too, use a sex pheromone for this purpose.[17]

Without homing, there could be no mycelium. Without mycelium, there could be no attraction between '-' and '+' mating types. Without sexual attraction there could be no sex. And without sex, there could be no truffle. However, the relationships between truffle fungi and their partner trees are just as important, and their chemical interactions must be intricately managed. The hyphae of young truffle fungi will soon die unless they find a plant to partner with. Plants must admit into their roots the fungal species that will form a mutually beneficial relationship, as opposed to the many that will cause disease. Both fungal hyphae and plant roots face the challenge of finding one another amid the chemical babble in the soil where countless other roots, fungi and microbes course and engage.[18]

It is another case of attraction and allure, of chemical call and response. Both plant and fungus use volatile chemicals to make themselves attractive to one another, just as truffles make themselves attractive to animals in a forest. Receptive plant roots produce plumes of volatile compounds that drift through the soil and cause spores to sprout and hyphae to branch and grow faster. Fungi produce plant growth hormones that manipulate roots, causing them to proliferate into masses of feathery branches – with a greater surface area, the chances of an encounter between root tips and fungal hyphae become more likely. (Many fungi produce plant and animal hormones to alter the physiology of their associates.)[19]

More than the architecture of roots has to change for a fungus to bond with a plant. In response to each other's distinctive chemical profiles, signalling cascades ripple through plant and fungal cells, activating suites of genes. Both reorient their metabolisms and developmental programmes. Fungi release chemicals that suspend their plant partners' immune responses, without which they can't get close enough to form symbiotic structures. Once established, mycorrhizal partnerships continue to develop. Connections between hyphae and roots are dynamic, formed and re-formed as root tips and fungal hyphae get old and die. These are relationships that ceaselessly remodel themselves. If you could place your olfactory epithelium into the soil, it would feel like the performance of a jazz group, with the players listening, interacting, responding to one another in real time.[20]

Piedmont white truffles and other prized mycorrhizal fungi, such as porcini, chanterelle and matsutake, have never been domesticated, in part because of the fluidity of their relationships with plants, and in part because of the intricacies of their sex lives. There are too many gaps in our understanding of how basic fungal communication happens. Some truffle species can be cultivated, such as the Périgord black, but trufficulture is immature by comparison with the venerable craft surrounding most human agricultural efforts, and even the success of seasoned cultivators can vary wildly. At Charles Lefevre's New World Truffieres, the proportion of seedlings that grow successfully with the mycelium of the Périgord black truffle hovers around 30 per cent. One year, with no deliberate change in method, he achieved a 100 per cent success rate. 'I have not been able to reproduce that result,' he told me. 'I don't know what I did right.'

To cultivate truffles effectively, you have to understand the quirks and needs not only of the fungi – with their idiosyncratic reproductive systems – but also of the trees and bacteria they live with. Moreover, you have to understand the importance of subtle variations in the surrounding soil, season and climate. 'It is an intellectually stimulating field because it's so interdisciplinary,' Ulf Büntgen, a professor of geography at Cambridge, and the first to report the fruiting of a Périgord black truffle in the British Isles, told me. 'It is microbiology, physiology, land management, agriculture, forestry, ecology, economy and climate change. You really have to take a holistic perspective.' Truffles' affairs quickly unspool into entire ecosystems. Scientific understanding hasn't yet caught up.[21]

For some, attracted by fungal chemical allure, the outcome is simpler: death.

Among the most impressive sensory feats are those performed by predatory fungi that trap and consume nematode worms. Hundreds of species of worm-hunting fungi can be found all over the world. Most spend their lives decomposing plant matter, and only start to hunt when there is insufficient material to eat. But they're subtle predators: unlike truffles, whose scent, once begun, is always on, nematode-eating fungi only produce worm-hunting organs and issue a chemical summons when they sense nematodes are close by. If there is plenty of material to rot, they don't bother, even if worms abound. To behave in this way, nematode-eating fungi must be able to detect the presence of worms with exquisite sensitivity. Nematodes all depend on the same class of molecule to serve a number of purposes, from regulating their development to attracting mates. In turn, fungi use these chemicals to eavesdrop on their prey.[22]

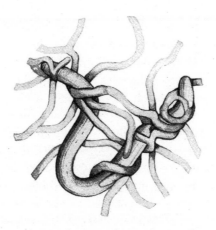

Nematode worm being devoured

The methods fungi use to hunt nematodes are grisly and diverse. It is a habit that has evolved multiple times – many fungal lineages have reached a similar conclusion but in different ways. Some fungi grow adhesive nets, or branches to which nematodes stick. Some use mechanical means, producing hyphal nooses which inflate in a tenth of a second when touched, ensnaring their prey. Some – including the commonly cultivated oyster mushroom, *Pleurotus ostreatus* – produce hyphal stalks capped with a single toxic droplet that paralyses nematodes, giving the hypha enough time to grow through their mouth and digest the worm from the inside. Others produce spores that can swim through the soil, chemically drawn towards nematodes, to which they bind. Once attached, the spores sprout and the fungus harpoons the worm with specialised hyphae known as 'gun cells'.[23]

Fungal worm-hunting is a variable behaviour: different individuals of a given species can respond idiosyncratically, producing different types of trap, or positioning traps in distinctive ways. One species – *Arthrobotrys oligospora* – behaves like a 'normal' decomposer in the presence of plenty of organic material and, if needs be, can produce nematode traps in its mycelium. It can also coil around the mycelium of other types of fungi, starving them, or develop specialised structures to penetrate and feed off plant roots. How it chooses between its many options remains unknown.[24]

How should we talk about fungal communication? In Italy, as we crowded around the hole in the muddy bank, peering in, I tried to imagine the scene from the truffle's point of view. In the excitement, Paride offered a lyrical interpretation. 'The truffle and its tree are like lovers, or husband and wife,' he crooned. 'If the threads are broken, there can be no going back. The bond is gone forever. The truffle was born from the root of the tree, defended by the wild rose.' He gestured to the brambles. 'It lay inside, protected by the thorns like Sleeping Beauty, waiting to be kissed by the dog.'

The prevailing scientific view is that it is a mistake to imagine that there is anything deliberate about most non-human interactions. Truffle fungi are not articulate. They don't speak. Like many of the animals and plants they depend on, truffle fungi react to their environment automatically, based on robotic routines that maximise their chance of survival. In stark contrast is the vivid experience of human life, where the quantity of a stimulus glides seamlessly into the quality of sensation; where stimuli are felt and arouse emotion; where we are affected.

I balanced on the muddy slope and suspended my nose over the pungent clod of fungus. No matter how hard I tried to reduce the truffle to an automaton, it kept springing to life in my mind.

When trying to understand the interactions of non-human organisms, it is easy to flip between these two perspectives: that of the inanimate behaviour of pre-programmed robots on the one hand, and that of rich, lived, human experience on the other. Framed as brainless organisms, lacking the basic apparatus required to have even a simple kind of 'experience', fungal interactions are no more than automatic responses to a series of biochemical triggers. Yet the mycelium of truffle fungi, like that of most fungal species, actively senses and responds to its surroundings in unpredictable ways. Their hyphae are chemically irritable, responsive, excitable. It is this ability to interpret the chemical emissions of others that allows fungi to negotiate a series of complex trading relationships with trees; to knead away at stores of nutrients in the soil; to have sex; to hunt; or to fend off attackers.

Anthropomorphism is usually thought of as an illusion that arises like a blister in soft human minds: untrained, undisciplined, unhardened. There are good reasons for this: when we humanise the world, we may prevent ourselves from understanding the lives of other organisms on their own terms. But are there things this stance might lead us to pass over – or forget to notice?[25]

The biologist Robin Wall Kimmerer, a member of the Citizen Potawatomi Nation of the Great Plains region of the United States, observes that the indigenous Potawatomi language is rich in verb forms that attribute aliveness to the more-than-human world. The word for hill, for example, is a verb: 'to be a hill'. Hills are always in the process of hilling, they are actively *being* hills. Equipped with this 'grammar of animacy', it is possible to talk about the life of other organisms without either reducing them to an 'it', or borrowing concepts traditionally reserved for humans. By contrast, in English, writes Kimmerer, there is no way to recognise the 'simple existence of another living being'. If you're not a human *subject*, by default you're an inanimate *object*: an 'it', a 'mere thing'. If you repurpose a human concept to help make sense of the life of a non-human organism, you've tumbled into the trap of anthropomorphism. Use 'it', and you've objectified the organism, and fallen into a different kind of trap.[26]

Biological realities are never black and white. Why should the stories and metaphors we use to make sense of the world – our investigative tools – be so? Might we be able to expand some of our concepts, such that speaking might not always require a mouth, hearing might not always require ears, and interpreting might not always require a nervous system? Are we able to do this without smothering other life forms with prejudice and innuendo?

Daniele wrapped up the truffle and carefully filled in the hole, pulling the clump of brambles back over the turned earth. Paride explained that it was to avoid disturbing the fungus's relationship with its tree's roots. Daniele said that it was to prevent other truffle hunters from following in our tracks. We strolled back through the field. The truffle's smell was less vivid by the time we arrived at the car, and more muted still by the time we got back to the weighing room. I wondered how faint it would be by the time it was grated onto a plate in Los Angeles.

*

Some months later, in the wooded hills outside Eugene, Oregon, I went out truffle hunting with Charles Lefevre and his Lagotto Romagnolo, Dante. Dante is what Lefevre calls a diversity dog. Production dogs – like Kika and Diavolo – are trained to find large amounts of a particular species; diversity dogs are trained to go after anything that smells interesting. This allows them to find species of truffle they have never smelled before. As a result Dante sometimes chases things that aren't truffles – pungent millipedes, for instance – but he has also unearthed four undescribed species of truffle. This is not so uncommon. Mike Castellano, a renowned truffle expert with a species named after him – he has described two new orders, more than two dozen new genera and some 200 new species of truffle – reports routinely discovering new species of truffle when collecting in California, a reminder of how much remains unknown.

As we ambled through the Douglas firs and sword ferns, Lefevre explained that humans have been cultivating truffles unintentionally for centuries. Truffles thrive in the disturbed environments that humans make. In Europe, truffle production plummeted during the twentieth century as the truffle-growing heartlands of managed woodland were either cleared for agriculture, or abandoned and left to grow into mature forests. Neither are good for truffle production. For Lefevre, the resurgence of trufficulture is exciting because it is a way to produce a cash crop from a forested landscape and divert private capital into environmental restoration. To grow truffles, you have to grow trees. You have to acknowledge that the soil is full of life. You can't cultivate truffles without thinking at the level of the ecosystem.

Dante zigzagged around, sniffing. Lefevre told me about the theory that manna – the providential food that sustained the Israelites during their passage through the desert – was in fact the desert truffle, a delicacy that erupts without warning from arid ground across much of the Middle East. He told me about his unsuccessful attempts to cultivate the evasive white truffle, and how little we understand about its relationship to its host trees. I thought of the many ways in which fungi respond to changing environments and find new ways to live alongside the plants and animals on which they depend.[27]

Back in a forest, hunting for truffles, I found myself once again searching for language to describe the lives of these remarkable organisms. Perfumers and wine tasters use metaphor to articulate differences in aromas. A chemical becomes 'cut grass', 'sweaty mango', 'grapefruit and hot horses'. Without these references, we would be unable to imagine it. Cis-3-hexenol smells like cut grass. Oxane smells like sweaty mango. Gardamide smells like grapefruit and hot horses. This is not to say oxane *is* sweaty mango, but if I were to pass you an open vial you'd almost certainly recognise the smell. Correlating human language with an odour involves judgement and prejudice. Our descriptions warp and deform the phenomena we describe, but sometimes this is the only way to talk about features of the world: to say what they are like, but are not. Might this also be the case when we talk about other organisms?[28]

Boil it down and there aren't many other options. Fungi may not have brains, but their many options entail decisions. Their fickle environments entail improvisation. Their trials entail errors. Whether in the homing response of hyphae within a mycelial network, the sexual attraction between two hyphae in separate mycelial networks, the vital fascination between a mycorrhizal hypha and a plant root, or the fatal attraction of a nematode to a fungal toxic droplet, fungi actively sense and interpret their worlds, even if we have no way of knowing what it is *like* for a hypha to sense or interpret. Perhaps it isn't so strange to think of fungi as articulating themselves using a chemical vocabulary, arranged and rearranged in such a way that it might be interpreted by other organisms, whether nematode, tree root, truffle dog or New York restaurateur. Sometimes – as with truffles – these molecules might translate into a chemical language we can, in our way, understand. The vast majority will always pass over our heads, or under our feet.

Dante started digging furiously. 'It looks like a truffle,' Lefevre said, reading the dog's body language, 'but it's deep.' I asked whether he ever worried about Dante's nose or feet getting hurt from all the frantic digging. 'Oh, he does keep injuring his pads,' Lefevre admitted. 'I keep meaning to get him some booties.' Dante snorted and scraped, but to no avail. 'I feel bad not rewarding him for his efforts when he's unsuccessful' – Lefevre crouched down and ruffled his curls – 'but I haven't found a treat that's worth more to him than a truffle. Truffles trump everything.' He grinned up at me. 'For Dante, God lives just below the surface of the soil.'

2

Living Labyrinths

I am so happy in the silky damp dark of the labyrinth and there is no thread.

Hélène Cixous

Imagine that you could pass through two doors at once. It's inconceivable, yet fungi do it all the time. When faced with a forked path, fungal hyphae don't have to choose one or the other. They can branch and take both routes.

One can confront hyphae with microscopic labyrinths and watch how they nose their way around. If obstructed, they branch. After diverting themselves around an obstacle, the hyphal tips recover the original direction of their growth. They soon find the shortest path to the exit, just as my friend's puzzle-solving slime moulds were able to find the quickest way out of the IKEA maze. If one follows the growing tips as they explore, it does something peculiar to one's mind. One tip becomes two, becomes four, becomes eight – yet all remain connected in one mycelial network. Is this organism singular or plural, I find myself wondering, before I'm forced to admit that it is somehow, improbably, *both*.[1]

Watching a hypha explore a single clinical maze is bewildering, but scale up: imagine millions of hyphal tips, each navigating a different maze at the same time within a tablespoon of soil. Scale up again: imagine billions of hyphal tips exploring a patch of forest the size of a football field.

Mycelium is ecological connective tissue, the living seam by which much of the world is stitched into relation. In school classrooms children are shown anatomical charts, each depicting different aspects of the human body. One chart reveals the body as a skeleton, another the body as a network of blood vessels, another the nerves, another the muscles. If we made equivalent sets of diagrams to portray ecosystems, one of the layers would show the fungal mycelium that runs through them. We would see sprawling, interlaced webs strung through the soil, through sulphurous sediments hundreds of metres below the surface of the ocean, along coral reefs, through plant and animal bodies both alive and dead, in rubbish dumps, carpets, floorboards, old books in libraries, specks of house dust and in canvases of old master paintings hanging in museums. According to some estimates, if one teased apart the mycelium found in a gram of soil – about a teaspoon – and laid it end to end, it could stretch anywhere from a hundred metres to ten kilometres. In practice, it is impossible to measure the extent to which mycelium perfuses the Earth's structures, systems and inhabitants – its weave is too tight. Mycelium is a way of life that challenges our animal imaginations.[2]

Lynne Boddy, a professor of microbial ecology at Cardiff University, has spent decades studying the foraging behaviour of mycelium. Her elegant studies illustrate the problems that mycelial networks are able to solve. In one experiment, Boddy allowed a wood-rotting fungus to grow within a block of wood. She then placed the block on a dish. Mycelium spread radially outwards from the block in all directions, forming a fuzzy white circle. Eventually the growing network encountered a new block of wood. Only a small part of the fungus touched the wood, but the behaviour of the entire network changed. The mycelium stopped exploring in all directions. It withdrew the exploratory parts of its network, and thickened the connection with the newly discovered block. After a few days, the network was unrecognisable. It had completely remodelled itself.[3]

She then repeated the experiment, but with a twist. She let the fungus grow out from the original block and discover the new block of wood. However, this time, before the network had time to remodel itself, she removed the original block of wood from the dish, stripped away all of the hyphae growing out of it, and placed it onto a fresh dish. The fungus grew out from the original block in the direction of the newly discovered block. The mycelium appeared to possess a directional memory, although the basis of this memory remains unclear.[4]

Boddy has a no-nonsense manner and talks with quiet amazement about what these fungi are able to do. Their behaviour is a bit like that of slime moulds, and she has tested them in similar ways. However, rather than modelling the Tokyo underground network, Boddy encouraged mycelium to work out the most efficient routes between the cities of Great Britain. She arranged soil into the shape of the British land mass, and marked cities using blocks of wood colonised with a fungus (the sulphur tuft, or *Hypholoma fasciculare*). The size of the wood blocks was proportional to the population of the cities they represented. 'The fungi grew out from the "cities" and made the motorway network,' Boddy recounted. 'You could see the M5, M4, M1, M6. I thought it was quite fun.'

One way to think about mycelial networks is as swarms of hyphal tips. Insects form swarms. A murmuration of starlings is a swarm, as is a school of sardines. Swarms are patterns of collective behaviour. Without a leader or command centre, a swarm of ants can work out the shortest route to a source of food. A swarm of termites can build giant mounds with sophisticated architectural features. However, mycelium quickly outgrows the swarm analogy because all the hyphal tips in a network are connected to one another. A termite mound is made up of units of termites. A hyphal tip would be the closest we could come to defining the unit of a mycelial 'swarm', although one can't dismantle a mycelial network hypha by hypha once it has grown, as we could pick apart a swarm of termites. Mycelium is conceptually slippery. From the point of view of the network, mycelium is a single interconnected entity. From the point of view of a hyphal tip, mycelium is a multitude.[5]

'I think there's lots we can learn, as humans, from mycelium,' Boddy reflected. 'You can't just go and close a road to see how the traffic flow changes, but you can sever a connection in a mycelial network.' Researchers have begun to use network-based organisms like slime moulds and fungi to solve human problems. The researchers who modelled the Tokyo train network using slime moulds are working to incorporate slime mould behaviour into the design of urban transport networks. Researchers at the Unconventional Computing Lab at the University of the West of England have used slime moulds to calculate efficient fire evacuation routes from buildings. Some are applying the strategies that fungi and slime moulds use to navigate labyrinths to solve mathematical problems or to program robots.[6]

Solving mazes and complex routing problems are non-trivial exercises. This is why mazes have long been used to assess the problem-solving abilities of many organisms, from octopuses to bees to humans. Nonetheless, mycelial fungi are maze-dwellers, and solving spatial and geometrical problems is what they have evolved to do. How best to distribute their bodies is a question fungi face on a moment-to-moment basis. By growing a dense network, mycelium can increase its capacity for transport, but dense networks aren't good for exploring over large distances. Sparse networks are better for foraging over large areas, but have fewer interconnections and so are more vulnerable to damage. How do fungi juggle this kind of trade-off while exploring a crowded rot-scape in search of food?[7]

Boddy's experiment with two blocks of wood illustrates a typical sequence of events. The mycelium starts in an exploratory mode, proliferating in all directions. Setting out to find water in a desert, we'd have to pick one direction to explore. Fungi can choose all possible routes at once. If the fungus discovers something to eat, it reinforces the links that connect it with the food and prunes back the links that don't lead anywhere. One can think of it in terms of natural selection. Mycelium overproduces links. Some turn out to be more competitive than others. These links are thickened. Less competitive links are withdrawn, leaving a few mainline highways. By growing in one direction while pulling back from another, mycelial networks can even migrate through a landscape. The Latin root of the word 'extravagant' means 'to wander outside or beyond'. It is a good word for mycelium, which ceaselessly wanders outside and beyond its limits, none of which are pre-set, as they are in most animal bodies. Mycelium is a body without a body plan.[8]

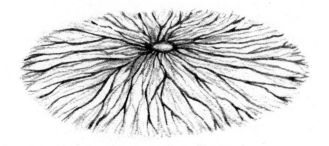

Mycelium exploring a flat surface

How does one part of a mycelial network 'know' what is happening in a distant part of the network? Mycelium sprawls, yet must somehow be able to stay in touch – with itself.

Stefan Olsson is a Swedish mycologist who has spent decades trying to understand how mycelial networks co-ordinate themselves and behave as integrated wholes. A number of years ago, he became interested in one of several species of fungus that produce bioluminescence, which causes their mushrooms and mycelium to glow in the dark and can help attract insects that disperse their spores. Coal miners in nineteenth-century England reported that bioluminescent fungi growing on wooden pit props were bright enough to 'see their hands by', and Benjamin Franklin proposed the use of the bioluminescent fungi known as 'foxfire' to illuminate the compass and depth gauge of the first submarine (the *Turtle* – developed in 1775 during the American Revolutionary War). The species Olsson had been studying was the bitter oyster, *Panellus stipticus*. 'You could read in the light of it when I grew it in jars,' he told me. 'It was like a little lamp standing on the shelf at home. My kids loved it.'[9]

To monitor the behaviour of *Panellus* mycelium, Olsson grew cultures in dishes in the lab, and placed two of them, glowing, in a perfectly dark box under constant conditions. He left them alone for a week with a camera sensitive enough to detect their bioluminescence taking pictures every few seconds. In the time-lapse video, two unconnected mycelial cultures grow outwards into the shape of irregular circles in their separate dishes, glowing more intensely in the middle than at their edges. After several days – about two minutes of video – there is a sudden shift. In one of the cultures, a wave of bioluminescence passes over the network from one edge to the other. A day later, a similar wave passes over the second culture. On mycelial time scales, it is high drama. In a matter of – mycelial – moments, each network flips into a different physiological state.[10]

'What the hell is that?' Olsson exclaimed to me. He jokes that left alone the fungus might have got bored, started playing, or become depressed. Although he left the cultures in the dark for several more weeks, the pulse never happened again. Years later, he still doesn't have a good explanation for what caused it. Nor for how the mycelium was able to co-ordinate its behaviour over such short timescales.[11]

Mycelial co-ordination is difficult to understand because there is no centre of control. If we cut off our head or stop our heart, we're finished. A mycelial network has no head and no brain. Fungi, like plants, are decentralised organisms. There are no operational centres, no capital cities, no seats of government. Control is dispersed: mycelial co-ordination takes place both everywhere at once and nowhere in particular. A fragment of mycelium can regenerate an entire network, meaning that a single mycelial individual – if you're brave enough to use that word – is potentially immortal.

Olsson was intrigued by the spontaneous waves of bioluminescence that he had recorded, and prepared another set of dishes for a follow-up experiment. He tried stabbing one side of a *Panellus* mycelium with the tip of a pipette. The wounded area lit up immediately. What confused him was that within ten minutes the light had spread a distance of nine centimetres across the whole network. This was far faster than a chemical signal could travel from one side to the other within the mycelium itself.

It occurred to Olsson that the wounded hyphae could have released a volatile chemical signal into the air that spread across the network in a gaseous cloud, thus avoiding the need to travel within the network. He tested this possibility by growing two genetically identical mycelia side by side. There were no direct connections between them, but they were close enough that chemicals drifting through the air would traverse the gap. Olsson stabbed one of the networks. The light propagated across the wounded network as it had before, but the signal did not spread to its neighbour. Some kind of rapid communication system had to be operating within the network itself. Olsson became increasingly preoccupied by the question of what this might be.

Mycelium is how fungi feed. Some organisms – such as plants that photosynthesise – make their own food. Some organisms, like most animals, find food in the world and put it inside their bodies, where it is digested and absorbed. Fungi have a differ- ent strategy. They digest the world where it is and then absorb it into their bodies. Their hyphae are long and branched, and only a single cell thick – between two and twenty microme- tres in diameter, over five times thinner than an average human hair. The more of their surroundings that hyphae can touch, the more they can consume. The difference between animals and fungi is simple: animals put food in their bodies, whereas fungi put their bodies in the food.[12]

However, the world is unpredictable. Most animals cope with uncertainty by moving. If food can be more easily found elsewhere, they move elsewhere. But to embed oneself in an irregular and unpredictable food supply as mycelium does, one must be able to shapeshift. Mycelium is a living, growing, opportunistic investigation – speculation in bodily form. This tendency is known as developmental 'indeterminism': no two mycelial networks are the same. What shape is mycelium? It's like asking what shape water is. We can only answer the question if we know where the mycelium happens to be growing. Compare this with humans, all of whom share a body plan and embark on similar developmental journeys. Short of an intervention, if we are born with two arms we will end up with two arms.

Mycelium decants itself into its surroundings, but its growth pattern isn't infinitely variable. Different fungal species form different kinds of mycelial network. Some species have thin hyphae, some thick. Some are picky about their food, others less so. Some grow into ephemeral puffs that don't range beyond their food source and could fit on a single speck of house dust. Other species form long-lived networks that roam over kilometres. Some tropical species don't forage for food at all. Instead, they behave like filter-feeding animals and grow nets out of thick strands of mycelium which they use to catch falling leaves.[13]

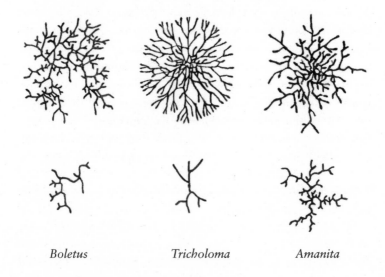

Boletus   Tricholoma   Amanita

Different mycelial types. Redrawn after Fries (1943)

No matter where fungi grow, they must be able to insinuate themselves within their source of food. To do so, they use pressure. In cases where mycelium has to break through particularly tough barriers, as disease-causing fungi do when infecting plants, they develop special penetrative hyphae which can reach pressures of fifty to eighty atmospheres and exert enough force to penetrate the tough plastics Mylar and Kevlar. One study estimated that if a hypha was as wide as a human hand, it would be able to lift an eight-tonne school bus.[14]

Most multicellular organisms grow by laying down new layers of cells. Cells divide to make more cells which then divide again. A liver is made by piling liver cells on top of liver cells. The same goes for a muscle or a carrot. Hyphae are different; they grow by getting longer. Under the right conditions, a hypha can prolong itself indefinitely.

At a molecular level, all cellular activity, whether fungal or not, is a blur of rapid activity. Even by these standards, hyphal tips are a commotion, busier than a court of 10,000 self-dribbling basketballs. The hyphae of some species grow so fast that one can watch them extend in real time. Hyphal tips must lay down new material as they advance. Small bladders filled with cellular building materials arrive at the tip from within, and fuse with it at a rate of up to 600 a second.[15]

In 1995, the artist Francis Alÿs walked around São Paulo carrying a can of blue paint with a hole punched in the bottom. Over many days, as he moved through the city, a continuous stream of paint dribbled onto the ground in a trail behind him. The line of blue paint made a map of his journey, a portrait of time. Alÿs' performance illustrates hyphal growth. Alÿs himself is the growing tip. The winding trail he leaves behind him is the body of the hypha. Growth happens at the tip; if one paused Alÿs as he walked around with his can of paint, the line would cease to grow. You can think of your life like this. The growing tip is the present moment – your lived experience of now – which gnaws into the future as it advances. The history of your life is the rest of the hypha, the blue lines that you've left in a tangled trail behind you. A mycelial network is a map of a fungus's recent history, and is a helpful reminder that all life forms are in fact *processes*, not *things*. The 'you' of five years ago was made from different stuff than the 'you' of today. Nature is an event that never stops. As William Bateson, who coined the word 'genetics', observed, 'We commonly think of animals and plants as matter, but they are really systems through which matter is continually passing.' When we see an organism, from a fungus to a pine tree, we catch a single moment in its continual development.[16]

Mycelium usually grows from hyphal tips, but not always. When hyphae felt together to make mushrooms, they rapidly inflate with water, which they must absorb from their surroundings – the reason why mushrooms tend to appear after rain. Mushroom growth can generate an explosive force. When a stinkhorn mushroom crunches through an asphalt road, it produces enough force to lift an object weighing 130 kilograms. In a popular fungal guidebook published in the 1860s, Mordecai Cooke reported that

> some years ago the [English] town of Basingstoke was paved; and not many months afterwards the pavement was observed to exhibit an unevenness which could not readily be accounted for. In a short time after, the mystery was explained, for some of the heaviest stones were completely lifted out of their beds by the growth of large toadstools beneath them. One of the stones measured twenty-two inches by twenty-one, and weighed eighty-three pounds.[17]

If I think about mycelial growth for more than a minute my mind starts to stretch.

In the mid 1980s, the American musicologist Louis Sarno recorded the music of the Aka people, living in the forests of the Central African Republic. One of these recordings is called 'Women Gathering Mushrooms'. As they wander around collecting mushrooms, their steps tracing the underground form of a mycelial network, the women sing amid the sounds of the animals in the forest. Each woman sings a different melody; each voice tells a different musical story. Many melodies intertwine without ceasing to be many. Voices flow around other voices, twisting into and beside one another.[18]

'Women Gathering Mushrooms' is an example of musical polyphony. Polyphony is singing more than one part, or telling more than one story, at the same time. Unlike the harmonies in a barbershop quartet, the voices of the women never weld into a unified front. No voice surrenders its individual identity. Nor does any one voice steal the show. There is no front woman, no soloist, no leader. If the recording was played to ten people and they were asked to sing the tune back, each would sing something different.[19]

Mycelium is polyphony in bodily form. Each of the women's voices is a hyphal tip, exploring a soundscape for itself. Although each is free to wander, their wanderings can't be seen as separate from the others. There is no main voice. There is no lead tune. There is no central planning. Nonetheless, a form emerges.

Whenever I listen to 'Women Gathering Mushrooms', my ears find their way into the music by choosing a single voice and riding with it, as if I were in the forest and could walk up to one of the women and stand next to her. To follow more than one line at a time is hard. It is like trying to listen to many conversations at once without flicking from one to another. Several streams of consciousness have to commingle in the mind. My attention has to become less focused and more distributed. I fail every time, but when I soften my hearing, something else happens. The many songs coalesce to make one song that doesn't exist in any one of the voices alone. It is an emergent song that I can't find by unravelling the music into its separate strands.

Mycelium is what happens when fungal hyphae – streams of embodiment rather than streams of consciousness – commingle.

However, as Alan Rayner, a mycologist specialising in mycelial development, reminded me, 'mycelium is not just amorphous cotton wool'. Hyphae can come together to form elaborate structures.

When you look at mushrooms, you're looking at fruit. Imagine bunches of grapes growing out of the ground in their place. Then imagine the vine that produced them, twisting and branching below the surface of the soil. Grapes and woody grape vines are made of different types of cell. Cut up a mushroom and you'll see that it is made of the same type of cell as mycelium: hyphae.

Hyphae grow into other structures besides mushrooms. Many species of fungus form hollow cables of hyphae known as 'cords' or 'rhizomorphs'. These range from slim filaments to strands several millimetres thick that can stretch for hundreds of metres. Given that individual hyphae are tubes, not threads – it is easy to forget about the fluid-filled space within the hyphae – cords and rhizomorphs are large pipes formed from

Mushrooms, like mycelium, are made of hyphae

many small tubes. They can conduct flow many times faster than through individual hyphae – nearly 1.5 metres per hour in one report – and allow mycelial networks to transport nutrients and water over large distances. Stefan Olsson told me about a forest in Sweden where he had observed a large *Armillaria* network that fruited over an area the size of two football fields. A small footbridge crossed a stream that flowed through the area. 'I started looking more closely at the bridge,' he remembered, 'and saw that the fungus had started to wind its cords under the bridge. It was actually crossing the stream using the bridge.' How fungi co-ordinate the growth of these structures remains a mystery.[20]

Cords and rhizomorphs are a good reminder that mycelial networks are transport networks. Lynne Boddy's mycelial road map is another good illustration. Mushroom growth is another: to push its way through asphalt, a mushroom must inflate with water. For this to happen, water must travel rapidly through the network from one place to another, and flow into a developing mushroom in a carefully directed pulse.

Over short distances, substances can be transported through mycelial networks on a network of microtubules – dynamic filaments of protein that behave like a cross between scaffolding and escalators. Transport using microtubule 'motors' is energetically costly, however, and over larger distances the contents of hyphae travel on a river of cellular fluid. Both approaches allow rapid transport across mycelial networks. Efficient transport allows different parts of a mycelial network to engage in different activities. When the English country house Haddon Hall was being renovated, a fruiting body of the dry rot fungus *Serpula* was found in a disused stone oven. Its mycelial connections wound back through 8 metres of stonework to a rotting floor elsewhere in the building. The floor was where it fed, the oven was where it fruited.[21]

The best way to appreciate flow within mycelium is to watch its contents shuttle around the network. In 2013, a group of researchers at the University of California, Los Angeles treated mycelium so that they could visualise cellular structures moving within the hyphae. Their videos show hordes of nuclei surging along. In some hyphae they travel faster than in others, in some they travel in different directions. Sometimes traffic jams form and nuclear traffic is rerouted on hyphal slip channels. Streams of nuclei merge with each other. Rhythmic pulses of nuclei – 'nuclear comets' – rush along, branching at junctions and darting down side ducts. It is a scene of 'nuclear anarchy', as one of the researchers wryly observed.[22]

Flow helps to explain how traffic circulates within a mycelial network, but it can't explain why fungi might grow in one direction rather than another. Hyphae are sensitive to stimuli, and at any one moment are confronted with a world of possibilities. Rather than extending in a straight line at a constant rate, hyphae steer themselves towards appealing prospects, and away from unappealing ones. How?

In the 1950s, the Nobel Prize-winning biophysicist Max Delbrück became interested in sensory behaviour. He chose as his model organism the fungus *Phycomyces blakesleeanus*. Delbrück was fascinated by *Phycomyces*' remarkable perceptual abilities. Its fruiting structures – essentially giant vertical hyphae – have a sensitivity to light similar to that of the human eye, and adapt to bright or low light as our eyes do. They can detect light at levels as low as that provided by a single star, and only become dazzled when exposed to full sunlight on a bright day. To provoke a response in a plant, one would have to expose it to light levels hundreds of times higher.[23]

At the end of his career, Delbrück wrote that he was still convinced that *Phycomyces* was 'the most intelligent' of the simpler multicellular organisms. Besides its exquisite sensitivity to touch – *Phycomyces* preferentially grows into wind at speeds as low as one centimetre per second, or 0.036 kilometres per hour – *Phycomyces* is able to detect the presence of nearby objects, a phenomenon known as the 'avoidance response'. Despite decades of painstaking investigation, the avoidance response remains an enigma. Objects within a few millimetres cause the fruiting body of *Phycomyces* to bend away without ever making contact. Regardless of the object – opaque or transparent, smooth or rough – *Phycomyces* starts to bend away after about two minutes. Electrostatic fields, humidity, mechanical cues and temperature have all been ruled out. Some hypothesise that *Phycomyces* uses a volatile chemical signal that deflects around the obstacle with tiny air currents, but this is far from proven.[24]

Although *Phycomyces* is an unusually sensitive species, most fungi are able to detect and respond to light (its direction, intensity or colour) temperature, moisture, nutrients, toxins and electrical fields. Like plants, fungi can 'see' colour across the spectrum using receptors sensitive to blue light and red light – unlike plants, fungi also have opsins, the light-sensitive pigments present in the rods and cones of animal eyes. Hyphae can also sense the texture of surfaces; one study reports that young hyphae of the bean rust fungus can detect grooves half a micrometre deep in artificial surfaces, three times shallower than the gap between the laser tracks on a CD. When hyphae felt together to make mushrooms, they acquire an acute sensitivity to gravity. And as we've seen, fungi maintain countless channels of chemical communication with other organisms and with themselves: when they fuse or have sex, hyphae distinguish 'self' from 'other', and between different kinds of 'other'.[25]

Fungal lives are lived in a flood of sensory information. And somehow, hyphae – piloted by their tips – are able to *integrate* these many data streams and determine a suitable trajectory for growth. Humans, like most animals, use brains to integrate sensory data and decide on the best course of action. Accordingly, we tend to look for particular places where integration might take place. We like a *where*, but with plants and fungi, asking 'where' only gets us so far. There are different parts of a mycelial network or a plant, but they aren't unique. There are many of everything. How, then, do sensory data streams come together within a mycelial network? How do brainless organisms link perception with action?

Plant scientists have wrestled with these questions for more than a century. In 1880, Charles Darwin and his son Francis published a book called *The Power of Movement in Plants*. In the final paragraph they suggest that, since root tips determine the trajectory for growth, it must be at the root tips that signals from different parts of the organism are integrated. Root tips, the Darwins write, act 'like the brain of one of the lower animals ... receiving impressions from the sense-organs, and directing the several movements'. The Darwins' conjecture has come to be known as the 'root-brain' hypothesis and is controversial, to put it mildly. This is not because anyone disputes their observations: it is clear that root tips do direct the movement of roots, just as growing tips direct the movement of shoots above ground. What divides plant scientists is their use of the word 'brain'. For some, it is a proposition that can draw us towards a richer understanding of plant life. For others, it is preposterous to suggest that plants have anything even like a brain.[26]

In some sense, the word 'brain' is a distraction. The Darwins' main point is that growing tips – which pilot roots and shoots – must be the place where information comes together to link perception and action, and determine a suitable course for growth. The same applies to fungal hyphae. Hyphal tips are the parts of the mycelium that grow, change direction, branch and fuse. They are the part of the mycelium that do the most. And they are numerous. A given mycelial network might have anywhere between hundreds and billions of hyphal tips, all integrating and processing information on a massively parallel basis.[27]

Hyphal tips may be the places where data streams come together to determine the speed and direction of growth, but how do tips in one part of the network 'know' what tips are doing in other,

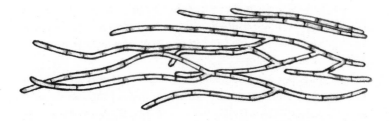

more distant parts of the network? We stumble back into Stefan Olsson's conundrum. His bioluminescent *Panellus* cultures were able to co-ordinate their behaviour over time periods too short to be caused by chemicals moving from A to B through the network. The mycelium of some fungal species grows into 'fairy rings' that stretch across hundreds of metres, reach hundreds of years in age, and then somehow produce a circle of mushrooms in a synchronised flush. In Lynne Boddy's experiments with foraging mycelium, only one part of the network discovered the new block of wood, but the behaviour of the entire mycelium changed, and changed rapidly. How are mycelial networks able to communicate with themselves? How does information travel across mycelial networks so quickly?[28]

There are a number of possibilities. Some researchers suggest that mycelial networks might transmit developmental cues using changes in pressure or flow – because mycelium is a continuous hydraulic network like a car's braking system, a sudden change in pressure in one part could, in principle, be felt rapidly everywhere else. Some have observed that metabolic activity – such as the accumulation and release of compounds within hyphal compartments – can take place in regular pulses that could help to synchronise behaviour across a network. Olsson, for his part, turned his attention to one of the few other options that remained: electricity.[29]

It has long been known that animals use electrical impulses, or 'action potentials', to communicate between different parts of their bodies. Neurones – the long, electrically excitable nerve cells that co-ordinate animal behaviour – have their own field of study: neuroscience. Although electrical signalling is normally thought of as an animal talent, animals aren't alone in producing action potentials. Plants and algae produce them, and it has been known since the 1970s that some types of fungi do also. Bacteria, too, are electrically excitable. 'Cable bacteria' form long electrically conductive filaments, known as nanowires. And it has been known since 2015 that bacterial colonies can co-ordinate their activity using action potential-like waves of electrical activity. Nonetheless, few mycologists imagined that it could play an important role in fungal lives. [30]

In the mid 1990s in Olsson's department at Lund University in Sweden, there was a research group working on insect neurobiology. In their experiments, they measured the activity of neurones by inserting fine glass microelectrodes into moth brains. Olsson approached them and asked if he could use their rig to ask a simple question: what would happen if he replaced the moth brains with fungal mycelium? The neuroscientists were intrigued. In principle, fungal hyphae should be well adapted to conduct electrical impulses. They are coated with proteins that insulate them, which would allow waves of electrical activity to travel long distances without dissipating – animal nerve cells have an analogous insulating sheath. Moreover, the cells in a mycelium are continuous with one another, possibly allowing impulses initiated in one part of the network to reach another part without interruption.

Olsson chose the species of fungus carefully. He surmised that if electrical communication systems did exist in fungi, they would be easier to detect in species with a greater need for communication over long distances. Just to be safe, he chose a honey fungus, or *Armillaria* – the species that forms the record-holding mycelial networks that stretch over kilometres and reach thousands of years in age.

When Olsson inserted the microelectrodes into *Armillaria's* hyphal strands, he detected regular action potential-like impulses, firing at a rate very close to that of animals' sensory neurons – around four impulses per second, which travelled along hyphae at a speed of at least half a millimetre per second, some ten times faster than the fastest rate of fluid flow measured in a fungal hypha. This caught his attention, but in itself it didn't suggest that impulses formed the basis of a rapid signalling system. Electrical activity can only play a role in fungal communication if it is sensitive to stimulation. Olsson decided to measure the response of the fungus to blocks of wood, which is food for this species.[31]

Olsson set up the rig and placed a block of wood onto the mycelium several centimetres from the electrodes. What he found was extraordinary. When the wood came into contact with the mycelium, the firing rate of the impulses doubled. When he removed the block of wood, the firing rate returned to normal. To make sure that the fungi weren't responding to the weight of the wooden block, he placed an inedible plastic block of the same size and weight onto the mycelium. The fungus didn't respond.

Olsson went on to test a range of other species of fungus, including a mycorrhizal fungus growing on the root system of a plant, *Pleurotus* (or oyster mushroom mycelium), and *Serpula* (the dry rot found fruiting in the oven at Haddon Hall). They all generated action potential-like impulses and were sensitive to a range of different stimuli. Olsson hypothesised that electrical signalling was a realistic way for a wide variety of fungi to send messages between different parts of themselves, messages that conveyed information about 'food sources, injury, local conditions within the fungus, or the presence of other individuals around it'.[32]

Many of the neurobiologists Olsson was working with became excited that mycelial networks could be behaving like brains. 'It was the first reaction from all the insect people,' Olsson recalled. 'They were thinking of these big mycelial networks in the forest sending electric signals around themselves. They imagined that maybe they were just big brains lying there.' I admit that I hadn't been able to ignore the superficial resemblance either. Olsson's findings suggested that mycelium might form fantastically complex networks of electrically excitable cells. Brains, too, are fantastically complex networks of electrically excitable cells.

'I don't think they're brains,' Olsson explained to me. 'I had to hold back the brain concept. As soon as one says it, people start thinking of brains like ours where we have language, and process thoughts to make decisions.' His caution is well placed. 'Brain' is a trigger word, burdened with concepts that spend most of their time in the animal world. 'When we say brain', Olsson continued, 'all associations are with animal brains.' Besides, as he pointed out, brains behave like brains because of the way they're built. The architecture of animal brains is very different from that of fungal networks. In animal brains, neurones connect with other neurones at junctions called synapses. At synapses, signals can combine with other signals. Neurotransmitter molecules pass across synapses, and allow different neurones to behave in different ways – some excite other neurones, some inhibit them. Mycelial networks don't share any of these features.

But if fungi did use waves of electrical activity to transmit signals around a network, wouldn't we think of mycelium as at least a brain-*like* phenomenon? In Olsson's view, there could be other ways to regulate electrical impulses in mycelial networks to create 'brain-like circuits, gates and oscillators'. In some fungi, hyphae are divided into compartments by pores, which can be sensitively regulated. Opening or closing a pore changes the strength of the signal that passes from one compartment to another, whether chemical, pressure or electrical. If sudden changes in the electrical charge within a hyphal compartment could open or close a pore, Olsson mused, a burst of impulses could change the way subsequent signals passed along the hypha, and form a simple learning loop. What's more, hyphae branch. If two impulses converged on one spot, they would both influence pore conductivity, integrating signals from different branches. 'You do not need much knowledge of how computers work to realise that such systems can create decision gates,' Olsson told me. 'If you combine these systems in a flexible and adaptable network we have the possibility for "a brain" that could learn and remember.' He held the word 'brain' at a safe distance, clamped in the forceps of inverted commas to emphasise that a metaphor was in play.[33]

That fungi could use electrical signalling as a basis for rapid communication has not been lost on Andrew Adamatzky, the director of the Unconventional Computing Lab. In 2018, he inserted electrodes into whole oyster mushrooms sprouting in clusters from blocks of mycelium and detected spontaneous waves of electrical activity. When he held a flame up to a mushroom, different mushrooms within the cluster responded with a sharp electrical spike. Shortly afterwards, he published a paper called 'Towards fungal computer'. In it, he proposed that mycelial networks 'compute' information encoded in spikes of electrical activity. If we knew how a mycelial network would respond to a given stimulus, Adamatzky argues, we could treat it like a living circuit board. By stimulating the mycelium, for example using a flame or a chemical, we could input data into the fungal computer.[34]

A fungal computer may sound fantastical, but 'biocomputing' is a fast-growing field. Adamatzky has spent years developing ways to use slime moulds as sensors and computers. These prototype biocomputers use slime moulds to solve a range of geometrical problems. The slime mould networks can be modified – for instance, by cutting a connection – to alter the set of 'logical functions' implemented by the network. Adamatzky's idea of a 'fungal computer' is just an application of slime mould computing to another type of network-based organism.[35]

As Adamatzky observes, the mycelial networks of some species of fungus are more convenient for computing than slime moulds. They form longer-lived networks and don't morph into new shapes quite so quickly. They are also larger, with more junctions between hyphae. It is at these junctions – what Olsson described as 'decision gates', and what Adamatzky describes as 'elementary processors' – that signals from different branches of the network would interact and combine. Adamatzky estimates that a network of honey fungus stretching over fifteen hectares would have nearly a trillion such processing units.

For Adamatzky, the point of fungal computers is not to replace silicon chips. Fungal reactions are too slow for that. Rather, he thinks humans could use mycelium growing in an ecosystem as a 'large-scale environmental sensor'. Fungal networks, he reasons, are monitoring a large number of data streams as part of their everyday existence. If we could plug into mycelial networks and interpret the signals they use to process information, we could learn more about what was happening in an ecosystem. Fungi could report changes in soil quality, water purity, pollution or any other features of the environment that they are sensitive to.[36]

We're some way off. Computing with living network-based organisms is in its infancy and many questions remain unanswered. Olsson and Adamatzky have shown that mycelium can be electrically sensitive, but they haven't shown that electrical impulses can link a stimulus to a response. It is as if you had stuck a pin in your toe, detected the nerve impulse that travelled through your body, but hadn't been able to measure your reaction to the pain.[37]

This is a challenge for the future. In the twenty-three years between Olsson's study on mycelium and Adamatzky's study on oyster mushrooms, no further research was conducted on electrical signalling in fungi. If he had the resources to pursue this line of enquiry, Olsson told me that he would try to demonstrate a clear physiological response to changes in electrical activity, and decode the patterns of electrical impulses. His dream is to 'hook up a fungus with a computer and communicate with it', to use electrical signals to get the fungus to change its behaviour. 'All sorts of weird and wonderful experiments could be done if this turns out to be right.'[38]

These studies raise a storm of questions. Are network-based life forms like fungi or slime moulds capable of a form of cognition? Can we think of their behaviour as intelligent? If other organisms' intelligence didn't look like ours then how might it appear? Would we even notice it?

Among biologists, opinion is divided. Traditionally, intelligence and cognition have been defined in human terms as something that requires at least a brain and, more usually, a mind. Cognitive science emerged from the study of humans and so naturally placed the human mind at the centre of its enquiry. Without a mind, the classical examples of cognitive processes – language, logic, reasoning, recognising oneself in a mirror – seem impossible. All require high-level mental functioning. But how we define intelligence and cognition is a question of taste. For many, the brain-centric view is too limited. The idea that a neat line can be drawn that separates non-humans from humans with 'real minds' and 'real comprehension' has been curtly dismissed by the philosopher Daniel Dennett as an 'archaic myth'. Brains didn't evolve their tricks from scratch, and many of their characteristics reflect more ancient processes that existed long before recognisable brains arose.[39]

A number of contemporary biologists and philosophers take a pragmatic perspective. The Latin root of the word 'intelligence' means 'to choose between'. Many types of brainless organism – plants, fungi, and slime moulds included – respond to their environments in flexible ways, solve problems and make decisions between alternative courses of action. Complex information-processing is evidently not restricted to the inner workings of brains. Some use the term 'swarm intelligence' to describe the problem-solving behaviour of brainless systems. Others suggest that the behaviour of these network-based life forms can be thought of as arising from 'minimal' or 'basal' cognition, and argue that the question we should ask is not whether an organism has cognition or not. Rather, we should assess the *degree* to which an organism might be cognisant. In all these views, intelligent behaviours can arise without brains. A dynamic and responsive network is all that's needed.[40]

The brain has long been thought of as a dynamic network. In 1940, the Nobel Prize-winning neurobiologist Charles Sherrington described the human brain as 'an enchanted loom where millions of flashing shuttles weave a dissolving pattern'. Today, 'network neuroscience' is the name given to the discipline that attempts to understand how the brain's activity emerges from the interlinked activity of millions of neurones. A single neuronal circuit within one's brain can't give rise to intelligent behaviour, just as the behaviour of a single termite can't give rise to the intricate architecture of a termite mound. No single neuronal circuit 'knows' what's going on any more than a single termite 'knows' the structure of the mound, but large numbers of neurones can build a network from which surprising phenomena can emerge. In this view, complex behaviours – including minds and the nuanced textures of lived, conscious experience – arise out of complex networks of neurones flexibly remodelling themselves.[41]

Brains are just one such network, one way of processing information. Even in animals, there is a lot that can take place without them. Researchers at Tufts University have illustrated this in striking experiments using flatworms. Flatworms are well-studied model organisms because of their ability to regenerate. If the head of a flatworm is cut off, it sprouts another head, brain and all. Flatworms can also be trained. The researchers wondered whether, if they trained a flatworm to remember features of its environment and then cut off its head, it would retain the memory when it had grown a new head and brain. Remarkably, the answer is yes. The flatworms' memory appeared to reside in a part of their body outside the brain. These experiments suggest that even within the body of brain-dependent animals, the flexible networks that underpin complex behaviours need not be limited to a small region inside the head. There are other examples. Most nerves in octopuses are not found in the brain, for instance, but are distributed throughout their bodies. A large number are found in the tentacles, which can explore and taste their surroundings without involving the brain. Even when amputated, tentacles are able to reach and grasp.[42]

Many types of organism, then, have evolved flexible networks to help solve the problems that life presents. Mycelial organisms appear to be some of the first to do so. In 2017, researchers at the Swedish Royal Museum of Natural History published a report in which they describe fossilised mycelium, preserved in the fractures of ancient lava flows. The fossils show branching filaments that 'touch and entangle each other'. The 'tangled network' they form, the dimensions of the hyphae, the dimensions of spore-like structures and the pattern of its growth all closely resemble modern-day fungal mycelium. It is an extraordinary discovery because the fossils date from 2.4 billion years ago, more than a billion years before fungi were thought to have branched off the tree of life. There is no way to identify the organism with certainty, but whether or not it was a true fungus, it clearly had a mycelial habit. It is a finding that makes mycelium one of the earliest known gestures towards complex multicellular life, an original tangle, one of the first living networks. Remarkably unchanged, mycelium has persisted for more than half of the four billion years of life's history, through countless cataclysms and catastrophic global transformations.[43]

Barbara McClintock, who won the Nobel Prize for her work on maize genetics, described plants as extraordinary 'beyond our wildest expectations'. Not because they have found ways to do what humans can do, but because a life lived rooted to one spot has coaxed them to evolve countless 'ingenious mechanisms' to deal with challenges that animals might avoid by simply running away. We could say the same of fungi. Mycelium is one such ingenious solution, a brilliant reply to some of life's most basic challenges. Mycelial fungi don't do as we do, and contain flexible networks that ceaselessly remodel themselves. They *are* flexible networks that ceaselessly remodel themselves.[44]

McClintock emphasises how important it is to acquire 'a feeling for the organism', to develop the patience to 'hear what the material has to say to you'. When it comes to fungi, do we really stand a chance? Mycelial lives are so *other*, their possibilities so strange. But perhaps they aren't quite so remote as they seem at first glance. Many traditional cultures understand life to be an entangled whole. Today, the idea that all things are interconnected has been so well used that it has collapsed into a cliché. The idea of the 'web of life' underpins modern scientific conceptions of nature; the school of 'systems theory', which arose during the twentieth century, understands all systems – from traffic flows to governments to ecosystems – to be dynamic networks of interaction; the field of 'artificial intelligence' solves problems using artificial neural networks; many aspects of human life are continuous with the digital networks of the Internet; network neuroscience invites us to understand *ourselves* as dynamic networks. Like a well-exercised muscle, 'network' has hypertrophied into a master concept. It is hard to think of a subject that networks aren't used to make sense of.[45]

Yet we still struggle to make sense of mycelium. I asked Lynne Boddy what aspects of mycelial lives remain most mysterious. 'Ah … that's a good question.' She faltered. 'I really don't know. There are just *so many things*. How do mycelial fungi work *as networks*? How do they sense their environment? How do they send messages back to other parts of themselves? How are those signals then integrated? These are all huge questions which hardly anyone seems to be thinking about. Yet understanding these things is crucial to understanding how fungi do almost everything that they do. We have techniques to do this work, but who is looking at basic fungal biology? Not many people.

I think it's a very worrying situation. We haven't put together many of the things we've found into an overall understanding.' She laughed. 'The field is ripe for picking! But I don't think there are many people out there doing the picking.'

In 1845, Alexander von Humboldt observed that 'Each step that we make in the more intimate knowledge of nature leads us to the entrance of new labyrinths.' Polyphonic songs like 'Women Gathering Mushrooms' emerge from the entangling of voices; mycelium emerges from the entangling of hyphae. A sophisticated understanding of mycelium is yet to emerge. We are standing at the entrance to one of the oldest of life's labyrinths.[46]

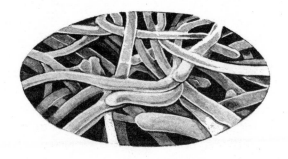

3

The Intimacy of Strangers

The problem was that we did not know whom we meant when we said 'we.'

Adrienne Rich

On 18 June 2016, the descent module of a Soyuz spacecraft landed on a bleak steppe in Kazakhstan. Three people were pulled safely from the scorched capsule following a stint at the International Space Station (ISS). The astronauts weren't alone as they plummeted to Earth. Under their seats were hundreds of living organisms packed tightly in a box.

Among the samples were several species of lichen that had been sent into space for one and a half years as part of the Biology and Mars Experiment. BIOMEX is an international consortium of astrobiologists who use trays mounted on the outside of the ISS – a piece of apparatus known as the EXPOSE facility – to incubate biological specimens in extraterrestrial conditions. 'Let's hope they have a safe return,' Natuschka Lee, one of the BIOMEX lichen team, remarked to me a few days before the landing was scheduled. I wasn't sure who she meant by 'they', but soon afterwards Lee got in touch to say that all was well. She had received an email from a lead researcher at the German Aerospace Center in Berlin, and read out the subject line, relieved: 'EXPOSE trays back on Earth ...' 'Soon', Lee smiled, 'we will have our samples back.'[1]

A number of organisms with extreme tolerances have been sent into orbit, from bacterial spores to free-living algae, to rock-dwelling fungi, to tardigrades – microscopic animals known as 'water bears'. Some can survive if shielded from the damaging effects of solar radiation. But few, apart from a handful of lichen species, are able to survive in full space conditions, drenched in unfiltered cosmic rays. So remarkable are these lichens' abilities that they have become model life forms for astrobiological research, ideal organisms 'to discern', as one researcher writes, 'the limits and limitations of terrestrial life'.[2]

It isn't the first time lichens have helped humans to fathom the limits of life as we know it. Lichens are living riddles. Since the nineteenth century, they have provoked fierce debate about what constitutes an autonomous individual. The closer we get to lichens, the stranger they seem. To this day, lichens confuse our concept of identity and force us to question where one organism stops and another begins.

In his lavishly illustrated book *Art Forms of Nature* (1904), the biologist and artist Ernst Haeckel vividly portrays a variety of lichen forms. His lichens sprout and layer deliriously. Veined ridges give way to smooth bubbles; stalks elaborate into prongs and dishes. Rugged coastlines meet unearthly pavilions, their forms lined with nooks and crannies. It was Haeckel who, in 1866, had coined the word 'ecology'. Ecology describes the study of the relationships between organisms and their environments: both the places where they live and the thicket of relationships that sustain them. Inspired by the work of Alexander von Humboldt, the study of ecology emerged from the idea that nature is an interconnected whole, 'a system of active forces'. Organisms could not be understood in isolation.[3]

Three years later, in 1869, the Swiss botanist Simon Schwendener published a paper advancing the 'dual hypothesis of lichens'. In it, he presented the radical notion that lichens were not a single organism, as had long been assumed. Instead, he argued that they were composed of two quite different entities: a fungus and an alga. Schwendener proposed that the lichen fungus (known today as the mycobiont) offered physical protection and acquired nutrients for itself and for the algal cells. The algal partner (known today as the photobiont, a role sometimes played by photosynthetic bacteria) harvested light and carbon dioxide to make sugars that provided energy. In Schwendener's view, the fungal partners were 'parasites, although with the wisdom of statesmen'. The algal partners were 'its slaves ... which it has sought out ... and forced into its service'. Together they grew into the visible body of the lichen. In their relationship, both partners were able to make a life in places where neither could survive alone.[4]

Schwendener's suggestion was vehemently opposed by his fellow lichenologists. The idea that two different species could come together in the building of a new organism with its own separate identity was shocking to many. 'A useful and invigorating parasitism?' one contemporary snorted. 'Who ever before heard of such a thing?' Others dismissed it as a 'sensational romance', an 'unnatural union between a captive Algal damsel and a tyrant Fungal master'. Some were more moderate. 'You see', wrote the English mycologist Beatrix Potter, best known for her children's books, 'we do not believe in Schwendener's theory.'[5]

Most worrisome for taxonomists – working hard to order life into neat lines of descent – was the prospect that a single organism could contain two separate lineages. Following Charles Darwin's theory of evolution by natural selection, first published in 1859, species were understood to arise by *diverging* from one another. Their evolutionary lineages forked, like the branches of a tree. The trunk of the tree forked into branches, which forked into smaller branches, which forked into twigs. Species were the leaves on the twigs of the tree of life. However, the dual hypothesis suggested that lichens were bodies composed of organisms with quite different origins. Within lichens, branches of the tree of life that had been diverging for hundreds of millions of years were doing something entirely unexpected: *converging.*[6]

Over the following decades a growing number of biologists adopted the dual hypothesis, but many disagreed with Schwendener's portrayal of the relationship. These were not sentimental concerns: Schwendener's choice of metaphor obstructed the larger questions raised by the dual hypothesis. In 1877, the German botanist Albert Frank coined the word 'symbiosis' to describe the living together of fungal and algal partners. In his study of lichens, it had become clear to him that a new word was required, one that didn't prejudice the relationship it described. Shortly afterwards, the biologist Heinrich Anton de Bary adopted Frank's term and generalised it to refer to the full spectrum of interactions between any type of organism, stretching from parasitism at one pole to mutually beneficial relationships at the other.[7]

Scientists made a number of major new symbiotic claims in the years that followed, including startling suggestions from Frank that fungi might help plants obtain nutrients from the soil (1885). All cited the dual hypothesis of lichens in support of their ideas. When algae were found living inside corals, sponges and green sea slugs, they were described by one researcher as 'animal lichens'. Several years later, when viruses were first observed within bacteria, their discoverer described them as 'microlichens'.[8]

Lichens, in other words, quickly grew into a biological principle. They were a gateway organism to the idea of symbiosis, an idea that ran against the prevailing currents in evolutionary thought in the late nineteenth and early twentieth centuries, best summed up in Thomas Henry Huxley's portrayal of life as a 'gladiator's show ... whereby the strongest, the swiftest and the cunningest live to fight another day'. In the wake of the dual hypothesis, evolution could no longer be thought of solely in terms of competition and conflict. Lichens had become a type case of inter-kingdom collaboration.[9]

Lichen: *Niebla*

Lichens encrust as much as 8 per cent of the planet's surface, an area larger than that covered by tropical rainforests. They clad rocks, trees, roofs, fences, cliffs and the surface of deserts. Some are a drab camouflage. Some are lime green or electric yellow. Some look like stains, others like small shrubs, others like antlers. Some leather and droop like bat wings, others, as the poet Brenda Hillman writes, are 'hung in hashtags'. Some live on beetles, whose lives depend on the camouflage the lichens provide. Untethered lichens – known as 'vagrants' or 'erratics' – blow around and don't live their lives *on* anything in particular. Against the 'plain story' of their surroundings, observes Kerry Knudsen, the curator of lichens at the herbarium at the University of California, Riverside, lichens 'look like fairy tales'.[10]

I have been most captivated by lichens on the islands off the coast of British Columbia, on the west coast of Canada. Seen from above, the coastline frays into the ocean. There is no hard edge. The land comes undone gradually into inlets and sounds, and then into channels and passages. Hundreds of islands scatter off the coast. Some are no bigger than a whale; the largest, Vancouver Island, is half the length of Britain. Most of the islands are solid granitic rock, the tops of submarine hills and valleys worn smooth by glaciers.

For a few days every year, I and a handful of friends pile onto a twenty-eight-foot sailing boat and set off around the islands. The boat, the *Caper,* has a dark green hull, no keel and one red sail. Making our way from the *Caper* onto the land is tricky. We paddle in an unstable dinghy with oars that slip their rowlocks at every other stroke. Pulling up to the shore is an art form. Waves shrug the dinghy onto the rocks and tug it away from our feet as we clamber out. But once on shore, the lichens begin. I've spent hours absorbed in the worlds they make – islands of life in a sea of rock. The names used to describe lichens sound like afflictions, words that get stuck in your teeth: crustose (crusty), foliose (leafy), squamulose (scaly), leprose (dusty), fruticose (branched). Fruticose lichens drape and tuft; crustose and squamulose lichens creep and seep; foliose lichens layer and flake. Some prefer to live on east-facing surfaces, some on the west. Some choose to live on exposed ledges, others in damp grooves. Some wage slow wars, repelling or disrupting their neighbours. Some inhabit the surfaces left exposed when other lichens have died and flaked off. They come to resemble the archipelagos and continents of an unfamiliar atlas, which is how *Rhizocarpon geographicum*, or the map lichen, got its name. The oldest surfaces are pitted by centuries of lichenous life and death.

Lichen's fondness for rock has changed the face of the planet, and continues to do so, sometimes literally. In 2006, the faces of the presidents carved into Mount Rushmore were pressure-hosed, removing more than sixty years of lichenous growth in the hope of extending the lifetime of the memorial. The presidents aren't alone. 'Every monument', writes the poet Drew Milne, 'has a lichen lining.' In 2019, the residents of Easter Island launched a campaign to scrub lichens off hundreds of monumental stone heads, or moai. Described by locals as 'leprosy', lichens are deforming the features of the statues, and softening the rock to a 'clay-like' consistency.[11]

Lichens mine minerals from rock in a two-fold process known as 'weathering'. First, they physically break up surfaces by the force of their growth. Second, they deploy an arsenal of powerful acids and mineral-binding compounds to digest the rock. Lichens' ability to weather makes them a geological force, yet they do more than dissolve the physical features of the world. When lichens die and decompose, they give rise to the first soils in new ecosystems. Lichens are how the inanimate mineral mass within rocks is able to cross over into the metabolic cycles of the living. A portion of the minerals in your body is likely to have passed through a lichen at some point. Whether on tombstones in a graveyard or encased within slabs of Antarctic granite, lichens are go-betweens that inhabit the boundary dividing life and non-life. Looking out from the *Caper* at the rocky Canadian coastline this becomes clear. Above the tideline, it is only after several metres of lichens and mosses that larger trees start to appear, rooted in crevices well beyond the water's reach where young soils have been able to form.[12]

Lichen: *Ramalina*

The question of what is and isn't an island is fundamental to the study of ecology and evolution. It is no less important for astrobiologists, including those on the BIOMEX team, many of whom wrestle with the question of 'panspermia', from the Greek *pan*, meaning all, and *sperma*, meaning seed. Panspermia deals with the question of whether planets, too, are islands, and whether life can travel through space between celestial bodies. It is an idea that has circulated since antiquity, although it didn't take on the form of a scientific hypothesis until the early twentieth century. Some advocates argue that life itself arrived from other planets. Some propose instead that life evolved on Earth *and* elsewhere, and periods of dramatic evolutionary novelty on Earth were triggered by the arrival of fragments of life from space. Others argue instead for a 'soft panspermia', where life itself evolved on Earth, but the chemical building blocks required for life arrived from space. There are many hypotheses as to how interplanetary transport might take place. Most are variations on a theme: organisms get trapped within asteroids or other debris ejected from planets during collisions with meteorites, and hurtle through space before colliding with another planetary body on which they may or may not be able to make a life.[13]

In the late 1950s, as the United States prepared to send rockets into space, the biologist Joshua Lederberg became concerned about the prospect of celestial contamination (it was Lederberg who in 2001 coined the word 'microbiome'). Humans were now able to spread earthly organisms to other parts of the solar system. More worrying was the thought that humans could bring back to Earth alien organisms that could cause ecological disruption – or worse, wreak havoc as diseases. Lederberg wrote urgent letters to the National Academy of Sciences to warn them of the possible 'cosmic catastrophe'. They paid attention and released an official statement of concern. There was still no word to describe the science of extraterrestrial life, so Lederberg coined one: 'exobiology'. It was the first version of the field now known as astrobiology.[14]

Lederberg was a prodigy. He enrolled at Columbia University at the age of fifteen, and in his early twenties made a discovery that helped transform our understanding of the history of life. He found that bacteria could trade genes with each other. One bacterium could acquire a trait from another bacterium 'horizontally'. Characteristics acquired horizontally are those that aren't inherited 'vertically' from one's parents. One picks them up along the way. We're used to the principle. When we learn or teach something, we're part of a horizontal exchange of information. Much of human culture and behaviour is transmitted in this fashion. However, for humans to engage in horizontal gene transfer as bacteria do is a fantastical prospect, even though it has taken place on occasion, deep in our evolutionary history. Horizontal gene transfer means that genes – and the traits they encode – are infectious. It is as if we noticed an unmarked trait lying by the side of the road, tried it on, and found that we had acquired a pair of dimples. Or perhaps we met someone on the street and swapped our straight hair for their curly hair. Or maybe we just picked up their eye colour. Or brushed up against a wolfhound quite by accident and developed an urge to run fast for several hours a day.[15]

Lederberg's discovery won him a Nobel Prize at the age of thirty-three. Before horizontal gene transfer was discovered, bacteria, like all other organisms, were understood to be biological islands. Genomes were closed systems. There was no way to take on new DNA midway through a lifetime, to acquire genes that had evolved 'off site'. Horizontal gene transfer changes this picture, and shows bacterial genomes to be cosmopolitan places, made up of genes that had evolved separately for millions of years. Horizontal gene transfer implied, as lichens had before, that branches of the evolutionary tree that had long since diverged were able to converge within the body of a single organism.

For bacteria, horizontal gene transfer is the norm – most of the genes in any given bacterium do not share an evolutionary history but are acquired piecemeal, just as objects accumulate in a home. In this way, a bacterium can acquire characteristics 'ready-made', speeding up evolution many times over. By exchanging DNA, a harmless bacterium can acquire antibiotic resistance and metamorphose into a virulent superbug in a single move. Over the last few decades it has become clear that bacteria aren't alone in this ability, although they remain its most adept practitioners: genetic material has been exchanged horizontally between all the domains of life.[16]

Lederberg's ideas were tinged with Cold War paranoia. In his hands, panspermia came to resemble horizontal gene transfer on a cosmic scale. For the first time in history, humans were capable – in theory – of infecting Earth and other planets with organisms that had not evolved on site. Life on Earth could no longer be considered a genetically closed system, a planetary island in an uncrossable sea. Just as bacteria could fast-forward evolution by picking up DNA horizontally, so the arrival of foreign DNA on Earth could 'short-circuit' the otherwise 'tortuous' process of evolution, with potentially catastrophic consequences.[17]

One of the main objectives of BIOMEX is to find out whether life forms can indeed survive a journey through space. Conditions outside the protective skin of the Earth's atmosphere are hostile. Among the many hazards are massive levels of radiation from the sun and other stars; a vacuum that causes biological material, lichens included, to dry out almost immediately; and rapid cycles of freezing, thawing and heating, with temperatures that swing from −120 to +120 degrees Centigrade and back again within twenty-four hours.[18]

The first attempt to send lichens into space didn't end well. In 2002, an unmanned Soyuz rocket carrying the samples exploded and crashed seconds after lift-off from a Russian spaceport. Months after the accident, when the snow had melted, the remains of the cargo were recovered. 'Curiously enough', the lead researchers reported, 'the LICHENS experiment was one of the few identifiable pieces of wreckage, and we discovered that despite the circumstances, the lichen[s] ... still showed some degree of biological activity.'[19]

The capacity of lichens to survive in space has since been demonstrated in a number of studies, and the findings are broadly the same. The hardiest lichen species can recover their metabolic activity in full within twenty-four hours of being rehydrated, and are able to repair much of the 'space-induced' damage they may have sustained. In fact, the toughest species – *Circinaria gyrosa* – has such high survival rates that three recent studies decided to expose samples to even higher levels of radiation than they receive in space, to test them to their 'uttermost limits of survival'. Sure enough, a dose of radiation could kill the lichens, but the amount required to disrupt their cells was enormous. Lichen samples exposed to 6 kilograys of gamma irradiation – six times the standard dose for food sterilisation in the United States, and 12,000 times the lethal dose for humans – were entirely untroubled. When the dose was doubled to 12 kilograys – 2.5 times the lethal dose for tardigrades – the lichens' ability to reproduce was impaired, although they survived and continued to photosynthesise with no apparent problems.[20]

For Trevor Goward, the curator of the lichen collection at the University of British Columbia, the extreme tolerances of lichens are an example of what he calls the 'lichening rod effect'. Lichens invite flashes of insight, or 'supercharged understanding', in Goward's words. The lichening rod effect describes what happens when lichens strike familiar concepts, splintering them into new forms. The idea of symbiosis is one such example. Survival in space is another, as is the threat that lichens posed to systems of biological classification. 'Lichens tell us things about *life*,' Goward exclaimed to me. '*They inform us*.'[21]

Goward is foremost a lichen obsessive (he has contributed around 30,000 lichen specimens to the university collection), and is no less a lichen taxonomist (he has named three genera and described thirty-six new lichen species). But he has the feel of a mystic about him. 'I like to say that lichens colonised the surface of my mind many years ago,' he told me with a chuckle. He lives on the edge of a large wilderness in British Columbia and runs a website called Ways of Enlichenment. For Goward, thinking deeply about lichens changes the way we apprehend life; they are organisms that can lure us towards new questions, and into new answers. 'What is our relationship to the world? What are we *about*?' Astrobiology pitches these questions at a cosmic scale. No wonder that lichens loom – if not large, then certainly vivid – at the front and centre of the panspermia debate.

However, it is closer to home that lichens and the concept of symbiosis they embody have triggered the most profound existential questions. Over the twentieth century, the concept of inter-kingdom collaboration transformed scientific understanding of how complex life forms evolved. Goward's questions may sound theatrical, but it is precisely our relationship to the world that lichens and their symbiotic way of life have led us to re-examine.

Life is divided into three domains. Bacteria make up one. Archaea – single-celled microbes that resemble bacteria, but which build their membranes differently – make up another. Eukarya make up the third. We're eukaryotes (pronounced *you-KA-ree-otes*), as are all other multicellular organisms, whether animal, plant, alga or fungus. The cells of eukaryotes are larger than bacterial and archaeal cells, and organise themselves around a number of specialised structures. One such structure is the nucleus, which contains most of the DNA in a cell. Mitochondria – the places where energy is produced – are another. Plants and algae have a further structure: chloroplasts, where photosynthesis happens.[22]

In 1967, the visionary American biologist Lynn Margulis became a vocal proponent of a controversial theory that gave symbiosis a central role in the evolution of early life. Margulis argued that some of the most significant moments in evolution had resulted from the coming together – and staying together – of different organisms. Eukaryotes arose when a single-celled organism engulfed a bacterium, which continued to live symbiotically inside it. Mitochondria were the descendants of these bacteria. Chloroplasts were the descendants of photosynthetic bacteria that had been engulfed by an early eukaryotic cell. All complex life that followed, human life included, was a story of the long-lasting 'intimacy of strangers'.[23]

The idea that eukaryotes had arisen 'by fusion and merger' had drifted in and out of biological thought since the start of the twentieth century, but it had remained at the margins of 'polite biological society'. By 1967, little had changed, and Margulis' manuscript was rejected fifteen times before it was finally accepted. After publication, her ideas were vigorously opposed, as similar suggestions had been before. (In 1970, the microbiologist Roger Stanier waspishly remarked that Margulis' 'evolutionary speculation ... can be considered a relatively harmless habit, like eating peanuts, unless it assumes the form of an obsession; then it becomes a vice.') However, in the 1970s Margulis was proved correct. New genetic tools revealed that mitochondria and chloroplasts had indeed started off as free-living bacteria. Since then, other examples of endosymbiosis have been found. The cells of some insects, for example, are inhabited by bacteria that themselves contain bacteria.[24]

Margulis' proposition amounted to a dual hypothesis of early eukaryotic life. No surprise, then, that she mobilised lichens to fight her cause – so too had the earliest proponents of her view at the turn of the twentieth century. The earliest eukaryotic cells could be thought of as 'quite analogous' to lichens, she argued. Lichens continued to figure prominently in her work over the following decades. 'Lichens are remarkable examples of innovation emerging from partnership,' she later wrote. 'The association is far more than the sum of its parts.'[25]

The endosymbiotic theory, as it came to be known, rewrote the history of life. It was one of the twentieth century's most dramatic shifts in biological consensus. The evolutionary biologist Richard Dawkins went on to congratulate Margulis on 'sticking by' the theory, 'from unorthodoxy to orthodoxy'. 'It is one of the great achievements of twentieth-century evolutionary biology,' Dawkins continued, 'and I greatly admire Lynn Margulis's sheer courage and stamina.' The philosopher Daniel Dennett described Margulis' theory as 'one of the most beautiful ideas [he'd] ever encountered', and Margulis as 'one of the heroes of twentieth-century biology'.[26]

Among the biggest implications of the endosymbiotic theory is that whole suites of abilities have been acquired in a flash, in evolutionary terms, ready-evolved, from organisms that are not one's parents, nor one's species, kingdom or even domain. Lederberg demonstrated that bacteria can horizontally acquire genes. The endosymbiotic theory proposed that single-celled organisms had horizontally acquired entire bacteria. Horizontal gene transfer transformed bacterial genomes into cosmopolitan places; endosymbiosis transformed cells into cosmopolitan places. The ancestors of all modern eukaryotes horizontally acquired a bacterium with a pre-existing ability to release energy using oxygen. Likewise, the ancestors of today's plants horizontally acquired bacteria with the ability to photosynthesise, ready-evolved.

In fact, this wording doesn't get it quite right. The ancestors of today's plants didn't acquire a bacterium with the ability to photosynthesise: they emerged from the combination of organisms that could photosynthesise with organisms that couldn't. In the two billion years they have lived together, both have become increasingly dependent on each other to the point we find ourselves in today, where neither can live without the other. Within eukaryotic cells, distant branches of the tree of life entwine and melt into an inseparable new lineage; they fuse, or anastomose, as fungal hyphae do.[27]

Lichens don't re-enact the origin of the eukaryotic cell exactly but, as Goward remarks, they certainly 'rhyme' with it. Lichens are cosmopolitan bodies, a place where lives meet. A fungus can't photosynthesise by itself, but by partnering with an alga or photosynthetic bacterium it can acquire this ability horizontally. Similarly, an alga or photosynthetic bacterium can't grow tough layers of protective tissue or digest rock, but by partnering with a fungus it gains access to these capabilities – suddenly. Together, these taxonomically remote organisms build composite life forms capable of entirely new possibilities. By comparison with plant cells which can't be parted from their chloroplasts, lichens' relationships are open. This gives them flexibility. In some situations, lichens reproduce without breaking up their relationship – fragments of a lichen containing all the symbiotic partners can travel as one to a new location and grow into a new lichen. In other situations, lichen fungi produce spores that travel alone. Upon arrival in a new place, the fungus must meet a compatible photobiont and form their relationship afresh.[28]

In joining forces, the fungal partners became part photobiont, and the photobionts part fungus. Yet lichens resemble neither. Just as the chemical elements of hydrogen and oxygen combine to make water, a compound entirely unlike either of its constituent elements, so lichens are emergent phenomena, entirely more than the sum of their parts. As Goward emphasises, it is a point so simple that it is hard to grasp. 'I often say that the only people who *can't* see a lichen are lichenologists. It's because they look at the parts, as scientists are trained to do. The trouble is that if you look at the parts of the lichen, *you don't see the lichen itself.*'[29]

It is exactly the emergent forms of lichens that are interesting from an astrobiological point of view. In the words of one study, 'it is hard to imagine a biological system that better summarises the characteristics of life on Earth.' Lichens are small biospheres that include both photosynthetic and non-photosynthetic organisms, thus combining the Earth's main metabolic processes. Lichens are, in some sense, micro-planets – worlds writ small.[30]

But what, exactly, do lichens do while in orbit around the Earth? To get around the problem of monitoring biological samples while they're in space, members of the BIOMEX team harvested specimens of the hardy species *Circinaria gyrosa* from the arid highlands of central Spain and took them to a Mars simulation facility. By exposing the lichens to space-like conditions on Earth, they hoped to be able to measure the lichens' activity in real time. It turned out there wasn't much to measure. Within an hour of 'turning on' Mars, the lichens had reduced their photosynthetic activity to near zero. They remained dormant for the rest of their time in the simulator and resumed their normal activity when rehydrated thirty days later.[31]

It is well known that the ability of lichens to survive extreme conditions depends on them entering a state of suspended animation – some studies have found that they can be successfully resuscitated after ten years of dehydration. If their tissues are dehydrated then freezing, thawing and heating don't cause much damage. Dehydration also protects them from the most hazardous consequence of cosmic rays: highly reactive free radicals, produced when radiation cleaves water molecules in two, that damage the structure of DNA.

Dormancy appears to be the most important survival strategy for lichens, but they have others. The hardiest lichen species have thick layers of tissue that block damaging rays. Lichens also produce more than a thousand chemicals that are not found in any other life forms, some of which act as sunscreens. A product of their innovative metabolisms, these chemicals have led lichens into all sorts of relationships with humans over the years: from medicines (antibiotics) to perfumes (oak moss), to dyes (tweeds, tartan, the pH indicator litmus), to foods – a lichen is one of the principal ingredients in the spice mix garam masala. Many fungi that produce compounds of importance to humans – including penicillin moulds – lived as lichens earlier in their evolutionary history but have since ceased to do so. Some researchers suggest that a number of these compounds, penicillin included, may have originally evolved as defensive strategies in ancestral lichens and persist today as metabolic legacies of the relationship.[32]

Lichens are 'extremophiles', organisms able to live, from our point of view, in other worlds. The tolerances of extremophiles are inconceivable. Collect samples in volcanic springs or superheated hydrothermal vents on the ocean floor and you'll find extremophilic microbes living apparently unfazed. Recent findings from the Deep Carbon Observatory report that more than half of all Earth's bacteria and archaea – so called 'infra-terrestrials' – exist kilometres below the planet's surface, where they live under intense pressure and extreme heat. These subsurface worlds are as diverse as the Amazon rainforest and contain billions of tonnes of microbes, hundreds of times the collective weight of all the humans on the planet. Some specimens are thousands of years old.[33]

Lichens are no less impressive. Indeed, their ability to survive many different types of extreme qualify them as 'polyextremophiles'. In the hottest, driest parts of the world's deserts, you'll find lichens prospering as crusts on the scorched ground. Lichens play a critical ecological role in these environments, stabilising the sandy surface of deserts, reducing dust storms and preventing further desertification. Some lichens grow inside cracks or pores within solid rock. The authors of one study, reporting the presence of lichens within chunks of granite, confess that they have no idea how these lichens got there in the first place. Several species of lichen are able to make a runaway success of life in the Antarctic Dry Valleys – an ecosystem so fierce it is used to approximate conditions on Mars. Long periods of freezing temperatures, irradiation with high levels of UV and near absence of water don't seem to trouble them. Even after immersion in liquid nitrogen at −195 degrees Centigrade lichens revive rapidly. And they live far longer than most organisms. The record-holding lichen lives in Swedish Lapland and is over 9,000 years old.[34]

In the already curious world of extremophiles, lichens are unusual for two reasons. First, they are complex multicellular organisms. Second, they arise from a symbiosis. Most extremophiles don't develop such sophisticated forms and enduring relationships. This is part of what makes lichens so interesting for astrobiologists. A lichen moving through space is a neat bundle of life – a whole ecosystem travelling as one. What better organisms to make interplanetary journeys?[35]

Although a number of studies have shown that lichens are capable of surviving in outer space, to be transported between planets, they would have to survive two additional challenges. First, the shock of their ejection from a planet by a meteorite. Second, the re-entry into a planetary atmosphere. Both present considerable hazards. Nonetheless, the shock of ejection is unlikely to be too much for them. In 2007, researchers demonstrated that lichens could withstand shock waves with a pressure of 10–50 gigapascals, 100–500 times greater than the pressure at the bottom of the Mariana Trench, the deepest place on Earth. This is well within the range of shock pressures experienced by rocks catapulted by meteorites into escape velocity from the surface of Mars. Re-entry into a planetary atmosphere might present more of a problem. In 2007, samples of bacteria and a rock-dwelling lichen were attached to the heat shield of a re-entry capsule. As the capsule scorched through the Earth's atmosphere, the samples were exposed to temperatures of over 2,000 degrees Centigrade for thirty seconds. In the process, the rocks partially melted and crystallised into new forms. When the remains were examined, there was no sign of any living cell whatsoever.[36]

This finding hasn't disheartened astrobiologists. Some argue that life forms encased deep within large meteorites would be protected from these extremes. Others point out that most of the material that arrives on Earth from space does so in the form of micro-meteorites, a type of cosmic dust. These small particles experience less friction and lower temperatures as they enter the atmosphere, and may be more likely to carry life forms safely to Earth than rocket capsules. As a number of researchers cheerily announce, the question remains open.[37]

No-one knows when lichens first evolved. The earliest fossils date from just over 400 million years ago, but it's possible that lichen-like organisms occurred before this. Lichens have evolved independently between nine and twelve times since. Today, one in five of all known fungal species form lichens, or 'lichenise'. Some fungi (such as *Penicillium* moulds) used to lichenise but don't any more; they have de-lichenised. Some fungi have switched to different types of photosynthetic partner – or re-lichenised – over the course of their evolutionary histories. For some fungi, lichenisation remains a lifestyle choice; they can live as lichens or not depending on their circumstances.[38]

It turns out that fungi and algae come together at the slightest provocation. Grow many types of free-living fungi and algae together, and they'll develop into a mutually beneficial symbiosis in a matter of days. Different species of fungi, different species of algae – it doesn't seem to matter. Completely new symbiotic relationships emerge in less time than it takes for a scab to heal. These remarkable findings, rare glimpses of the 'birth' of new symbiotic relationships, were published in 2014 by researchers at Harvard University. When fungi were grown with algae, they coalesced into visible forms that looked like soft green balls. They weren't the elaborate lichen forms depicted by Ernst Haeckel and Beatrix Potter. But then they hadn't spent millions of years in each other's company.[39]

Not just any fungus could partner with any alga, however. One critical condition had to be fulfilled for a symbiotic relationship to arise. Each partner had to be able to do something that the other couldn't achieve on its own. The identity of the partners didn't matter so much as their ecological fit. In the words of the evolutionary theorist W. Ford Doolittle, it was the 'song, not the singer' that appeared to be important. This finding sheds light on lichens' ability to survive in extreme conditions. As Trevor Goward points out, lichens by their nature are a kind of 'shotgun marriage' that arises in conditions too severe for either partner to survive alone. Whenever it was that lichens occurred for the first time, their very existence implies that life outside the lichen was less bearable, that together they were able to sing a metabolic 'song' neither can sing in isolation. Viewed in this way, lichens' extremophilia, their ability to live life on the edge, is as old as lichens themselves, and a direct consequence of their symbiotic way of life.[40]

There is no need to go to the Antarctic Dry Valleys or a Mars simulation facility to see lichen extremophilia in action. Most shorelines will do just fine. It is on the rocky coast of British Columbia that I've found lichens' tenacity most eye-catching. A foot or so above the barnacles, just beyond the furthest reach of the water, is a black smear that stretches across the rock in a band about two feet high. Close up, it looks like cracked tar on a dock. It forms a ribbon that traces the line of the shore, which becomes important when we're sailing around the islands. We use it when we anchor, to help us bet against the tide; it is a sure indicator of the limits of the water's reach. The dry land mark.

The black streak is a type of lichen, though one might never guess it was a living organism. It certainly doesn't grow into elaborate structures. Nonetheless, along much of the upper west coast of North America this species, *Hydropunctaria maura* ('water speckled midnight'), is the first organism to live beyond the reach of the waves. Look at high tide lines around the world and you'll see something similar. Most rocky shorelines are rimmed with lichen. Lichens start where the seaweeds stop, and some extend down into the water. When a volcano creates a new island in the middle of the Pacific Ocean, the first things to grow on the bare rock are lichens, which arrive as spores or fragments carried by the wind or birds. Likewise when a glacier retreats. The growth of lichens on freshly exposed rock is a variation on the theme of panspermia. These bare surfaces are inhospitable islands, remote possibilities for most organisms. Barren, seared by intense radiation and exposed to wild storms and temperature fluctuations, they may as well be other planets.[41]

Lichens are places where an organism unravels into an ecosystem and where an ecosystem congeals into an organism. They flicker between 'wholes' and 'collections of parts'. Shuttling between the two perspectives is a confusing experience. The word 'individual' comes from the Latin, meaning un-dividable. Is the whole lichen the individual? Or are its constituent members, the parts, the individuals? Is this even the right question to ask? Lichens are a product less of their parts than of the exchanges between those parts. Lichens are stabilised networks of relationships; they never stop lichenising; they are verbs as well as nouns.[42]

One of the people worrying these categories is a lichenologist from Montana called Toby Spribille. In 2016, Spribille and his colleagues published a paper in the journal *Science* that pulled the rug from underneath the dual hypothesis. Spribille described a new fungal participant in one of the major evolutionary lineages of lichens, a partner that had gone entirely undetected despite one and a half centuries of painstaking scrutiny.[43]

Spribille's discovery was an accident. A friend challenged him to grind up a lichen and sequence the DNA of all its participant organisms. He expected the results to be straightforward. 'The textbooks were clear,' he told me. 'There could only be two partners.' However, the more Spribille looked, the less this appeared to be the case. Each time he analysed a lichen of this type, he found additional organisms besides the expected fungus and alga. 'I dealt with these "contaminant" organisms for a long time,' he recalled, 'until I convinced myself that there was no such thing as lichens without "contamination", and we found that the "contaminants" were remarkably consistent. The more we dug in, the more they seemed to be the rule, not the exception.'

Researchers have long hypothesised that lichens might involve additional symbiotic partners. After all, lichens don't contain microbiomes. They *are* microbiomes, packed with fungi and bacteria besides the two established players. Nonetheless, until 2016, no new stable partnerships had been described. One of the 'contaminants' Spribille discovered – a single-celled yeast – turned out to be more than a temporary resident. It is found in lichens across six continents and can make such a substantial contribution to lichens' physiology as to give them the appearance of an entirely different species. This yeast was a crucial third partner in the symbiosis. Spribille's bombshell finding was only the beginning. Two years later, he and his team found that wolf lichens – some of the best-studied species – contain yet another fungal species, a *fourth* fungal partner. Lichens' identity splintered into even smaller shards. Yet this is still an oversimplification, Spribille told me. 'The situation is infinitely more complex than anything we've published. The "basic set" of partners is different for every lichen group. Some have more bacteria, some fewer; some have one yeast species, some have two, or none. Interestingly, we have yet to find any lichen that matches the traditional definition of one fungus and one alga.'[44]

What do the new fungal partners actually *do* in the lichen, I asked.

'We're not yet sure,' Spribille replied. 'Every time we go in and try to find out who's doing what, we get confounded. Instead of finding out the roles of the players, we bump into yet more players. The deeper we dig, the more we find.'

Spribille's findings are troubling for some researchers because they suggest that the lichen symbiosis is not as 'locked in' as it has been thought to be. 'Some people think about symbiosis as being like a package from IKEA,' Spribille explained, 'with clearly identified parts, and functions and order in which it's assembled.' His findings suggest instead that a broad range of different players might be able to form a lichen, and that they just need to 'tickle each other in the right way'. It's less about the identity of the 'singers' in the lichen, and more about what they do – the metabolic 'song' that each of them sings. In this view, lichens are dynamic *systems*, rather than a catalogue of interacting components.

It's a very different picture from the dual hypothesis. Since Schwendener's portrayal of the fungus and alga as master and slave, biologists have argued about which of the two partners is in control of the other. But now a duet has become a trio, the trio has become a quartet, and the quartet sounds more like a choir. Spribille seems unperturbed by the fact that it isn't possible to provide a single, stable definition of what a lichen actually is. It is a point Goward often returns to, relishing the absurdity: '*There is an entire discipline that can't define what it is that they study ... ?*' 'It doesn't matter what you call it,' writes Brenda Hillman on lichens. 'Anything so radical & ordinary stands for something.' For more than a hundred years, lichens have stood for many things, and will probably continue to challenge our understanding of what living organisms are.[45]

Meanwhile, Spribille is pursuing a number of promising new leads. 'Lichens are completely jam-packed full of bacteria,' he told me. In fact, lichens contain so many bacteria that some researchers hypothesise – in another twist on the panspermia theme – that they act as microbial reservoirs that seed barren habitats with crucial bacterial strains. Within lichens, some bacteria provide defence; others make vitamins and hormones. Spribille suspects that they might be doing more. 'I think a few of these bacteria might be necessary to tie the lichen system together and get it to form something other than a blob on a dish.'[46]

Spribille told me about a paper called 'Queer theory for lichens' ('It comes up as the first thing in Google when you enter "queer" and "lichen"'). Its author argues that lichens are queer beings that present ways for humans to think beyond a rigid binary framework: the identity of lichens is a question, rather than an answer known in advance. In turn, Spribille has found queer theory a helpful framework to apply to lichens. 'The human binary view has made it difficult to ask questions that aren't binary,' he explained. 'Our strictures about sexuality make it difficult to ask questions about sexuality, and so on. We ask questions from the perspective of our cultural context. And this makes it extremely difficult to ask questions about complex symbioses like lichens because we think of ourselves as autonomous individuals and so find it hard to relate.'[47]

Spribille describes lichens as the most 'extroverted' of all symbioses. Yet it is no longer possible to conceive of any organism – humans included – as distinct from the microbial communities they share a body with. The biological identity of most organisms can't be prised apart from the life of their microbial symbionts. The word 'ecology' has its roots in the Greek word *oikos*, meaning 'house', 'household', or 'dwelling place'. Our bodies, like those of all other organisms, are dwelling places. Life is nested biomes all the way down.

We can't be defined on anatomical grounds because our bodies are shared with microbes, and consist of more microbial cells than our 'own' – cows can't eat grass, for example, but their microbial populations can, and cows' bodies have evolved to house the microbes that sustain them. Neither can we be defined developmentally, as the organism that proceeds from the fertilisation of an animal egg, because we depend, like all mammals, on our symbiotic partners to direct parts of our developmental programmes. Nor is it possible to define us genetically, as bodies made up of cells that share an identical genome – many of our symbiotic microbial partners are inherited from our mothers alongside our 'own' DNA, and at points in our evolutionary history, microbial associates have permanently insinuated themselves into the cells of their hosts: our mitochondria have their own genome, as do plants' chloroplasts, and at least 8 per cent of the human genome originated in viruses (we can even swap cells with other humans when we grow into 'chimeras', formed when mothers and foetuses exchange cells or genetic material *in utero)*. Nor can our immune systems be taken as a measure of individuality, although our immune cells are often thought of as answering this question for us by distinguishing 'self' from 'non-self'. Immune systems are as concerned with managing our relationships with our resident microbes as fighting off external attackers, and appear to have evolved to enable colonisation by microbes rather than prevent it. Where does this leave you? Or perhaps y'all?[48]

Some researchers use the term 'holobiont' to refer to an assemblage of different organisms that behaves as a unit. The word 'holobiont' derives from the Greek word *holos*, which means whole. Holobionts are the lichens of this world, the more-than-the-sums of their parts. Like 'symbiosis' and 'ecology', 'holobiont' is a word that does useful work. If we only have words that describe neatly bounded autonomous individuals, it is easy to think that they actually exist.[49]

The holobiont is not a utopian concept. Collaboration is always a blend of competition and co-operation. There are many instances where the interests of all the symbionts don't align. A bacterial species in our gut can make up a key part of our digestive system but cause a deadly infection if it gets into our blood. We're used to this idea. A family can function as a family, a touring jazz group can give a captivating performance, and both still be fraught with tension.[50]

Perhaps it isn't so hard for us to relate to lichens after all. This sort of relationship-building enacts one of the oldest evolutionary maxims. If the word 'cyborg' – short for cybernetic organism – describes the fusion between a living organism and a piece of technology, then we, like all other life forms, are symborgs, or symbiotic organisms. The authors of a seminal paper on the symbiotic view of life take a clear stance on this point. 'There have never been individuals,' they declare. 'We are all lichens.'[51]

*

Drifting around on the *Caper*, we spend a lot of time looking at sea charts. On these maps, the familiar role of sea and land are reversed. The land masses are blank, beige expanses. The water is busy with contours and indications, which pucker around the rocks. Faceless flakes of land are laced with branching, joining seaways. The ocean moves through the network of waterways unpredictably. Some passages can only be navigated at certain times of the day. When the tide rushes through one perilous, narrow channel, its currents course together into a five-foot-high wave that stands still, a self-supporting wall of water. In a particularly treacherous corridor between two islands, fifty-foot tidal whirlpools appear and suck down floating logs.

Many of these seaways are edged with rock. Granitic bluffs tumble down to the sea. Trees lean, toppling in slow motion. Along the shore, trees, moss and lichens are rinsed off by the tides, revealing boulders and ledges, many wearing glacial scratch marks. It is hard to forget that much of the land is solid rock, slowly falling to pieces. Uneven shelves slope their way into sheer drops. My brother and I often sleep on these ledges overnight. Lichens are everywhere, and I wake up with a face full of them. For days afterwards I find fragments lining my trouser pockets. I turn them out, feeling like a human meteorite, and wonder how many will make a life in the unexpected places where they now find themselves.

4

Mycelial Minds

There is a world beyond ours … That world talks. It has a language of its own. I report what it says. The sacred mushroom takes me by the hand and brings me to the world where everything is known … I ask them and they answer me.

María Sabina

On a scale of one to five – one being 'not at all', and five being 'extreme' – how would you rate the sense of loss of your usual identity? How would you rate your experience of pure Being? How would you rate your sense of fusion into a larger whole?

I lay on my bed in the clinical drug testing unit, towards the end of my LSD trip, and puzzled over these questions. The walls appeared to breathe gently, and I found it difficult to focus on the words on the screen. There was a soft murmuring around my stomach, and the willow trees outside swayed, green and vivid.

LSD, like psilocybin – the active ingredient in many species of 'magic' mushrooms – is classified as both a psychedelic (or 'mind-manifesting') and an entheogen (a substance that can elicit an experience of 'the divine within'). With effects ranging from auditory and visual hallucinations and dream-like, ecstatic states to powerful shifts in cognitive and emotional perspective and a dissolving of time and space, these chemicals loosen the grip of our everyday perceptions, reach into our consciousness and touch us somewhere deep. Many users report mystical experiences or a connection with divine beings or entities, a feeling of 'oneness' with the natural world and a loss of a neatly-bounded sense of self.[1]

The psychometric questionnaire I was struggling to complete had been designed to assess this kind of experience. But the more I tried to cram my sensations into a five-point scale on a page, the more confused I became. How can one measure the experience of timelessness? How can one measure the experience of unity with an ultimate reality? These are qualities, not quantities. Yet science deals in quantities.

I squirmed, took several deep breaths and tried to approach the questions from a different angle. *How do you rate your experience of amazement?* The bed seemed to sway gently, and a school of thoughts scattered through my mind like startled minnows. *How do you rate your experience of infinity?* I could feel the scientific procedure groaning under the strain of what seemed to be an impossible task. *How do you rate the loss of your usual sense of time?* I succumbed to a fit of uncontrollable laughter – a common effect of LSD, I had been warned in a preparatory risk assessment. *How do you rate the loss of your usual awareness about where you are?*

I recovered from my laughter and looked up at the ceiling. Come to think of it, how *had* I ended up here? A fungus had evolved a chemical that had been used to make a drug. Quite by accident, this drug had been discovered to alter human experience. For seven decades or so, LSD's peculiar effects on our minds had generated astonishment, confusion, evangelical zeal, moral panic and everything in between. As it filtered through the twentieth century, it had left an indelible cultural residue that we still struggle to make sense of. I was lying in this hospital room as part of a clinical trial because its effects remained as bewildering as they had always been.

No wonder I was baffled. LSD and psilocybin are fungal molecules that have found themselves entangled within human life in complicated ways exactly because they confound our concepts and structures, including the most fundamental concept of all: that of our selves. It is their ability to pull our minds into unexpected places that have caused psilocybin-producing 'magic' mushrooms to be enveloped within the ritual and spiritual doctrines of human societies since antiquity. It is their ability to soften the rigid habits of our minds that makes these chemicals powerful medicines capable of relieving severe addictive behaviours, otherwise incurable depression and the existential distress that can follow the diagnosis of terminal illness. And it is their ability to modify the inner experience of our minds that has helped to change the way that the very nature of mind is understood within modern scientific frameworks. Yet why certain fungal species evolved these abilities remains a source of puzzlement and speculation.

I rubbed my eyes, rolled over and plucked up the courage to look once more at the words on the screen. *How do you rate your sense that the experience cannot be described adequately in words?*

The most prolific and inventive manipulators of animal behaviour are a group of fungi that live within the bodies of insects. These 'zombie fungi' are able to modify their host's behaviour in ways that bring a clear benefit: by hijacking an insect, the fungus is able to disperse its spores and complete its life cycle.

One of the best-studied cases is that of the fungus *Ophiocordyceps unilateralis*, which organises its life around carpenter ants. Once infected by the fungus, ants are stripped of their instinctive fear of heights, leave the relative safety of their nests and climb up the nearest plant – a syndrome known as 'summit disease'. In due course the fungus forces the ant to clamp its jaws around the plant in a 'death grip'. Mycelium grows from the ant's feet and stitches them to the plant's surface. The fungus then digests the ant's body and sprouts a stalk out of its head, from which spores shower down on ants passing below. If the spores miss their targets, they produce secondary sticky spores which extend outwards on threads that act like trip wires.[2]

Zombie fungi control the behaviour of their insect hosts with exquisite precision. *Ophiocordyceps* compels ants to perform the death grip in a zone with just the right temperature and humidity to allow the fungus to fruit: a height of twenty-five centimetres above the forest floor. The fungus orients ants according to the direction of the sun, and infected ants bite in synchrony, at noon. They don't bite any old spot on the leaf's underside. Ninety-eight per cent of the time, the ants clamp onto a major vein.[3]

How zombie fungi are able to control the minds of their insect hosts has long puzzled researchers. In 2017, a team headed by David Hughes, a leading expert on fungal manipulative behaviours, infected ants with *Ophiocordyceps* in the lab. The researchers preserved the ants' bodies at the moment of their death bite, sliced them into thin pieces and reconstructed a three-dimensional picture of the fungus living within their tissues. They found that the fungus becomes, to an unsettling degree, a prosthetic organ of ants' bodies. As much as 40 per cent of the biomass of an infected ant is fungus. Hyphae wind through their body cavities, from heads to legs, enmesh their muscle fibres, and co-ordinate their activity via an interconnected mycelial network. However, in the ants' brains, the fungus is conspicuous by its absence. To Hughes and his team, this was unexpected. They anticipated that the fungus would have to be present in the brain to exert such fine control over the ants' behaviour.[4]

Instead, the fungus's approach appears to be pharmacological. The researchers suspect that the fungus is able to puppeteer the ants' movements by secreting chemicals that act on their muscles and central nervous system even if the fungus does not have a physical presence in their brains. Exactly what chemicals these are isn't known. Nor is it known whether the fungus is able to cut the ant's brain off from its body and co-ordinate its muscle contractions directly. However, *Ophiocordyceps* is closely related to the ergot fungi from which the Swiss chemist Albert Hofmann originally isolated the compounds used to make LSD, and is able to produce the family of chemicals that LSD derives from – a group known as 'ergot alkaloids'. Inside infected ants the parts of the *Ophiocordyceps* genome responsible for the production of these alkaloids are activated, suggesting that they might have a role to play in the manipulation of ant behaviour.[5]

However they do it, these fungal interventions are remarkable by any human standard. After decades of research, and many billions of dollars of investment, the ability to regulate human behaviour using drugs is anything but fine-tuned. Antipsychotic drugs, for example, don't target specific behaviours; they just tranquillise. Compare this with the 98 per cent success rate of *Ophiocordyceps* in causing an ant not just to climb upwards, or perform its death bite – these always happen – but to bite onto the specific part of the leaf with the best conditions for the fungus to fruit. To be fair, *Ophiocordyceps*, like many zombie fungi, have had a long time to fine-tune their methods. The behaviours of infected ants don't pass without a trace. Ants' death grips leave distinctive scars on leaf veins, and fossilised scars push the origins of this behaviour back into the Eocene epoch, 48 million years ago. It is likely that fungi have been manipulating animal minds for much of the time that there have been minds to manipulate.[6]

Ophiocordyceps sprouting from an ant

I was seven when I discovered that humans can alter their minds by eating other organisms. My parents took me and my brother to stay in Hawaii with a friend of theirs, the eccentric author, philosopher and ethnobotanist Terence McKenna. His great passion was mind-altering plants and fungi. He had been a hashish smuggler in Bombay, a butterfly collector in Indonesia and a psilocybin mushroom grower in Northern California. Now he lived in an offbeat bolthole called Botanical Dimensions, several kilometres up a potholed road on the slopes of the volcano Mauna Loa. He had set up the land in Hawaii as a forest garden, a living library of rare and not-so-rare psychoactive and medicinal plants harvested from many corners of the tropical world. To reach the outhouse, one had to walk along a winding trail through the forest, ducking under dripping leaves and lianas. A few kilometres down the road, streams of lava flowed into the sea and made it froth and boil.

McKenna reserved his greatest enthusiasm for psilocybin mushrooms. He had first eaten them while travelling in the Colombian Amazon with his brother Dennis in the early 1970s. In the years that followed, fuelled by regular 'heroic' doses of mushrooms, McKenna discovered a rare gift of the gab and flair for public speaking. 'I realised that my innate Irish ability to rave had been turbo-charged by years of psilocybin mushroom use,' he recalled. 'I could talk to small groups of people with what appeared to be electrifying effect about ... peculiarly transcendental matters.' McKenna's bardic musings – eloquent and widely broadcast – remain celebrated and denounced in more or less equal measure.[7]

After a few days at Botanical Dimensions, I came down with a fever. I remember lying under a mosquito net, watching as McKenna ground up a preparation in a large pestle and mortar. I assumed it was a remedy for my sickness and asked what he was doing. In his zany metallic drawl, he explained that it was no such thing. This plant, like some types of mushroom, could make us dream. If we were lucky these organisms could even speak to us. These were powerful medicines that humans had used for a long time, but they could also be scary. He grinned a languorous smile. When I was older, he said, I could try some of the preparation – a mind-altering cousin of sage called *Salvia divinorum,* as it turned out. But not now. I was transfixed.

There are many examples of intoxication in the animal world – birds eat inebriating berries, lemurs lick millipedes, moths drink the nectar of psychoactive flowers – and it is likely that we have been using mind-altering drugs for longer than we have been human. The effects of these substances are 'frequently inexplicable, and indeed uncanny', wrote Richard Evans Schultes, a professor of biology at Harvard, and a leading authority on psychoactive plants and fungi. 'Without any doubt, [these compounds] have been known and employed in human experience since earliest man's experimentation with his ambient vegetation.' Many have 'strange, mystical and confounding' effects and, like psilocybin mushrooms, are intimately bound up within human cultures and spiritual practices.[8]

A number of fungi have mind-altering properties. The iconic red and white spotted mushroom *Amanita muscaria*, eaten by shamans in parts of Siberia, elicits euphoria and hallucinatory dreams. Ergot fungi induce a grisly portfolio of effects from hallucinations to convulsions to a sensation of unbearable burning. Involuntary muscle twitching is one of the primary symptoms of ergotism, and the ability of ergot alkaloids to induce muscle contractions in humans may mirror their role in ants infected by *Ophiocordyceps*. A number of the horrors depicted by the Renaissance painter Hieronymus Bosch are thought to have been inspired by the symptoms of ergot poisoning, and some hypothesise that the numerous outbreaks of 'dancing mania' between the fourteenth and seventeenth centuries, in which hundreds of townspeople took to dancing for days without rest, were caused by convulsive ergotism.[9]

The longest well-documented use of psilocybin mushrooms is in Mexico. The Dominican friar Diego Durán reported that mind-altering mushrooms – known as 'flesh of the gods' – were served at the coronation of the Aztec emperor in 1486.

Dr Francisco Hernández, the physician to the King of Spain, described mushrooms that 'when eaten cause not death but madness that on occasion is lasting, of which the symptom is a kind of uncontrolled laughter ... There are others again which, without inducing laughter, bring before the eyes all kinds of visions, such as wars and the likeness of demons.' The Franciscan friar Bernardino de Sahagún (1499–1590) provided one of the most vivid accounts of mushroom use:[10]

> They ate these little mushrooms with honey, and when they began to be excited by them, they began to dance, some singing, others weeping ... Some did not want to sing but sat down in their quarters and remained there as if in a meditative mood. Some saw themselves dying in a vision and wept; others saw themselves being eaten by a wild beast ... When the intoxication from the little mushrooms had passed, they talked over among themselves the visions which they had seen.

Unequivocal records of mushroom consumption in Central America stretch back to the fifteenth century, but the use of psilocybin mushrooms in the region almost certainly predates this. Hundreds of mushroom-shaped statues have been found, dating from the second millennium BCE, and codices from before the Spanish conquest depict mushrooms being eaten and held aloft by feathered deities.[11]

In McKenna's view, human consumption of psilocybin mushrooms was an even more ancient phenomenon and lay at the root of human biological, cultural and spiritual evolution. Evidence of religion, complex social organisation, commerce and the earliest art arises within a relatively short period in human history around fifty to seventy thousand years ago. What triggered these developments is not known. Some scholars attribute them to the invention of complex language. Others hypothesise that genetic mutations brought about changes in brain structure. For McKenna, it was psilocybin mushrooms that had ignited the first flickerings of human self-reflection, language and spirituality, somewhere in the proto-cultural fog of the Palaeolithic. Mushrooms were the original tree of knowledge.

Cave paintings preserved by the dry heat of the Sahara Desert in southern Algeria provided McKenna with the most impressive evidence for ancient mushroom consumption. Dating from between nine and seven thousand years BCE, the Tassili paintings include a figure of a deity with an animal's head, and mushroom-like forms sprouting from its shoulders and arms. As our ancestors roamed the 'mushroom-dotted grasslands and savannahs of tropical and subtropical Africa', McKenna conjectured, 'the psilocybin-containing mushrooms were encountered, consumed and deified. Language, poetry, ritual, and thought emerged from the darkness of the hominid mind.'[12]

There are many variations on the 'stoned ape' hypothesis but, as with most origin stories, it is difficult to prove either way. A rich bloom of speculation proliferates around psilocybin mushrooms wherever they are eaten. Surviving texts and artefacts are patchy, and almost always ambiguous. Does the Tassili painting represent a mushroom deity? It might. Then again, it might well not. The evidence from Neanderthal tooth plaque, the 'Iceman' and other well-preserved corpses provides proof that humans' knowledge of mushrooms as food and medicine stretches back many thousands of years. However, none of these bodies have been found with traces of psilocybin mushrooms. A number of primate species are known to seek out and consume mushrooms as food, and there are anecdotal accounts of primates consuming psilocybin mushrooms, but no well-documented instances. Some suspect that ancient Eurasian populations used psilocybin mushrooms as part of religious ceremonies, the best known being the Eleusinian Mysteries, secretive rites celebrated in ancient Greece and thought to have been attended by many luminaries including Plato. But once again there's no definitive record. And yet the absence of evidence does not provide evidence of absence. This makes speculation inevitable. And McKenna, turbo-charged by psilocybin, was a master of the art.[13]

Psilocybe cubensis

Ophiocordyceps has been the inspiration for at least two fictional monsters: the cannibals in the video game *The Last of Us* and the zombies in the book *The Girl with All the Gifts*. It sounds like a strange-but-true special case – one of evolution's left-field outcomes. However, *Ophiocordyceps* is just one well-studied example. This type of manipulative behaviour is not exceptional. It has evolved multiple times across the fungal kingdom in unrelated lineages, and there are numerous non-fungal parasites that are also able to manipulate the minds of their hosts.[14]

Fungi use a variety of approaches to tweak the biochemical dials that regulate their hosts' behaviour. Some use immunosuppressants to over-ride the insects' defensive responses. Two such compounds have found their way into mainstream medicine for these very reasons. Cyclosporine is an immunosuppressant drug that makes organ transplants possible. Myriocine has become the blockbuster multiple sclerosis drug fingolimod and was originally extracted from fungus-infested wasps that are eaten in parts of China as a nostrum for eternal youth.[15]

In 2018, researchers at the University of California, Berkeley, published a study documenting a startling technique used by *Entomophthora,* a mind-manipulating fungus that infects flies. There are parallels with *Ophiocordyceps.* Infected flies climb up high. When they extend their mouth parts to feed, a glue produced by the fungus sticks them to whatever surface they touch. When the fungus has consumed the fly's body, starting with the fatty parts and finishing with the vital organs, it pushes a stalk out of the fly's back and ejects spores into the air.

The researchers were surprised to find that the *Entomophthora* fungus carries around a type of virus that infects insects, not fungi. The lead author of the study reported it to be 'one of the whackiest discoveries' of his time in science. What's whacky is the implication: that the fungus uses the virus to manipulate the mind of insects. It's still a hypothesis, but it's plausible. A number of related viruses specialise in modifying insect behaviour. One such virus is injected by parasitic wasps into ladybirds, which tremble, remain rooted to the spot and become guardians for the wasp's eggs. Another similar virus makes honeybees more aggressive. By harnessing a mind-manipulating virus, the fungus wouldn't have to evolve the ability to modify the mind of its insect host.[16]

One of the more surprising twists in the story of zombie fungi came from research carried out by Matt Kasson and his team at West Virginia University. Kasson studies the fungus *Massospora,* which infects cicadas and causes the rear third of their bodies to disintegrate, allowing it to discharge its spores out of their ruptured back ends. Infected male cicadas – 'flying saltshakers of death', in Kasson's words – become hyperactive and hypersexual despite the fact that their genitals have long since crumbled away, a testament to how expertly the fungus is able to arrange their deterioration. Within their decaying bodies, their central nervous systems remain intact.[17]

In 2018, Kasson and his team analysed the chemical profile of the 'plugs' of fungus that sprout from the cicadas' broken bodies. They were amazed to find that the fungus produced cathinone, an amphetamine in the same class as the recreational drug mephedrone. Cathinone naturally occurs in the leaves of *khat* (*Catha edulis*), a plant cultivated in the Horn of Africa and the Middle East, which has been chewed for centuries by humans for its stimulant effects. Cathinone had never before been found outside plants. More astonishing was the presence of psilocybin, which was one of the most abundant chemicals in the fungal plugs – although one would have to eat several hundred infected cicadas to notice any effect. It's surprising because *Massospora* sits in an entirely different division of the fungal kingdom from the species known to produce psilocybin, separated by a gulf of hundreds of millions of years. Few suspected that psilocybin would show up in such a distant part of the fungal evolutionary tree, playing a behaviour-modifying role in a very different story.[18]

What exactly is *Massospora* able to accomplish by drugging its hosts with a psychedelic and an amphetamine? The researchers presume that these drugs play a part in the fungal manipulation of the insect. But how, exactly, isn't known.[19]

Accounts of psychedelic experiences frequently involve hybrid beings and inter-species transformations. Myths and fairy tales, too, are full of composite animals from werewolves and centaurs to sphinxes and chimeras. Ovid's *Metamorphoses* is a catalogue of transformations from one creature into another, and even includes a land where 'men grew from rainswept fungus'. In many traditional cultures, it is believed that composite creatures exist, and that the boundaries between organisms are fluid. The anthropologist Eduardo Viveiros de Castro reports that shamans in indigenous Amazonian societies believe they can temporarily inhabit the mind and body of other animals and plants. Among the Yukaghir people in northern Siberia, writes the anthropologist Rane Willerslev, humans dress and behave like elk when they hunt elk.[20]

These accounts seem to stretch the limits of biological possibility, and are rarely taken seriously within modern scientific circles. However, the study of symbiosis reveals that life is full of hybrid life forms, such as lichens, which are composed of several different organisms. Indeed, all plants, fungi and animals, including ourselves, are composite beings to some extent: eukaryotic cells are hybrids, and we all inhabit bodies that we share with a multitude of microbes without which we could not grow, behave and reproduce as we do. It's possible that many of these 'beneficial' microbes share some of the manipulative abilities of parasites like *Ophiocordyceps*. A growing number of studies have made a link between animal behaviour and the trillions of bacteria and fungi that live in their guts, many of which produce chemicals that influence animal nervous systems. The interaction between gut microbes and brains – the 'microbiome–gut–brain axis' – is far-reaching enough to have birthed a new field: neuromicrobiology. However, mind-manipulating fungi remain some of the most dramatic examples of composite organisms. In the words of David Hughes, an infected ant is a 'fungus in ant's clothing'.[21]

It's possible to make sense of this sort of shape-shifting within a scientific framework. In *The Extended Phenotype*, Richard Dawkins points out that genes don't just provide the instructions to build the body of an organism. They also provide instructions to build certain behaviours. A bird's nest is part of the outward expression of the bird's genome. A beaver's dam is part of the outward expression of a beaver's genome. And an ant's death grip is part of the outward expression of the genome of *Ophiocordyceps* fungi. Through inherited behaviours, Dawkins argues, the outward expression of an organism's genes – known as its 'phenotype' – extends into the world.

Dawkins was careful to place 'stringent requirements' on the idea of the extended phenotype. Although it is a speculative concept, he dutifully reminds us, it is a 'tightly limited speculation'. There are three crucial criteria that have to be met to prevent phenotypes becoming *too* extended (if a beaver's dam is an expression of the beaver's genome, then what about the pond that forms upstream of the dam, and the fish that live in the pond, and ...?).[22]

First, extended traits must be inherited – *Ophiocordyceps*, for example, inherits a pharmacological talent for infecting and manipulating ants. Second, extended traits must vary from generation to generation – some *Ophiocordyceps* are more precise manipulators of ant behaviour than others. Third, and most important, variation must affect an organism's ability to survive and reproduce, a quality known as its 'fitness' – *Ophiocordyceps* that can more precisely control their insect's movements are better able to spread their spores. Provided these three conditions are met – traits must be inherited, they must vary, and their variation must affect an organism's fitness – extended characteristics will be subject to natural selection and will evolve in an analogous way to their bodily characteristics. Beavers that make better dams are more likely to survive and pass on the ability to make better dams. But human dams – or any human building for that matter – don't count as part of our extended phenotype because we aren't born with an instinct to build specific structures that directly affect our fitness.

Summit disease and the death grip, on the other hand, fully qualify as fungal behaviours, not ant behaviours. The fungus doesn't have a twitchy, muscular, animal body with a centralised nervous system or an ability to walk, bite or fly. So it commandeers one. It is a strategy that works so well that it has lost the ability to survive without it. For part of its life, *Ophiocordyceps* must wear an ant's body. In nineteenth-century spiritualist circles, human mediums were understood to become possessed by the spirits of the dead. Lacking their own bodies or voices, spirits were said to borrow a human body to speak and act through. In an analogous way, mind-manipulating fungi possess the insects that they infect. Infected ants stop behaving like ants and become mediums for the fungi. It is in this sense that Hughes referred to an ant infected with *Ophiocordyceps* as a fungus in ant's clothing. Impelled by the fungus, the ant veers off the tracks of its own evolutionary story – tracks that guide its behaviours and relationships to the world and other ants – and onto the tracks of the evolutionary story of *Ophiocordyceps*. In physiological, behavioural and evolutionary terms, the ant *becomes fungus*.

Ophiocordyceps and other insect-manipulating fungi have evolved a remarkable ability to cause harm to the animals they influence. Psilocybin mushrooms, as a growing number of studies report, have evolved an astonishing ability to cure a wide range of human problems. In one sense, this is news: since the 2000s, rigorously controlled trials and the latest brain-scanning techniques have helped researchers interpret psychedelic experiences using the language of modern science – it was this new wave of psychedelic research that brought me into the hospital for the LSD study. These recent findings have broadly confirmed the opinions of many researchers in the 1950s and 1960s, who came to regard LSD and psilocybin as miracle cures for a wide range of psychiatric conditions. In another sense, however, much of the research that has taken place in modern scientific contexts broadly confirms what is well known to the traditional cultures who have used psychoactive plants and fungi as medicines and psycho-spiritual tools for an unknowably long time. From this point of view, modern science is simply catching up.[23]

Many recent findings are extraordinary by the standards of conventional pharmaceutical interventions. In 2016, two sister studies at New York University and Johns Hopkins University administered psilocybin alongside a course of psychotherapy to patients suffering from anxiety, depression and 'existential distress' following diagnoses with terminal cancer. After a single dose of psilocybin, 80 per cent of patients showed substantial reductions in their psychological symptoms, reductions that persisted for at least six months after the dose. Psilocybin reduced 'demoralisation and hopelessness, improved spiritual well-being and increased quality of life'. Participants described 'exalted feelings of joy, bliss and love', and 'a movement from feelings of separateness to interconnectedness'. More than 70 per cent of participants rated their experiences as one of the top five most meaningful experiences in their lives. 'You may say, what does that mean?' Roland Griffiths, a senior researcher on the study, remarked in an interview. 'Initially I wondered if they had pretty dull lives. But no.' Participants compared their experiences to the birth of their first child or the death of a parent. These studies are considered to be some of the most effective psychiatric interventions in the history of modern medicine.[24]

Profound changes in people's minds and personalities are rare; that they should happen over the course of such a short experience is striking. Nonetheless, these aren't anomalous findings. Several recent studies report the dramatic effects of psilocybin on people's minds, outlooks and perspectives. Using some of the psychometric questionnaires that I had contended with, many of these studies have found that psilocybin can reliably induce experiences classified as 'mystical'. Mystical experiences include feelings of awe; of everything being inter-connected; of transcending time and space; of profound intuitive understanding about the nature of reality; and of deeply felt love, peace or joy. They often include the loss of a clearly defined sense of self.[25]

Psilocybin can leave a lasting impression on people's minds, like the grin on the Cheshire Cat in *Alice in Wonderland*, which 'remained some time after the rest of it had gone'. In one study, researchers found that a single high dose of psilocybin increased the openness to new experiences, psychological well-being and life satisfaction of healthy volunteers, a change that persisted in most cases for more than a year. Some studies have found that experiences with psilocybin have helped smokers or alcoholics break their addictions. Other studies have reported enduring increases in subjects' sense of connection with the natural world.[26]

Out of the flurry of recent research into psilocybin some themes are starting to emerge. One of the most interesting is the way in which participants in psilocybin trials make sense of their experiences. As Michael Pollan reports in *How to Change Your Mind*, most of the people who take psilocybin don't interpret their experiences in modern biology's mechanistic terms, of molecules moving around their brains. Quite the opposite. Pollan found that many of those he interviewed had 'started out stone-cold materialists or atheists ... and yet several had had "mystical experiences" that left them with the unshakable conviction that there was something more than we know – a "beyond" of some kind that transcended the physical universe.' These effects pose a riddle. That a chemical can induce a profound mystical experience appears to support the prevailing scientific view that our subjective worlds are underpinned by the chemical activity of our brains; that the world of spiritual beliefs and experience of the divine can spring from a material, biochemical phenomenon. However, as Pollan points out, the very same experiences are so powerful as to convince people that a non-material reality – the raw ingredient of religious belief – exists.[27]

Ophiocordyceps and gut-dwelling microbes influence animal minds by living inside their bodies, fine-tuning their chemical secretions in real time. This is not the case with psilocybin mushrooms. One can inject a person with synthetic psilocybin and elicit the full range of psycho-spiritual effects. How does it work?

Once inside the body, psilocybin is converted to the chemical psilocin. Psilocin slips into the workings of the brain by stimulating receptors normally stimulated by the neurotransmitter serotonin. By mimicking one of our most widely used chemical messengers, psilocybin, like LSD, infiltrates our nervous systems, intervenes directly in the passage of electrical signals around our bodies, and can even change the growth and structure of neurones.[28]

How exactly psilocybin changes patterns of neuronal activity wasn't known until the late 2000s, when researchers from the Beckley/Imperial Psychedelic Research Program gave subjects psilocybin and monitored the activity of their brains. Their findings were surprising. The scans revealed that psilocybin didn't increase the activity of the brain as one might expect, given its dramatic effects on people's minds and cognition. Rather, it reduced the activity of certain key areas.

The type of brain activity reduced by psilocybin forms the basis of what is termed the default mode network (DMN). When we're not focusing on much, when our minds are wandering idly, when we're self-reflecting, when we're thinking of the past or making plans about the future, it's our DMN that's active. The DMN has been described by researchers as the 'capital city' or 'corporate executive' of the brain. In the riot of cerebral processes going on at any one time, the DMN is understood to keep a kind of order – a schoolteacher in a chaotic classroom.

The study showed that subjects who reported the strongest sense of 'ego dissolution', or loss of a sense of self on psilocybin, had the most dramatic reductions in the activity of their DMNs. Shut down the DMN, and the brain is let off the leash. Networks of activity previously distant from one another link up. In the terms of the metaphor used by Aldous Huxley in his seminal exploration of psychedelic experience, *The Doors of Perception*, psilocybin appears to shut down a 'reducing valve' in our consciousness. The outcome? An 'unconstrained style of cognition'. The authors conclude that psilocybin's ability to change people's minds is related to these states of cerebral flux.[29]

Brain-imaging studies provide an important description of the way that psychedelics act on our bodies, but they don't do much to explain participants' feelings. After all, it is people who have experiences, not brains. And it is exactly people's experiences that seem to underpin the therapeutic effects of psilocybin. In the studies that measured the effects of psilocybin on terminally ill cancer patients, it was those who had the strongest mystical experiences that showed the most pronounced reductions in symptoms of depression and anxiety. Similarly, in a study of psilocybin and tobacco addiction, the patients with the best results were those who had undergone the most powerful mystical experiences. Psilocybin appears to take effect not by pushing a set of biochemical buttons, but by opening patients' minds to new ways of thinking about their lives and behaviours.

It is a finding that echoes much of the research into LSD and psilocybin that took place during the first wave of modern psychedelic research in the mid twentieth century. Abram Hoffer, a Canadian psychiatrist and researcher into the effects of LSD in the 1950s, remarked that 'from the first, we considered not the chemical, but the experience as a key factor in therapy.' This may sound like common sense, but from the standpoint of mechanistic medicine at the time it was a radical notion. The conventional approach was – and remains to a large degree – to use *stuff*, whether drugs or a surgical tool, to treat the *stuff* that the body is made out of, just as we might use tools to repair a machine. Drugs are normally understood to work through a pharmacological circuit that bypasses the conscious mind entirely: a drug affects a receptor, which triggers a change in symptoms. By contrast, psilocybin – like LSD and other psychedelics – appears to act on symptoms of mental illness *via the mind*. The standard circuit is enlarged: a drug affects a receptor, which triggers a change of mind, which triggers a change in symptoms. Patients' psychedelic experiences themselves appear to be the cure.[30]

In the words of Matthew Johnson, a psychiatrist and researcher at Johns Hopkins, psychedelics like psilocybin 'dope-slap people out of their story. It's literally a reboot of the system ... Psychedelics open a window of mental flexibility in which people can let go of the mental models we use to organise reality.' Toughened habits, such as those that give rise to substance addiction, or those that add up to the 'rigid pessimism' of depression, become more pliable. By softening the categories that organise human experience, psilocybin and other psychedelics are able to open up new cognitive possibilities.[31]

One of our most robust mental models is that of the *self*. It is exactly this sense of self that psilocybin and other psychedelics seem to disrupt. Some call it ego dissolution. Some simply report that they lost track of where they ended and their surroundings began. The well-defended 'I' that humans depend on for so much can vanish entirely, or just dwindle, shading off into otherness gradually. The result? Feelings of merging with something greater, and a reimagined sense of one's relationship to the world.[32] In many instances – from lichens to the boundary-stretching behaviour of mycelium – fungi challenge our well-worn concepts of identity and individuality. Psilocybin-producing mushrooms, like LSD, do so too, but in the most intimate possible setting: the inside of our own minds.

In the case of *Ophiocordyceps,* an infected ant's behaviour can be thought of as fungal behaviour. The death grip, summit disease, these are extended characteristics of the fungus, part of its extended phenotype. Can the alterations in human consciousness and behaviour brought about by psilocybin mushrooms be thought of as part of the extended phenotype of the fungus? The extended behaviour of *Ophiocordyceps* leaves an imprint in the world in the form of fossilised scars on the underside of leaves. Can the extended behaviour of psilocybin mushrooms be thought of as leaving an imprint in the world in the form of ceremonies, rituals, chants and the other cultural and technological outgrowths of our altered states? Do psilocybin fungi wear our minds, as *Ophiocordyceps* and *Massospora* wear insect bodies?

Terence McKenna was a great advocate of this view. Given a sufficiently large dose, he asserted, the mushroom could be expected to speak, plainly and clearly, talking 'eloquently of itself in the cool night of the mind'. Fungi have no hands with which to manipulate the world but with psilocybin as a chemical messenger, they could borrow a human body, and use its brain and senses to think and speak through. McKenna thought fungi could wear our minds, occupy our senses and, most important, impart knowledge about the world out there. Among other things, fungi could use psilocybin to influence humans in an attempt to deflect our destructive habits as a species. For McKenna, this was a symbiotic partnership that presented possibilities 'richer and even more baroque' than those available to humans or fungi alone.[33]

As Dawkins reminds us, how far we're willing to go depends on how far we're willing to speculate. How we speculate in turn depends on how we arrange our biases. 'You think the world is what it looks like in fine weather at noon day,' the philosopher Alfred North Whitehead once observed to his former student Bertrand Russell. 'I think it is what it seems like in the early morning when one first wakes from deep sleep.' In Whitehead's terms, Dawkins speculates in fine weather at noon day. He takes pains to ensure that his speculation about extended phenotypes remains 'disciplined' and 'tightly limited'. He is clear that phenotypes can extend beyond the body, but they can't be *too* extended. By contrast, McKenna speculates at dawn. His requirements are less stringent, his explanations less tightly limited. Between the two poles lies a continent of possible opinion.[34]

How do psilocybin mushrooms stand up to Dawkins' three 'stringent requirements'?

A mushroom's ability to produce psilocybin is certainly inherited. It is also an ability that varies, from mushroom species to mushroom species, and between individual mushrooms.

However, for the be-mushroomed state – the visions, the mystical experiences, the ego dissolution, the loss of a sense of self – to count as part of the extended phenotype of the fungi, the final key condition must be met. Fungi that orchestrate 'better' altered states – whatever that means – must pass on their genes more successfully. Fungi must differ in the ability to influence humans, and the ones that provide more fulsome and desirable experiences must benefit at the expense of those providing less desirable experiences.

At a glance, this third requirement seems to decide the issue. Psilocybin-producing fungi may influence human behaviour but unlike *Ophiocordyceps,* they don't live on or within our bodies. Moreover, McKenna's speculation is hard to reconcile with the fact that humans are latecomers to the psilocybin story. Psilocybin was produced by fungi for tens of millions of years before the genus *Homo* evolved – the current best estimate puts the origin of the first 'magic' mushroom at around 75 million years ago. For more than 90 per cent of their evolutionary history, psilocybin-producing fungi have lived on a human-free planet and have done just fine. If the fungi do indeed benefit from our altered states, they can't have done so for very long.[35]

Then what *did* psilocybin do for those fungi that evolved an ability to produce it? Why bother to make it in the first place? It's a question that has been pored over for decades by mycologists and magic mushroom enthusiasts alike.

It's possible that psilocybin didn't do much at all for the fungi that made it until humans came along. There are lots of compounds in fungi and plants that accumulate in biochemical backwaters playing Z-list roles as incidental metabolic byproducts. Sometimes these 'secondary compounds' encounter an animal that they attract, confound or kill, at which point they might start to benefit the fungus and become an evolutionary adaptation. However, sometimes they don't do much more than provide variations on a biochemical theme that might one day prove useful, or not.

Two studies published in 2018 suggest that psilocybin did provide a benefit to the fungi that could make it. Analysis of the DNA of psilocybin-producing fungal species reveals that the ability to make psilocybin evolved more than once. More surprising was the finding that the cluster of genes needed to make psilocybin has jumped between fungal lineages by horizontal gene transfer several times over the course of its history. As we've seen, horizontal gene transfer is the process by which genes and the characteristics they underpin move between organisms without the need to have sex and produce offspring. It is an everyday occurrence in bacteria – and how antibiotic resistance can spread rapidly through bacterial populations – but it is rare in mushroom-forming fungi. It is even more rare for complex clusters of metabolic genes to remain intact as they jump between species. The fact that the psilocybin gene cluster remained in one piece as it moved around suggests that it provided a significant advantage to any fungi who expressed it. If it didn't, the trait would have quickly degenerated.[36]

But what could this advantage have been? The psilocybin gene cluster jumped between species of fungus that lived similar lifestyles in rotting wood and animal dung. These habitats are also the home of numerous insects that 'eat or compete' with fungi, all of whom should be sensitive to the potent neurological activity of psilocybin. It seems probable that the evolutionary value of psilocybin lay in its ability to influence animal behaviour. But how, exactly, isn't clear. Fungi and insects share a long and complicated history. Some fungi, like *Ophiocordyceps* or *Massospora*, kill. Some co-operate over immense tracts of evolutionary time, like those that live with leafcutter ants and termites. In either case, fungi use chemicals to change insect behaviour. *Massospora* even goes so far as to use psilocybin to accomplish its purpose. Which way did psilocybin swing? Opinion is divided. Monitoring the effects of psilocybin on the organisms that consume it isn't straightforward even with humans, who can at least attempt to talk about their experience and fill out psychometric questionnaires. What chance do we have of finding out what psilocybin might do to the mind of an insect? Animal studies on the subject are scarce, which makes matters worse.[37]

Could psilocybin be a deterrent produced by fungi to fuddle the wits of their insect pests? If so, it doesn't seem to be very effective. There are species of gnat and fly that routinely make their homes within magic mushrooms. Snails and slugs devour them without apparent ill effect. And leafcutter ants have been observed to actively forage for a certain type of psilocybin mushroom, carrying them back to their nests in one piece. These findings have led some to suppose that, far from being a deterrent, psilocybin served as a lure, somehow changing insect behaviour in ways that benefited the fungus.[38]

The answer probably falls somewhere in between. Psilocybin mushrooms that are toxic to some animals could still make a good meal for those able to develop resistance. Some species of fly are resistant to the poisons produced by the death cap mushroom, for instance, and have near exclusive access as a result. Could these psilocybin-tolerant insects serve the fungus by helping to spread its spores? By defending it from other pests? Once again, we're left speculating.

We might not know how psilocybin served fungal interests for the first several million years of its existence. But from our current vantage point, it is clear that the interaction of psilocybin with human minds has transformed the evolutionary fortunes of those mushrooms that produce it. Psilocybin-producing fungi develop an easy rapport with humans. Far from acting as a repellent – to stand a chance of overdosing, a human would have to eat a thousand times more mushrooms than required for an average trip – psilocybin has caused humans to seek out the mushrooms, carry them from place to place and develop methods to cultivate them. In doing so, we have helped to spread their spores, which are both light enough to travel over great distances in the air, and numerous: left on any surface for just a few hours, a single mushroom will eject enough spores to leave a thick black smear. In colliding with a new type of animal, a chemical that might once have served to baffle and deter pests has been transformed into a glittering lure in a few swift moves. The passage of magic mushrooms from obscurity to international stardom over a few decades in the twentieth century is one of the most dramatic stories in the long history of human relationships with fungi.[39]

In the 1930s, the Harvard botanist Richard Evans Schultes read the fifteenth-century accounts of the 'flesh of the gods' written by the Spanish friars and became intrigued. From the few sources that survived, it was clear that in parts of Central America psilocybin mushrooms had grown into cultural and spiritual centres of gravity. They had found their way into the hands of the local deities, and their consumption had fuelled a conception of the divine in which the mushrooms themselves featured heavily.

Could these mushrooms still be found growing in modern-day Mexico? Schultes received a tip-off from a Mexican botanist, and in 1938 set off for the remote valleys in northeastern Oaxaca to find out. (This was the same year that Albert Hofmann first isolated LSD from ergot fungi in a pharmaceutical lab in Switzerland.) Schultes found mushroom use among the Mazatec people to be alive and well. *Curanderos*, or healers, held regular mushroom vigils to heal the sick, locate lost property and give advice. Mushrooms were common in the pastures surrounding the villages. Schultes collected specimens and published his findings. He reported that consumption of these mushrooms resulted in 'hilarity, incoherent talking, and … fantastic visions in brilliant colors'.[40]

In 1952, Gordon Wasson, an amateur mycologist and a vice president of the bank J. P. Morgan, received a letter from the poet and scholar Robert Graves describing Schultes' report. Wasson was fascinated by Graves' news of the mind-altering 'flesh of the gods' and travelled to Oaxaca in search of the mushrooms. There, Wasson met a *curandera* called

María Sabina, who invited him to a mushroom vigil. Wasson described his experience as 'soul-shattering'. In 1957, he published an account of his experience in *Life* magazine. The article was titled 'Seeking the magic mushroom: a New York banker goes to Mexico's mountains to participate in the age-old rituals of Indians who chew strange growths that produce visions'.[41]

Wasson's article was a sensation and read by millions. By this time, the mind-altering properties of LSD had been known about for fourteen years, and there was an active community of researchers conducting studies into its effects. Nonetheless, Wasson's was among the very first accounts of a psychedelic mind-altering substance to reach the general public. 'Magic mushrooms' became a household term – and gateway concept – more or less overnight. In his autobiography, Dennis McKenna remembers his brother Terence, then a precocious ten-year-old, 'trailing our mother as she did her housework, waving the magazine demanding to know more. But of course she had nothing to add.'[42]

Things moved quickly. Hofmann was sent a sample of the 'magic' mushrooms by a member of Wasson's expedition, and had soon identified, synthesised and named the active ingredient: psilocybin. In 1960, the well-respected Harvard academic Timothy Leary heard of the magic mushrooms through a friend and went to Mexico to try them. His experience, a 'visionary voyage', had a profound impact on him, and he returned 'a changed man'. Back at Harvard, inspired by his experience with the mushrooms, Leary abandoned his research programme and set up the Harvard Psilocybin Project. 'Since eating seven mushrooms in a garden in Mexico', he later wrote of his gateway experience, 'I have devoted all my time and energy to the exploration and description of these strange deep realms.'[43]

Leary's methods proved controversial. He left Harvard and began in earnest to promote his vision that cultural revolution and spiritual enlightenment could be attained via the consumption of psychedelics, and soon became notorious. In numerous TV and radio appearances, he evangelised about LSD and its many benefits. In an interview with *Playboy* he advised that on an average acid trip women could expect to have a thousand orgasms. He ran against Ronald Reagan for governor of California and lost. Fuelled in part by Leary's proselytising, the countercultural movement of the 1960s picked up momentum. In 1967, in San Francisco, Leary, now 'High Priest' of the psychedelic movement, addressed the 'Human Be-In' that was attended by tens of thousands. Soon afterwards, in a haze of backlash and scandal, LSD and psilocybin were made illegal. By the end of the decade, almost all of the research taking place into the effects of psychedelics had been shut down or driven underground.[44]

The outlawing of psilocybin and LSD marked the start of a new chapter in the evolutionary history of psilocybin mushrooms. Most of the psychedelic research of the 1950s and 60s had taken place with LSD, or synthetic psilocybin in pill form, much of it produced by Hofmann in Switzerland. But by the early 1970s, in part because of the legal risks associated with pure psilocybin and LSD, and in part due to their scarcity, interest in magic mushrooms grew. By the mid 1970s, species of psilocybin mushrooms had been discovered growing in many parts of the world, from the United States to Australia. However, the supply of wild mushrooms is limited by seasonal conditions and location. When they returned from Colombia in the early 1970s, Terence and Dennis McKenna sought a steadier supply. Their solution was radical. In 1976, the McKennas published a short book entitled *Psilocybin: Magic Mushroom Growers' Guide*. Armed with this slim volume, the brothers advised, with little more than jars and a pressure cooker, anyone could produce unlimited quantities of a powerful psychedelic from the comfort of their garden shed. The process was only a little bit more complicated than making jam, and even a novice could soon find themselves, in Terence's words, 'neck deep in alchemical gold'.[45]

The McKennas were not the first to cultivate psilocybin mushrooms, but they were the first to publish a reliable method for growing large quantities of mushrooms without specialist laboratory equipment. The *Growers' Guide* was a runaway success and went on to sell more than 100,000 copies in the five years following its release. It kick-started a new field of DIY mycology, and influenced a young mycologist named Paul Stamets, the discoverer of four new species of psilocybin mushroom and author of a guidebook to psilocybin mushroom identification.

Stamets was already working on new ways to cultivate a range of 'gourmet and medicinal' mushrooms, and in 1983 he published *The Mushroom Cultivator*, which simplified growing techniques even further. In the 1990s, as online forums for magic mushroom growers sprang up, Dutch entrepreneurs spotted a loophole in the law that allowed them to sell psilocybin mushrooms openly, and many Dutch growers of edible supermarket mushrooms switched over to psychedelic mushroom production. By the early 2000s, the craze had spread to England, and crates of fresh psilocybin mushrooms were being sold on London high streets. By 2004, the Camden Mushroom Company alone was shifting 100 kilograms of fresh mushrooms a *week*, equivalent to about 25,000 trips. Fresh psilocybin mushrooms were made illegal soon afterwards, but the secret was out. Today, just-add-water kits are readily available online. Crosses between fungal strains are producing new varieties, from 'Golden Teacher' to 'Mc Kennai', each with subtly different effects.[46]

For as long as humans have sought out psilocybin mushrooms – thus serving as enthusiastic agents of spore dispersal – the fungi have benefited from their ability to tinker with our consciousness. Since the 1930s, these benefits have multiplied many times over. Before Wasson's trip to Mexico, few people outside indigenous communities in Central America knew of the existence of psilocybin mushrooms. Yet within two decades of their arrival in North America, a new story of domestication was under way. In cupboards, bedrooms and warehouses, a handful of tropical fungal species found new lives in otherwise inhospitable temperate climates.[47]

What's more, since Schultes' first paper in the late 1930s, more than 200 new species of psilocybin-producing fungi have been described, including a psilocybin-producing lichen which grows in the Ecuadorian rainforest. It turns out that there are few environments where these mushrooms don't grow, given sufficient rainfall. As one researcher observes, psilocybin mushrooms 'occur in abundance wherever mycologists abound'. Guidebooks make it possible for humans to find, identify and pick – and thus disperse – psilocybin mushrooms that would have been off the radar a few decades ago. Several of these species seem to have a fondness for disturbed habitats and find an easy home in our messy wake. As Stamets wryly confides, many have an affection for public spaces, including 'parks, housing developments, schools, churches, golf courses, industrial complexes, nurseries, gardens, freeway rest areas and government buildings – including county and state courthouses and jails'.[48]

Do the events of the last several decades bring us any closer to satisfying Dawkins' third criterion? Can these fungi be thought of as borrowing a human brain to think with, a human consciousness to experience with? Does a human under the influence of mushrooms really fall under their influence, as an infected ant falls under the influence of *Ophiocordyceps*?

For our altered states to count as an extended phenotype of the fungi the be-mushroomed human would need to serve the reproductive interests of the very fungi they had eaten. However, this doesn't seem to be the case. Only a small number of species are cultivated, and for a large part, the decision of which fungal strains to grow is made on the basis of which are the easiest to cultivate and provide the biggest yields – it's not clear that 'better' mind alterers are selected over 'worse' mind alterers. More problematic is that if all humans were made extinct in a single instant, most species of psilocybin mushroom would carry on untroubled. Psilocybin-producing fungi don't depend entirely on our altered states, as *Ophiocordyceps* depends entirely on the altered behaviour of ants. For tens of millions of years, they have grown and reproduced perfectly well without humans, and would probably continue to do so.

Does it really matter? 'One might think that with the isolation … of psilocybin and psilocin, the mushrooms of Mexico had lost their magic,' wrote Schultes and Hofmann in 1992. With the domestication of psilocybin-producing fungi, hundreds of kilograms of mushrooms can be grown in warehouses in Amsterdam. With the isolation of psilocybin, the default mode network can be disabled on demand in brain scanners. Mystical experiences, awe and a loss of one's sense of self can be elicited in a hospital bed. How much closer do these advances propel us towards an understanding of the way that psilocybin influences human minds?

For Schultes and Hofmann, the answer was 'not very'. Mystical experiences are those by definition resistant to rational explanation. They don't readily fit within numbered scales on psychometric questionnaires. They confound and enthral. And they undoubtedly occur. As Schultes and Hofmann observe, scientific investigation into the identity and structure of psilocybin and psilocin had 'merely shown that the magic properties of the mushrooms are the properties of two crystalline compounds'. It is a finding that does little more than kick the question down the road. 'Their effect on the human mind is just as inexplicable, and just as magical, as that of the mushrooms themselves.'[49]

The effects of psilocybin mushrooms might not count as an extended phenotype in a strict sense, but does this mean we should dismiss the speculation of Terence McKenna? Perhaps we shouldn't be too hasty. 'Our normal waking consciousness', wrote the philosopher and psychologist William James in 1902, 'is but one special type of consciousness, whilst all about it, parted from it by the filmiest of screens, there lie potential forms of consciousness entirely different.' For reasons that are poorly understood, certain fungi lead humans out of familiar stories into forms of consciousness that are entirely different, and towards the edge of new questions. 'No account of the universe in its totality can be final which leaves these other forms of consciousness quite disregarded,' James concluded.[50]

Whether for a researcher, a patient or just an interested bystander, the curious thing about these fungal chemicals is exactly the *experiences* that they elicit. McKenna's mushroom-fuelled speculation may stretch the limits of mental and biological possibility. But that is precisely the point: the effects of psilocybin on human minds stretch the limits of what seems possible. In Mazatec culture, it is self-evident that mushrooms speak; anyone who takes them can experience this for themselves. Theirs is a view shared by many traditional cultures that ritually use entheogenic plants or fungi. And it is a view commonly reported by contemporary users in non-traditional settings, many of whom report a thinning of the boundaries between 'self' and 'other', and an experience of 'merging' with other organisms.

Is the world what it looks like in fine weather at noon day? Or is it what it seems to be at dawn when we first wake from sleep? Perhaps there are things that everyone can agree on. Whether or not fungi actually speak through humans and occupy our senses, the impact of psilocybin mushrooms on our thoughts and beliefs is real enough. If we imagined that a fungus could wear our minds and enjoyed splashing around in our consciousness, what would we expect to see? There might be songs sung about mushrooms, statues of mushrooms, paintings of mushrooms, myths and stories in which mushrooms play leading roles, ceremonies built around the celebration of mushrooms, a global community of DIY mycologists developing new ways to cultivate mushrooms in their homes, mycological evangelists like Paul Stamets talking to large audiences about how mushrooms can save the world. And people like Terence McKenna who claim to be able to speak English for fungi.

Psilocybe semilanceata, or the 'liberty cap'

5

Before Roots

You'll never be free of me
He'll make a tree from me
Don't say goodbye to me
Describe the sky to me

Kathleen Brennan & Tom Waits

Some time around 600 million years ago, green algae began to move out of shallow fresh waters and onto the land. These were the ancestors of all land plants. The evolution of plants transformed the planet and its atmosphere and was one of the pivotal transitions in the history of life – a profound breakthrough in biological possibility. Today, plants make up 80 per cent of the mass of all life on Earth and are the base of the food chains that support nearly all terrestrial organisms.[1]

Before plants, land was scorched and desolate. Conditions were extreme. Temperatures fluctuated wildly and landscapes were rocky and dusty. There was nothing that we would recognise as soil. Nutrients were locked up in solid rocks and minerals, and the climate was dry. This isn't to say that land was completely devoid of life. Crusts made up of photosynthetic bacteria, extremophile algae and fungi were able to make a living in the open air. But the harsh conditions meant that life on Earth was overwhelmingly an aquatic event. Warm, shallow seas and lagoons teemed with algae and animals. Sea scorpions several metres long ranged the ocean floor. Trilobites ploughed silty seabeds using spade-like snouts. Solitary corals started to form reefs. Molluscs thrived.[2]

Despite its comparatively inhospitable conditions, land provided considerable opportunities for any photosynthetic organisms that could cope. Light was unfiltered by water, and carbon dioxide was more accessible – no small incentives for organisms that make a living by eating light and carbon dioxide. But the algal ancestors of land plants had no roots, no way to store or transport water, and no experience in extracting nutrients from solid ground. How did they manage the fraught passage onto dry land?

When it comes to piecing together origin stories it's difficult to find agreement among scholars. Evidence is usually sparse, and what fragments there are can often be mobilised to support different points of view. And yet, amid the slow-burning disputes that surround the early history of life, one piece of academic consensus stands out: it was only by striking up new relationships with fungi that algae were able to make it onto land.[3]

These early alliances evolved into what we now call mycorrhizal relationships. Today, more than 90 per cent of all plant species depend on mycorrhizal fungi. They are the rule, not the exception: a more fundamental part of planthood than fruit, flowers, leaves, wood or even roots. Out of this intimate partnership – complete with co-operation, conflict and competition – plants and mycorrhizal fungi enact a collective flourishing that underpins our past, present and future. We are unthinkable without them, yet seldom do we think about them. The cost of our neglect has never been more apparent. It is an attitude we can't afford to sustain.[4]

As we've seen, algae and fungi have a tendency to partner with one another. Their association can take many forms. Lichens are one example. Seaweeds – also algae – are another; many seaweeds washed up on shorelines depend on fungi to nourish them and prevent them from drying out. And then there are the soft green balls produced in days by the Harvard researchers when they introduced free-living fungi and algae to one another. As long as fungi and algae have a good ecological fit – as long as they sing a metabolic 'song' together that neither can sing alone – they will coalesce into entirely new symbiotic relationships. In this sense, the union of fungi and algae that gave rise to plants is part of a larger story, an evolutionary refrain.[5]

Whereas the partners in lichens come together to make a body altogether unlike those of their individual members, the partners in a mycorrhizal relationship do not: plants stay recognisable as plants, and mycorrhizal fungi stay recognisable as fungi. This makes for a very different, more promiscuous type of symbiosis, in which a single plant can be coupled to many fungi at once, and a single fungus can be coupled to many plants.

For the relationship to thrive, both plant and fungus must make a good metabolic match. It is a familiar pact. In photosynthesis, plants harvest carbon from the atmosphere and forge the energy-rich carbon compounds – sugars and lipids – on which much of the rest of life depends. By growing within plant roots, mycorrhizal fungi acquire privileged access to these sources of energy: they get fed. However, photosynthesis isn't enough to support life. Plants and fungi need more than a source of energy. Water and minerals must be scavenged from the ground – full of textures and micropores, electrically charged cavities and labyrinthine rot-scapes. Fungi are deft rangers in this wilderness and can forage in a way that plants can't. By hosting fungi within their roots, plants gain hugely improved access to these sources of nutrients. They, too, get fed. By partnering, plants gain a prosthetic fungus, and fungi gain a prosthetic plant. Both use the other to extend their reach. It is an example of Lynn Margulis' 'long-lasting intimacy of strangers'. Except that they're hardly strangers any more. Look inside a root, and this becomes clear.

Roots turn into worlds under a microscope. I've spent weeks immersed in them, sometimes enthralled, sometimes frustrated. Put fresh, fine roots in a dish of water and you'll see fungal hyphae stringing off them. Boil roots in dye, squash them onto a glass slide, and you'll see an intertwining. Fungal hyphae fork and fuse and erupt within plant cells in a riot of branching filaments. Plant and fungus clasp one another. It's difficult to imagine a more intimate set of poses.

The strangest thing I've seen under a microscope is germinating dust seeds. Dust seeds are the smallest plant seeds in the world. A single seed is just visible to the naked eye like a small hair or the tip of an eyelash. Orchids make them, as do some other plants. They weigh almost nothing and disperse easily with wind or rain. And they won't germinate until they've met a fungus. I spent a long time trying to catch them in the act. I buried thousands of dust seeds in small bags and dug them up after a few months hoping that some would have sprouted. Under the microscope I pushed seeds around a glass dish with a needle searching for signs of life. After several days, I found what I was looking for. Some seeds had swollen into fleshy clumps tangled up in fungal hyphae, sticky streamers that trailed out into the dish. Inside the developing roots, hyphae ravelled into knots and coils. This wasn't sex: fungal and plant cells hadn't fused and pooled their genetic information. But it was sexy: cells from two different creatures had met, incorporated each other, and were collaborating in the building of a new life. To imagine the future plant as separable from the fungus was absurd.

It isn't clear how mycorrhizal relationships first arose. Some venture that the earliest encounters were soggy, disorganised affairs: fungi seeking food and refuge within algae that washed up onto the muddy shores of lakes and rivers. Some propose instead that the algae arrived on land with their fungal partners already in tow. Either way, explained Katie Field, a professor at the University of Leeds, 'they soon became dependent on each other'.

Field is a brilliant experimentalist who has spent years studying the most ancient lineages of plants alive today. Using radioactive tracers, she measures the exchanges taking place between fungi and plants in growth chambers that simulate ancient climates. Their symbiotic manners provide clues about how plants and fungi behaved towards one another in the earliest stages of their migration onto land. Fossils, too, provide a striking glimpse of these early alliances. The finest specimens date from around 400 million years ago and bear the unmistakable imprint of mycorrhizal fungi within them: feathery lobes that look just as they do today. 'You can see the fungus actually living in the plant cells,' Field marvelled.[6]

The earliest plants were little more than puddles of green tissue, with no roots or other specialised structures. Over time, they evolved coarse fleshy organs to house their fungal associates, who scavenged the soil for nutrients and water. By the time the first roots evolved, the mycorrhizal association was already some 50 million years old. Mycorrhizal fungi are the roots of all subsequent life on land. The word 'mycorrhiza' has it right. Roots (*rhiza*) followed fungi (*mykes*) into being.[7]

Today, hundreds of millions of years later, plants have evolved thinner, faster-growing, opportunistic roots that behave more like fungi. But even these roots can't outmanoeuvre fungi when it comes to exploring the soil. Mycorrhizal hyphae are fifty times finer than the finest roots and can exceed the length of a plant's roots by as much as a hundred times. They came before roots, and range beyond roots. Some researchers take it further. 'Plants don't have roots,' one of my undergraduate professors confided to a class of astounded students. 'They have fungus-roots, myco-rhizas.'[8]

Mycorrhizal fungi are so prolific that their mycelium makes up between a third and a half of the living mass of soils. The numbers are astronomical. Globally, the total length of mycorrhizal hyphae in the top ten centimetres of soil is around half the width of our galaxy (4.5×10^{17} kilometres versus 9.5×10^{17} kilometres). If these hyphae were ironed into a flat sheet, their combined surface area would cover every inch of dry land on Earth two and a half times over. However, fungi don't stay still. Mycorrhizal hyphae die back and regrow so rapidly – between ten and sixty times per year – that over a million years their cumulative length would exceed the diameter of the known universe (4.8×10^{10} light years of hyphae, versus 9.1×10^{9} light years in the known universe). Given that mycorrhizal fungi have been around for some 500 million years and aren't restricted to the top ten centimetres of soil, these figures are certainly underestimates.[9]

In their relationship, plants and mycorrhizal fungi enact a polarity: plant shoots engage with the light and air, while the fungi and plant roots engage with the solid ground. Plants pack up light and carbon dioxide into sugars and lipids. Mycorrhizal fungi unpack nutrients bound up in rock and decomposing material. These are fungi with a dual niche: part of their life happens within the plant, part in the soil. They are stationed at the entry point of carbon into terrestrial life cycles and stitch the atmosphere into relation with the ground. To this day, mycorrhizal fungi help plants cope with drought, heat and the many other stresses life on land has presented from the very beginning, as do the symbiotic fungi that crowd into plant leaves and stems. What we call 'plants' are in fact fungi that have evolved to farm algae, and algae that have evolved to farm fungi.

The word 'mycorrhiza' was coined in 1885 by the German biologist Albert Frank – the same Albert Frank whose fascination with lichens had led him to coin the word 'symbiosis', eight

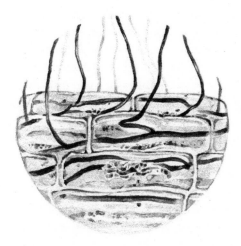

Mycorrhizal fungus inside a plant root

years earlier. He was subsequently employed by the Ministry of Agriculture, Domains and Forestry for the Kingdom of Prussia, to 'promote the possibility of truffle cultivation', a post that caused him to turn his attention towards the soil. As for many before and since, truffles were the lure that tugged him into a fungal underground.

Frank didn't have much success in cultivating truffles, but in his enquiries he documented in vivid detail the entanglement between tree roots and the mycelium of truffle fungi. His diagrams portray root tips entangled within a mycelial sleeve, with hyphae writhing outwards onto the page. Frank was struck by the intimacy of the association, and suggested that the relationship between plant roots and their companion fungi might be mutually beneficial rather than parasitic. As was common among scientists studying symbiosis, Frank used lichens as an analogy to make sense of the mycorrhizal association. In his view, plants and mycorrhizal fungi were bound in an 'intimate, reciprocal dependence'. Mycorrhizal mycelium behaved like a 'wet nurse', and enabled 'the entire nourishment of the tree from the soil'.[10]

Frank's ideas were fiercely attacked, as Simon Schwendener's dual hypothesis of lichens had been. For Frank's critics, the idea that the symbiosis could be of mutual benefit – a 'mutualism' – was a sentimental illusion. If one partner appeared to benefit, they did so at a price. Any symbiosis that appeared to be mutually beneficial was actually one of conflict and parasitism in disguise.[11]

Undeterred, Frank worked for ten years to understand plants' relationships with their fungal 'nurses'. He performed elegant experiments with pine seedlings. Some he grew in sterilised soil; some he grew in soil collected from a nearby pine forest. Those that grew in forest soil formed fungal relationships and developed into larger, healthier saplings than those grown in sterile conditions.[12]

Frank's findings caught the eye of J. R. R. Tolkien, who had a well-known fondness for plants, and trees in particular. Mycorrhizal fungi soon found their way into *The Lord of the Rings*.[13] 'For you, little gardener and lover of trees', said the elf Galadriel to the hobbit Sam Gamgee,

I have only a small gift … In this box there is earth from my orchard … if you keep it and see your home again at last, then perhaps it may reward you. Though you should find all barren and laid waste, there will be few gardens in Middle-earth that will bloom like your garden, if you sprinkle this earth there.

When he finally returned home to find a devastated Shire:

Sam Gamgee planted saplings in all the places where specially beautiful or beloved trees had been destroyed, and he put a grain of the precious dust from Galadriel in the soil at the root of each … All through the winter he remained as patient as he could, and tried to restrain himself from going round constantly to see if anything was happening. Spring surpassed his wildest hopes. His trees began to sprout and grow, as if time was in a hurry and wished to make one year do for twenty.

Tolkien could have been describing plant growth in the Devonian period, 300 to 400 million years ago. By now well established on land, and fuelled by high levels of light and carbon dioxide, plants spread across the world and evolved larger and more complex forms faster than at any time before. Metre-tall trees evolved into thirty-metre-tall trees in a few million years. Over this period, as plants boomed, the amount of carbon dioxide in the atmosphere dropped by 90 per cent, triggering a period of global cooling. Could plants and their fungal associates have played a part in this massive atmospheric transformation? A number of researchers, Field included, think it's probable.[14]

'The levels of carbon dioxide in the atmosphere drop off dramatically at the same time as land plants are evolving increasingly complex structures,' Field explained. The surge in plant productivity in turn depended on their mycorrhizal partners. It's a predictable sequence of events. One of the biggest limits to plant growth is a scarcity of the nutrient phosphorus. One of the things that mycorrhizal fungi do best – one of their most prominent metabolic 'songs' – is to mine phosphorus from the soil and transfer it to their plant partners. If plants are fertilised with phosphorus, they grow more. The more plants grow, the more they draw down carbon dioxide from the atmosphere. The more plants live, the more plants die, and the more carbon is buried in soils and sediments. The more carbon that is buried, the less there is in the atmosphere.

Phosphorus is only part of the story. Mycorrhizal fungi deploy acids and high pressure to burrow into solid rock. With their help, plants in the Devonian period were able to mine minerals like calcium and silica. Once unlocked, these minerals react with carbon dioxide, pulling it out of the atmosphere. The resulting compounds – carbonates and silicates – find their way into the oceans where they are used by marine organisms to make their shells. When the organisms die, the shells sink and pile up hundreds of metres thick on the ocean floor, which becomes an enormous burial ground for carbon. Add all of this up and climates start to change.[15]

Is there a way to measure the impact of mycorrhizal fungi on ancient global climates, I wondered.

'Yes and no,' Field replied. 'I recently tried.' To do so, she collaborated with the biogeochemist Benjamin Mills, a fellow researcher at the University of Leeds, who works with computer models that give predictions about the climate and the composition of the atmosphere.[16]

Lots of researchers build climate models. Weather forecasters and climate scientists depend on these digital simulations to predict future scenarios. So do researchers trying to reconstruct major transitions in the planet's past. By varying the numbers dialled into the model, one can test different hypotheses about the history of the Earth's climate. Turn up carbon dioxide, and what happens? Turn down the amount of phosphorus that plants can access, and what happens? The model can't say what actually occurred, but it can tell us which factors are capable of making a difference.

Before Field approached him, Mills hadn't included mycorrhizal fungi in the model. He could vary the amount of phosphorus that plants could obtain. However, without taking account of mycorrhizal fungi, there is no way to make realistic estimates of how much phosphorus plants were able to access. Field could help. In a series of experiments, she had found that the outcome of mycorrhizal relationships varied depending on the climatic conditions in her growth chambers. Sometimes plants benefited more from the relationship, and sometimes less, a trait she terms 'symbiotic efficiency'. If plants are hitched to an efficient mycorrhizal partner, they receive more phosphorus and grow more. Field was able to estimate how efficient mycorrhizal exchange would have been around 450 million years ago, when atmospheric carbon dioxide levels were several times higher than they are today.

When Mills added mycorrhizal fungi to the model using Field's measurements, he found that it was possible to change the entire global climate simply by turning the symbiotic efficiency up or down. The amount of carbon dioxide and oxygen in the atmosphere, and global temperatures – all varied according to the efficiency of mycorrhizal exchange. Based on Field's data, mycorrhizal fungi would have made a substantial contribution to the dramatic drawdown of carbon dioxide that followed the plant boom in the Devonian period. 'It's one of those moments where you think, wow, actually, hang on!' Field exclaimed. 'Our results suggest that mycorrhizal relationships have played a role in the evolution of much of life on Earth.'[17]

They continue to do so. The book of Isaiah in the Old Testament has it that 'all flesh is grass'. It is a logic that we might today describe as ecological: in animal bodies, grass becomes flesh. But why stop there? Grass only becomes grass when sustained by the fungi that live in its roots. Does this mean that all grass is fungus? If all grass is fungus, and all flesh is grass, does it follow that all flesh is fungus?

Maybe not all, but certainly some: mycorrhizal fungi can provide up to 80 per cent of a plant's nitrogen, and as much as 100 per cent of its phosphorus. Fungi supply other crucial nutrients to plants, such as zinc and copper. They also provide plants with water, and help them to survive drought as they've done since the earliest days of life on land. In return, plants allocate up to 30 per cent of the carbon they harvest to their mycorrhizal partners. Exactly what is taking place between a plant and a mycorrhizal fungus at any given moment depends on who's involved. There are many ways to be a plant, and many ways to be a fungus. And there are many ways to form a mycorrhizal relationship: it is a way of life that has evolved on over sixty separate occasions in different fungal lineages since algae first migrated onto land. As with many traits that have defied the odds to evolve more than once – whether the ability to hunt nematodes, form lichens or manipulate animal behaviour – it is hard to avoid the feeling that these fungi have stumbled upon a winning strategy.[18]

A plant's fungal partners can have a noticeable impact on its growth – and its flesh. A number of years ago, at a conference on mycorrhizal relationships, I met a researcher who had been growing strawberry plants with different communities of mycorrhizal fungus. The experiment was simple. If the same species of strawberry was grown with different species of fungus, would the flavour of the strawberries change? He conducted blind taste tests and found that different fungal communities did seem to change the flavour of the fruit. Some had more flavour, some were juicier, some were sweeter.

When he repeated the experiment a second year running, unpredictable weather swamped the effects of mycorrhizal fungi on the taste of the strawberries, but a number of other striking effects surfaced. Bumblebees were more attracted to the flowers of strawberry plants grown with some fungal species, and less attracted to others. Plants grown with some mycorrhizal species produced more berries than others. And the appearance of the berries changed depending which fungi they partnered with. Some mycorrhizal communities made the berries look more appealing, some less so.[19]

Strawberries aren't alone in being sensitive to the identity of their fungal partners. Most plants – from a potted snapdragon to a giant sequoia – will develop differently when grown with different communities of mycorrhizal fungus. Basil plants, for example, produce different profiles of the aromatic oils that make up their flavour when grown with different mycorrhizal strains. Some fungi have been found to make tomatoes sweeter than others; some change the essential oil profile of fennel, coriander and mint; some increase the concentration of iron and carotenoids in lettuce leaves, the antioxidant activity in artichoke heads, or the concentrations of medicinal compounds in St John's wort and echinacea. In 2013, a team of Italian researchers baked loaves of bread using wheat that had been grown with different mycorrhizal communities. The bread was subjected to testing with an electronic nose, and a tasting panel consisting of ten 'well-trained testers' at the University of Gastronomic Sciences in Bra, Italy. (Each tester, the authors assert reassuringly, 'had a minimum of two years' experience in sensory evaluation'.) Surprisingly, given how many stages occur between harvest and tasting – milling, mixing and baking, besides the addition of yeast – both the panel and the electronic nose were able to tell the loaves apart. The bread grown with an enhanced mycorrhizal fungal community had a higher 'flavour intensity', and improved 'elasticity and crumbliness'. By smelling a flower, by chewing on twigs, leaves or bark, by drinking a wine, how many other aspects of a plant's mycorrhizal underground might we be able to taste? I often wonder.[20]

'How delicate is the mechanism by which the balance of power is maintained among members of the soil population,' reflected the mycologist Mabel Rayner in *Trees and Toadstools,* a book on mycorrhizal relationships, published in 1945. Different species of mycorrhizal fungus might cause a basil leaf to taste different or a strawberry plant to produce more delicious-looking berries. But how? Are some fungal partners 'better'

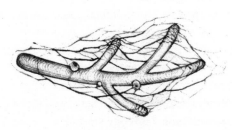

Mycorrhizal root tip

than others? Are some plant partners 'better' than others? Can plants and fungi tell the difference between alternative partners? Decades have elapsed since Rayner's remark, but we are only just beginning to understand the intricate behaviours that maintain a symbiotic balance between plants and mycorrhizal fungi.[21]

Social interactions are demanding. According to some evolutionary psychologists, humans' large brains and flexible intellects arose to allow us to navigate our way through complex social situations. Even the smallest interaction is embedded within a shifting social constellation. According to the *Chambers Dictionary of Etymology*, the word 'entangle' was originally used to describe such human interactions, or our involvement in 'complex affairs'. Not until later did the word take on other meanings. We humans became as clever as we are, so the argument goes, because we were entangled within a demanding flurry of interaction.[22]

Plants and mycorrhizal fungi don't have recognisable brains or intellects, but they certainly live entangled lives and have had to evolve ways to manage their complex affairs. Plants' actions are informed by what is happening in the sensory world of their fungal partners. Similarly, fungal behaviours are informed by what is happening in the sensory world of their plant partners. Using information from between fifteen and twenty different senses, a plant's shoots and leaves explore the air and adjust their behaviour based on continuous subtle changes in their surroundings. Anywhere from thousands to billions of root tips explore the soil, each able to form multiple connections to different fungal species. Meanwhile, a mycorrhizal fungus must sniff out sources of nutrients, proliferate within them, mingle with crowds of other microbes – whether fungal, bacterial or other – absorb the nutrients, and divert them around its rambling network of a body. Information must be integrated across an immense number of hyphal tips, which at any one moment can be strung between several different plants, and sprawl over tens of metres.

Toby Kiers, a professor at Vrije Universiteit Amsterdam, is one of the researchers who has done the most to investigate how plants and fungi maintain their 'balance of power'. Using radioactive labels, or by attaching light-emitting tags to molecules, she and her team are able to trace the carbon that moves from plant roots into fungal hyphae, and phosphorus that moves from fungi into plant roots. By carefully measuring these fluxes, she has been able to describe some of the ways in which both partners manage their exchange. How *do* plants and mycorrhizal fungi navigate their demanding social landscapes, I asked Kiers.

She laughed. 'We really want to get our hands on the complexity of what's happening. We know that trade is taking place. The question is whether we can predict how trading strategies change. It's overwhelming, but why not try?'

Kiers' findings are surprising because they suggest that neither plant nor fungus is in complete control of the relationship. Between them, they are able to strike compromises, resolve trade-offs and deploy sophisticated trading strategies. In one set of experiments, she found that plant roots were able to supply carbon preferentially to fungal strains that provided them with more phosphorus. In return, fungi that received more carbon from the plant supplied it with yet more phosphorus. Exchange was in some sense negotiated between the two depending on the availability of resources. Kiers hypothesised that these 'reciprocal rewards' have helped to keep plant and fungal associations stable over evolutionary time. Because both partners share control of the exchange, neither partner would be able to hijack the relationship for their own exclusive benefit.[23]

Although both plants and fungi tend to benefit from the relationship overall, different species of plant and fungus have different symbiotic manners. Some fungi make more co-operative partners; some are less co-operative and will 'hoard' phosphorus rather than exchange it with their plant partners. However, even a hoarder might not hoard all the time. Their behaviour is flexible, a set of ongoing negotiations that depend on what is taking place around them and in other parts of themselves. We don't know much about the workings of these behaviours, but it's clear that at any one moment plants and fungi face a number of options. And options entail choices, however those choices turn out to be made – whether in a conscious human mind, an unconscious computer algorithm or anything in between.[24]

Are plants and fungi making decisions, albeit brainless ones, I wondered.

'I use the word "decision" all the time,' Kiers told me. 'There's a set of options, and somehow information has to be integrated and one of the options has to be chosen. I think that a lot of what we are doing is studying micro-scale decisions.' There are many ways that these choices could unfold. 'Are there *absolute* decisions being made in every hyphal tip?' Kiers mused. 'Or is it all *relative*, in which case what happens would depend on what else is happening across the network.'

Intrigued by these questions, and having read Thomas Piketty's work on wealth inequality in human societies, Kiers began thinking about the role of inequality within fungal networks. She and her team exposed a single mycorrhizal fungus to an unequal supply of phosphorus. One part of the mycelium had access to a big patch of phosphorus. Another part had access to a small patch. She was interested in how this would affect the fungus's trading decisions in different parts of the same network. Some recognisable patterns emerged. In parts of a mycelial network where phosphorus was scarce, the plant paid a higher 'price', supplying more carbon to the fungus for every unit of phosphorus it received. Where phosphorus was more readily available, the fungus received a less favourable 'exchange rate'. The 'price' of phosphorus seemed to be governed by the familiar dynamics of supply and demand.[25]

Most surprising was the way that the fungus co-ordinated its trading behaviour across the network. Kiers identified a strategy of 'buy low, sell high'. The fungus actively transported phosphorus – using its dynamic microtubule 'motors' – from areas of abundance, where it fetched a low 'price' when exchanged with a plant root, to areas of scarcity, where it was in higher demand and fetched a higher 'price'. By doing so, the fungus was able to transfer a greater proportion of its phosphorus to the plant at the more favourable 'exchange rate', thus receiving larger quantities of carbon in return.[26]

How are these behaviours controlled? Can the fungus detect differences in 'exchange rate' across its network and actively transport phosphorus to 'play' the system? Or does it always transport phosphorus within its network from areas of abundance to areas of scarcity, sometimes receiving a 'pay-off' from the plant, and sometimes not? We still don't know. Nonetheless, Kiers' studies illuminate some of the intricacies of plant and fungal exchange, and show how solutions to complex challenges are able to emerge. All of these behaviours illustrate a general pattern. How a given plant or fungus behaves depends on *who* they find themselves partnering with and *where* they happen to be. One can think of mycorrhizal relationships as stretched along a continuum, with parasites at one pole and co-operative mutualists at the other. Some plants benefit from their fungal partners under some conditions and not under others. Grow plants with plenty of phosphorus, and they might become less picky about which fungal species they partner with. Grow co-operative fungi alongside other co-operative fungi, and they might become less co-operative. Same fungus, same plant, different setting, different outcome.[27]

One of my collaborators, a professor at the University of Marburg, told me about a sculpture he had seen as a child. *The Vertical Earth Kilometre* is a brass pole one kilometre long buried in the ground. The only visible part of it is the very end of the pole: a brass circle that lies flat on the floor and looks like a coin. He described the imaginative vertigo it had triggered in him, the sense of floating on the surface of an ocean of land, looking down into its depths. The experience inspired his lifelong fascination with roots and mycorrhizal fungi. I feel a similar sense of vertigo when I think about the complexity of mycorrhizal relationships – kilometres of entangled life – jostling beneath my feet.

The vertigo really sets in when I try to scale from the very small to the very large, from the microscopic trading decisions taking place at a cellular level, up to the entire planet, the atmosphere, the three trillion-odd trees that make their lives on land, and the quadrillions of kilometres of mycorrhizal fungi that weave them into relation with the soil. Our minds aren't good at keeping their balance when faced with numbers this big. Yet the story of mycorrhizal relationships makes many such dizzying swoops, from very large to very small and back again.[28]

Scale is an issue in the field of mycorrhizal research. Mycorrhizal relationships are conducted out of sight. It is hard to experience them, to see them or touch them. Their inaccessibility means that most knowledge of mycorrhizal behaviour comes from studies in controlled laboratory or greenhouse settings. Scaling up these findings to complex real-world ecosystems isn't always possible. Much of the time we only see a small part of the picture. The result is that researchers know more about what mycorrhizal fungi are capable of doing than what they're actually doing.[29]

Even in controlled settings, it's difficult to get a feel for how mycorrhizal fungi actually behave on a moment-to-moment basis. By contrast with Kiers' studies, there are situations in which plant and fungal exchanges don't seem to obey what we would recognise as rational trading strategies. Is something missing from our understanding? No-one can be sure. We have very little idea of exactly how the chemical exchange between plants and fungi takes place, and how it is controlled at a cellular level. 'We're trying to study how stuff moves within a network,' Kiers told me; 'we're trying to get videos of it. It's so crazy what's going on in there. But these studies are *hard*, and I can understand why people would want to work with different organisms.' Many mycologists share this combination of excitement and frustration.[30]

Are there other ways to think about these associations, other ways to quell the vertigo? Some of my colleagues find more intuitive outlets for their mycorrhizal enthusiasm. A number of them are passionate mushroom hunters. By foraging for mushrooms – from truffle to porcini to chanterelle to matsutake – they involve themselves with mycorrhizal relationships in a more spontaneous way. Others spend hours looking at mycorrhizal fungi under microscopes, which is almost the equivalent of a marine biologist going for a dive. Some of them spend hours sifting mycorrhizal spores from the soil, colourful orbs that under the microscope glisten like fish eggs. One of my colleagues in Panama was a skilled spore wrangler. Some evenings we made snacks from spores, fragments of cracker, and sour cream: tiny crumbs of mycorrhizal caviar that we had to prepare under the microscope and tweezer into our mouths. We didn't learn much, but that wasn't the point. It was an exercise that helped us to keep our balance as we careened from the small to the large. These were rare moments of unmediated contact with our experimental subjects, goofs to remind us that mycorrhizal fungi aren't mechanical schematic entities – one can't eat a machine or a concept – but living organisms engaged in lives that we still struggle to understand.

Plants remain the easiest way in. It is through plants that the mycorrhizal extravaganza below ground most commonly erupts into everyday human life. The countless microscopic interactions that occur between fungi and roots express themselves in the forms, growth, tastes and smells of plants. Sam Gamgee, like Albert Frank, could see the outcome of young trees' mycorrhizal relationships with his own eyes: the saplings began to sprout and grow 'as if time was in a hurry'. Eat a plant, and we taste the outgrowth of a mycorrhizal relationship. Cultivate plants – in a plant pot, flower bed, garden or city park – and we cultivate mycorrhizal relationships. Scale up yet further, and the microscopic trading decisions made by plants and fungi can shape the populations of forests across entire continents.

The last Ice Age ended around 11,000 years ago. As the vast Laurentide Ice Sheet retreated, it revealed millions of square kilometres of North America. Over a period of several thousand years forests expanded northwards. Using pollen records, it is possible to reconstruct the migration timelines of different species of tree. Some – beech, alder, pine, fir, maple – moved quickly, over a hundred metres per year. Some – plane, oak, birch, hickory – moved more slowly, around ten metres per year.[31]

What was it about these different species that determined their response to the changing climate? The relationship between fungi and the ancestors of plants allowed them to migrate onto dry land. Could mycorrhizal relationships have continued to play a part in plants' movements around the planet hundreds of millions of years later? It's possible. Neither plants nor fungi inherit each other. They inherit a tendency to associate, but they conduct what are, by the standards of many other ancient symbioses, open relationships. As in the earliest days of life on land, plants form their relationships depending on who's around. The same goes for fungi. Though this might be a limitation – a plant seed that finds no compatible fungi is unlikely to survive – the ability to reform their relationships, or evolve entirely new ones, can allow partners to respond to changing circumstances. A study published in 2018 by researchers at the University of British Columbia found that the speed of tree migration may indeed depend on their mycorrhizal proclivity. Some species of tree are more promiscuous than others and can enter into relationships with many different fungal species. As the Laurentide Ice Sheet retreated, the species that migrated faster were the more promiscuous ones, those that stood a better chance of meeting a compatible fungus when they arrived somewhere new.[32]

The fungi that live in plant leaves and shoots – known as 'endophytes' – can have similarly dramatic effects on a plant's ability to make a life in a new place. Take a grass from salty coastal soils, grow it without its fungal endophytes, and it won't be able to survive in its natural salty habitat. The same goes for grasses growing in hot geothermal soils. Researchers swapped the fungal endophytes that lived in each type of grass so that coastal grasses were grown with hot geothermal fungi and vice versa. The grasses' ability to survive in each habitat switched. Coastal grasses could no longer grow in salty coastal soils but thrived in hot geothermal soils. Hot geothermal grasses could no longer grow in the hot geothermal soils but thrived in the salty coastal soils.[33]

Fungi can determine which plants grow where; they can even drive the evolution of new species by isolating plant populations from one another. Lord Howe Island is 9 kilometres long, around a kilometre wide, and lies between Australia and New Zealand. On it grow two species of palm that have diverged from each other. One species, the Belmore sentry palm (*Howea belmoreana*), grows on acidic volcanic soils, while its sister species, the Kentia palm (*Howea forsteriana*), lives on alkaline chalky soils. What enabled the Kentia palm's radical switch of habitats has long puzzled botanists. A study published in 2017 by researchers at Imperial College London shows that mycorrhizal fungi are largely responsible. They found that the two palm species associate with different fungal communities. The Kentia palm is able to form relationships with fungi that allow it to live on the alkaline chalky soils. However, its ability to do so makes it difficult to form relationships with the mycorrhizal fungi in the ancestral volcanic soils. This means that the Kentia palm benefits only from the fungi present in the chalky soils, whereas the Belmore sentry palm benefits only from the fungi present in the volcanic soils. Over time, living on different mycorrhizal 'islands', though sharing the same tiny geographical island, one species became two.[34]

The ability of plants and mycorrhizal fungi to reshape their relationships has profound implications. We are familiar with the story: throughout human history, partnerships with other organisms have extended the reach of both humans and non-humans. Human relationships with corn brought about new forms of civilisation. Relationships with horses allowed new forms of transport. Relationships with yeast permitted new forms of alcohol production and distribution. In each case, humans and their non-human partners redefined their possibilities.

Horses and humans remain separate organisms, as do plants and mycorrhizal fungi, but both are echoes of an ancient tendency for organisms to associate. The anthropologists Natasha Myers and Carla Hustak argue that the word 'evolution', which literally means 'rolling outwards', doesn't capture the readiness of organisms to involve themselves in one another's lives. Myers and Hustak suggest that the word 'involution' – from the word 'involve' – better describes this tendency: a 'rolling, curling, turning inwards'. In their view, the concept of involution better captures the entangled pushing and pulling of 'organisms constantly inventing new ways to live with and alongside one another'. It was their tendency to involve themselves in the lives of others that enabled plants to borrow a root system for fifty million years while they evolved their own. Today, even with their own root systems, almost all plants still depend on mycorrhizal fungi to manage their underground lives. Their involutionary tendencies enabled fungi to borrow a photosynthesising alga to handle their atmospheric affairs. They still do. Mycorrhizal fungi are not built into plant seeds. Plants and fungi must constantly form and re-form their relationships. Involution is ongoing, and extravagant: by associating with one another, all participants wander outside and beyond their prior limits.[35]

Faced with catastrophic environmental change, much of life depends on the ability of plants and fungi to adapt to new conditions, whether in polluted or deforested landscapes or in newly created environments such as urban green roofs. Increases in atmospheric carbon dioxide, changes in climate and pollution all influence the microscopic trading decisions of plant roots and their fungal partners. As has long been the case, the influences of these trading decisions scale up and spill out over whole ecosystems and land masses. A large study published in 2018 suggested that the 'alarming deterioration' of the health of trees across Europe was caused by a disruption of their mycorrhizal relationships, brought about by nitrogen pollution. Mycorrhizal associations born of the Anthropocene will determine much of humans' ability to adapt to the worsening climate emergency. Nowhere are the possibilities – and pitfalls – more apparent than in agriculture.[36]

'On the efficiency of this mycorrhizal association the health and well-being of mankind must depend.' So wrote Albert Howard, a founding figure in the modern organic farming movement and a passionate spokesman for mycorrhizal fungi. In the 1940s, Howard argued that the widespread application of chemical fertilisers would disrupt mycorrhizal associations, the means by which 'the marriage of a fertile soil and the tree it nourishes … is arranged'. The consequences of such a breakdown would be far-reaching. To cut these 'living fungous threads' would be to reduce the health of the soil. In turn, the health and productivity of crops would suffer, as would the animals and people that consumed them. 'Can mankind regulate its affairs so that its chief possession – the fertility of the soil – is preserved?' Howard challenged. 'On the answer to this question the future of civilisation depends.'[37]

Howard's tone is dramatic, but eighty years on his questions cut deep. By some measures, modern industrial agriculture has been effective: crop production doubled over the second half of the twentieth century. But a single-minded focus on yield has incurred steep costs. Agriculture causes widespread environmental destruction and is responsible for a quarter of global greenhouse gas emissions. Between 20 and 40 per cent of crops are lost each year to pests and diseases, despite colossal applications of pesticide. Global agricultural yields have plateaued, despite a 700-fold increase in fertiliser use over the second half of the twentieth century. Worldwide, thirty football fields' worth of topsoil are lost to erosion every minute. Yet a third of food is wasted, and demand for crops will double by 2050. It is difficult to overstate the urgency of the crisis.[38]

Could mycorrhizal fungi form part of the solution? Perhaps it's a silly question. Mycorrhizal relationships are as old as plants and have been shaping Earth's future for hundreds of millions of years. They have forever featured in our efforts to feed ourselves, whether we've thought about them or not. For millennia in many parts of the world, traditional agricultural practices have attended to the health of the soil, and thus supported plants' fungal relationships implicitly. But over the course of the twentieth century, our neglect has led us into trouble. In 1940, Howard's greatest worry was that industrial agricultural techniques would develop without taking account of the 'life of the soil'. His concern was justified. In viewing soils as more or less lifeless places, agricultural practices have ravaged the underground communities that sustain the life we eat. There are parallels with much of twentieth-century medical science, which considered 'germ' and 'microbe' to mean the same thing. Of course some soil organisms, like some microbes that live on your body, can cause disease. Most do quite the opposite. Disrupt the ecology of microbes that live in your gut, and your health will suffer – a growing number of human diseases are known to arise because of efforts to rid ourselves of 'germs'. Disrupt the rich ecology of microbes that live in the soil – the guts of the planet – and the health of plants too will suffer.[39]

A study published in 2019 by researchers at Agroscope in Zurich measured the scale of the disruption by comparing the impact of organic and conventional 'intensive' farming practices on fungal communities in the roots of crops. By sequencing fungal DNA, the authors were able to compile networks showing which fungal species associated with one another. They found 'remarkable differences' between organic and conventionally managed fields. Not only was the abundance of mycorrhizal fungi higher in organically managed fields, but the fungal communities were also far more complex: twenty-seven species of fungi were identified as highly connected, or 'keystone species', compared with none in the conventionally managed fields. Many studies report similar findings. Intensive farming practices – through a combination of ploughing and application of chemical fertilisers or fungicides – reduce the abundance of mycorrhizal fungi and alter the structure of their communities. More sustainable farming practices, organic or otherwise, tend to result in more diverse mycorrhizal communities and a greater abundance of fungal mycelium in the soil.[40]

Does it matter? Much of the story of agriculture is one of ecological sacrifice. Forests are cleared to make way for fields. Hedgerows are cleared to make way for bigger fields. Surely it is the same with the communities of microbes in the soil? If humans feed crops by adding fertiliser to fields, don't we take over the job of mycorrhizal fungi? Why care about the fungi if we have made them redundant?

Mycorrhizal fungi do more than feed plants. The researchers at Agroscope describe them as keystone organisms, but some prefer the term 'ecosystem engineers'. Mycorrhizal mycelium is a sticky living seam that holds soil together; remove the fungi, and the ground washes away. Mycorrhizal fungi increase the volume of water that the soil can absorb, reducing the quantity of nutrients leached out of the soil by rainfall by as much as 50 per cent. Of the carbon that is found in soils – which, remarkably, amounts to twice the amount of carbon found in plants and the atmosphere combined – a substantial proportion is bound up in tough organic compounds produced by mycorrhizal fungi. The carbon that floods into the soil through mycorrhizal channels supports intricate food webs. Besides the hundreds or thousands of metres of fungal mycelium in a teaspoon of healthy soil, there are more bacteria, protists, insects and arthropods than the number of humans who have ever lived on Earth.[41]

Mycorrhizal fungi can increase the quality of a harvest, as the experiments with basil, strawberries, tomatoes and wheat illustrate. They can also increase the ability of crops to compete with weeds and enhance their resistance to diseases by priming plants' immune systems. They can make crops less susceptible to drought and heat, and more resistant to salinity and heavy metals. They even boost the ability of plants to fight off attacks from insect pests by stimulating the production of defensive chemicals. The list goes on: the literature is awash with examples of the benefits that mycorrhizal relationships provide to plants. However, putting this knowledge into practice is not always straightforward. For one thing, mycorrhizal associations don't always increase crop yields. In some cases, they can even reduce them.[42]

Katie Field is one of the many researchers being funded to develop mycorrhizal solutions to agricultural problems. 'The whole relationship is much more plastic and affected by the environment than we thought,' she told me. 'A lot of the time the fungi *aren't* helping the crops take up nutrients. The results are super-variable. It totally depends on the type of fungus, the type of plant and the environment in which it's growing.' A number of studies report similarly unpredictable outcomes. Most modern crop varieties have been developed with little thought for their ability to form high-functioning mycorrhizal relationships. We've bred strains of wheat to grow fast when they are given lots of fertiliser, and ended up with 'spoilt' plants that have almost lost the ability to co-operate with fungi. 'The fact that the fungi are colonising these cereal crops at all is a minor miracle,' Field pointed out.[43]

The subtleties of mycorrhizal relationships mean that the most obvious intervention – supplementing plants with mycorrhizal fungi and other microbes – can cut two ways. Sometimes, as Sam Gamgee found, introducing plants to a community of soil microbes can support the growth of crops and trees and help restore life to devastated soils. However, the success of this approach depends on the ecological 'fit'. Poorly matched mycorrhizal species might do more harm to plants than good. Worse, introducing opportunistic fungal species to new environments might displace local fungal strains with unknown ecological consequences. It is a fact not always taken into account by the fast-growing industry of commercial mycorrhizal products, often marketed as one-size-fits-all quick fixes. As in the ballooning market for human probiotics, many of the microbial strains sold are selected not because they are particularly suitable, but because they are easy to produce in manufacturing facilities. Even if done wisely, seeding an environment with microbial strains can only do so much. Like any organism, mycorrhizal fungi must be provided with the conditions to thrive. The soil's microbial communities live in a state of ongoing assembly and won't hold together for long in the face of continued disruption. For microbial interventions to be effective, more profound changes to agricultural practices are required, analogous to the changes in diet or lifestyle we might make in an effort to restore health to damaged gut flora.[44]

Other researchers are approaching the problem from a different angle. If humans have unthinkingly bred varieties of crops that form dysfunctional symbioses with fungi, surely we can turn around and breed crops that make high-functioning symbiotic partners? Field is taking this approach, and hopes to develop more co-operative plant varieties, 'a new generation of super-crops that can form amazing associations with fungi'. Toby Kiers, too, is interested in these possibilities but looks at the question from the fungal point of view. Rather than breed more co-operative plants, she is working on breeding fungi that behave more altruistically: strains that hoard less, and possibly even put the needs of plants above their own.[45]

In 1940, Albert Howard professed that we lacked a 'complete scientific explanation' of mycorrhizal relationships. Scientific explanations remain far from complete, but prospects for working with mycorrhizal fungi to transform agriculture and forestry and to restore barren environments have only increased as environmental crises have worsened. Mycorrhizal relationships evolved to deal with the challenges of a desolate and windswept world in the earliest days of life on land. Together, they evolved a form of agriculture, although it is not possible to say whether plants learned to farm fungi, or fungi learned to farm plants. Either way, we're faced with the challenge of altering our behaviour so that plants and fungi might better cultivate one another.[46]

It's unlikely we'll get far unless we question some of our categories. The view of plants as autonomous individuals with neat borders is causing destruction. 'Consider a blind man with a stick,' wrote the theorist Gregory Bateson. 'Where does the blind man's self begin? At the tip of the stick? At the handle of the stick? Or at some point halfway up the stick?' The philosopher Maurice Merleau-Ponty employed a similar thought experiment nearly thirty years earlier. He concluded that a person's stick was no longer just an object. The stick extends their senses, and becomes part of their sensory apparatus, a prosthetic organ of their body. Where the person's self begins and ends is not as straightforward a question as it might seem at first glance. Mycorrhizal relationships challenge us with a similar question. Can we think about a plant without also thinking about the mycorrhizal networks that lace outward – extravagantly – from its roots into the soil? If we follow the tangled sprawl of mycelium that emanates from its roots, then where do we stop? Do we think about the bacteria that surf through the soil along the slimy film that coats roots and fungal hyphae? Do we think about the neighbouring fungal networks that fuse with those of our plant? And – perhaps most perplexing of all – do we think about the other *plants* whose roots share the very same fungal network?[47]

6

Wood Wide Webs

Gradually, the observer realises that these organisms are connected to each other, not linearly, but in a net-like, entangled fabric.

Alexander von Humboldt

In the Pacific North-west of North America the forests are overwhelmingly green. So I am startled by clumps of bright white plants that push their way up through the drifts of fallen fir needles. These ghost plants don't have leaves. They look like clay tobacco pipes balanced on their ends. Small scales wrap around their stalks where their leaves ought to be. They sprout from deeply shaded parts of the forest floor where no other plant can grow and crowd in tight clusters as some mushrooms do. In fact, if they weren't so clearly flowers one would think they were mushrooms. Their name is *Monotropa uniflora*, and they are plants pretending not to be.

Monotropa – 'ghost pipes' – long ago gave up their ability to photosynthesise. With it, they abandoned leaves and their green colour. But how? Photosynthesis is one of the most ancient of plant habits. In most cases, it is a non-negotiable feature of planthood. Yet *Monotropa* have left it behind. Imagine discovering a species of monkey that doesn't eat, and instead harbours photosynthetic bacteria in its fur, which it uses to make energy from sunlight. It's a radical departure.

The solution is fungal. *Monotropa* – like the majority of green plants – depend on their mycorrhizal fungal partners to survive. However, their symbiotic manners differ. 'Normal' green plants supply energy-rich carbon compounds, whether sugars or lipids, to their fungal partners in exchange for mineral nutrients from the soil. *Monotropa* have worked out how to sidestep the exchange part. Instead, they receive *both* carbon and nutrients from mycorrhizal fungi, and don't appear to give anything back.

Then where does *Monotropa's* carbon come from? Mycorrhizal fungi obtain all their carbon from green plants. This means that the carbon that powers the life of *Monotropa* – the bulk of the stuff from which they are made – must ultimately come from other plants, via a shared mycorrhizal network: if carbon didn't flow from a green plant to *Monotropa* through shared fungal connections, *Monotropa* couldn't survive.

Monotropa have long puzzled biologists. In the late nineteenth century, a Russian botanist, wrestling with the question of how these strange plants were able to exist, was the first to suggest that substances could pass between plants via fungal connections. The idea didn't catch on. It was a passing conjecture buried in an obscure article and it sank more or less without a trace. The *Monotropa* riddle mouldered away for another seventy-five years before it was picked up by the Swedish botanist Erik Björkman, who in 1960 injected trees with radioactive sugars and was able to show that radioactivity accumulated in nearby *Monotropa* plants. It was the first demonstration that substances might pass between plants via a fungal pathway.[1]

Monotropa lured botanists into uncovering an entirely new biological possibility. Since the 1980s, it has become clear that *Monotropa* isn't an anomaly. Most plants are promiscuous and can engage with many mycorrhizal partners. Mycorrhizal fungi, too, are promiscuous in their relationships with plants. Separate fungal networks can fuse with each other. The result? Potentially vast, complex and collaborative systems of shared mycorrhizal networks.

*

Monotropa uniflora

'The fact that it's connected underground wherever we walk is just mind-blowing,' Toby Kiers enthused. 'It's huge. I can't believe everybody isn't studying it.' I share her sentiment. Lots of organisms interact. If one makes a map of who interacts with whom, one sees a network. However, fungal networks form physical connections between plants. It is the difference between having twenty acquaintances, and having twenty acquaintances with whom one shares a circulatory system. These shared mycorrhizal networks – known by researchers in the field as 'common mycorrhizal networks' – embody the most basic principle of ecology: that of the relationships between organisms. Humboldt's 'net-like, entangled fabric' was a metaphor he used to describe the 'living whole' of the natural world – a complex of relationships in which organisms are inextricably embedded. Mycorrhizal networks make the net and fabric real.[2]

One of the next people to pick up the *Monotropa* question and run with it was the English researcher David Read, who is among the most distinguished researchers in the history of mycorrhizal biology and a co-author of the definitive textbook on the subject. For his work on mycorrhizal associations he received a knighthood and was made a fellow of the Royal Society. Known by his colleagues in the United States as 'Sir Dude', Read is well known for his charm and fierce wit, and is most often described by fellow researchers as a 'character'.

In 1984, Read and his colleagues were the first to show conclusively that carbon could pass between normal green plants through fungal connections. Researchers had hypothesised that such a transfer could take place since the studies on *Monotropa* in the 1960s. But no-one had been able to demonstrate that the sugars hadn't leaked out of one plant's roots, passed into the open soil, and been taken up by the other plant's roots. No-one, in other words, had shown that the carbon moved between plants through a direct fungal channel.

Read devised an approach that allowed him to actually see the transfer of carbon from plant to plant. He grew 'donor' and 'receiver' plants next to each other, either with or without mycorrhizal fungi. After six weeks, he fed donor plants with radioactive carbon dioxide. He then harvested the plants and exposed their root systems to radiographic film. Where no mycorrhizal fungi are present, radioactivity is visible in the roots of the donor plant only. Where fungal networks are allowed to form, radioactivity is visible in the roots of the donor plant, the fungal hyphae and the roots of the receiver plants. Read's was a key advance. He had shown that the inter-plant transfer of carbon was not a habit unique to plants like *Monotropa*. However, larger questions remained. Read had conducted his experiments in a laboratory setting, and there was nothing to suggest that the inter-plant transfer of carbon could take place outside, in a natural environment.[3]

Thirteen years later, in 1997, a Canadian PhD student, Suzanne Simard, published the first study suggesting that carbon could pass between plants in a natural setting. Simard exposed pairs of tree seedlings growing in a forest to radioactive carbon dioxide. After two years, she found that carbon had passed from birch trees to fir trees, which shared a mycorrhizal network, but not between birch and cedar, which didn't. The amount of carbon obtained by the fir trees – on average 6 per cent of the labelled carbon taken up by the birch – was, by Simard's reckoning, a meaningful transfer: over time, one would expect this to make a difference to the life of the trees. What's more, when fir seedlings were shaded – which limited photosynthesis and deprived them of their supply of carbon – they received more carbon from their birch donors than when they were unshaded. Carbon seemed to flow 'downhill' between plants, from plenty to scarcity.[4]

Simard's finding turned heads. Her study was accepted by the journal *Nature*, and the editor asked David Read to write a commentary. In his piece, 'The ties that bind', Read suggested that Simard's study should 'stimulate us to examine forest ecosystems from a fresh standpoint'. Printed in large letters on the cover of the journal was a new phrase that Read had coined in his discussions with the editor of *Nature*: 'The Wood Wide Web'.[5]

Before the work of Read, Simard and others in the 1980s and 1990s, plants had been thought of as more or less distinct entities. Some species of tree have long been known to form root grafts, where the roots of one tree fuse with those of another. However, root grafts had been considered a marginal phenomenon, and most plant communities were understood to be made up of individuals that competed for resources. Simard's and Read's findings suggested that it might not be appropriate to think of plants as such neatly separable units. As Read wrote in his commentary in *Nature*, the possibility that resources could pass between plants suggested 'that we should place less emphasis on competition between plants, and more on the distribution of resources within the community.'[6]

Simard published her findings at a major moment in the development of modern network science. The network of cables and routers that comprises the Internet had been expanding since the 1970s. The World Wide Web – the system of information based on web pages and links between them, made possible by the hardware of the Internet – was invented in 1989 and became publicly available two years later. After the US National Science Foundation gave up its stewardship of the Internet in 1995, it began to expand in an uncontrolled and decentralised manner. As the network scientist Albert-László Barabási explained to me, 'it was in the mid 1990s that networks started to enter the public consciousness.'[7]

In 1998, Barabási and his colleagues embarked on a project to map the World Wide Web. Up to this point, scientists lacked the tools to analyse the structure and properties of complex networks, despite their prevalence in human life. The branch of mathematics that models networks – graph theory – was unable to describe the behaviour of most networks in the real world, and many questions remained unanswered. How could epidemics and computer viruses spread so quickly? Why could some networks continue to function despite massive disruption? Out of Barabási's study of the World Wide Web emerged new mathematical tools. A few key principles appeared to govern the behaviour of a wide range of networks, from human sexual relationships to biochemical interactions within organisms. The World Wide Web, Barabási remarked, appeared to have 'more in common with a cell or an ecological system than with a Swiss watch'. Today, network science is inescapable. Pick any field of study – from neuroscience to biochemistry, to economic systems, disease epidemics, web search engines, machine learning algorithms that underpin much of 'AI', to astronomy and the very structure of the universe itself, a cosmic web criss-crossed with filaments of gas and clusters of galaxies – the chances are that it makes sense of the phenomenon using a network model.[8]

As David Read explained to me, inspired by Simard's paper, and impelled by the catchy concept of the Wood Wide Web, 'the whole notion of shared mycorrhizal networks expanded prolifically' – eventually finding its way into James Cameron's *Avatar*, in the form of a glowing, living network that linked plants underground. Read's and Simard's studies had raised a number of exciting new questions. What, apart from carbon, might pass between plants? How common was this phenomenon in nature? Could the influence of these networks extend over whole forests or ecosystems? And what difference did they make?

No-one denies that shared mycorrhizal networks are widespread in nature. They are inevitable given the promiscuity of plants and fungi, and the readiness of mycelial networks to fuse with each other. However, not everyone is convinced that they do anything important.

On one hand, since Simard's 1997 paper in *Nature*, many studies have measured the transfer of substances between plants. Some have shown that not only carbon, but also nitrogen, phosphorus and water can pass between plants via fungal networks in meaningful quantities. A study published in 2016 found that 280 kilograms of carbon per hectare of forest could be transferred between trees via fungal connections. This is a substantial sum: 4 per cent of the total carbon pulled out of the atmosphere in a year by the same hectare of forest, and enough carbon to power an average home for a week. These findings imply that shared mycorrhizal networks have an important ecological role.[9]

On the other hand, a number of studies have failed to observe the transfer of substances between plants. In itself, this doesn't mean that shared mycorrhizal networks have no role to play: a germinating seedling that can plug into a large, pre-existing fungal network wouldn't have to supply the carbon needed to grow its own mycorrhizal network from scratch. Nonetheless, these findings suggest that it isn't straightforward to generalise from one ecosystem to another, or from one type of fungus to another. There are many situations where shared mycorrhizal networks don't seem to do much more for their plant partners than a single – 'private' – mycorrhizal partner would.[10]

One would expect the behaviour of shared mycorrhizal networks to be variable. There are many different types of mycorrhizal relationship, and different fungal groups can behave in quite different ways. Moreover, the symbiotic conduct of even a single plant and fungus can vary wildly depending on their circumstances. Nonetheless, the variety of experimental findings has given rise to a range of opinions within the research community. For some, the available evidence shows that shared mycorrhizal networks permit forms of interaction that aren't otherwise possible and can have a profound influence on the behaviour of ecosystems. Others interpret the evidence differently, and conclude that shared mycorrhizal networks don't enable unique ecological possibilities and are no more important to plants than sharing root space or air space with one another.[11]

Monotropa help to navigate the debate. In fact, they appear to settle it: their dependence on shared mycorrhizal networks is total. I brought up the subject with David Read, who took an unambiguous position: 'The idea that inter-plant transfer via a fungal pathway is never of any significance is patently absurd.' *Monotropa* plants are full-time receivers, vivid living testaments to the fact that shared mycorrhizal networks can support a unique way of life.

Monotropa are what's known as 'mycoheterotrophs'. 'Myco' because they depend on a fungus for their nutrition; 'heterotroph' (from 'hetero-', meaning 'other', and '-troph', meaning 'feeder') because they don't make their own energy from the sun and have to get it from somewhere else. It's an unlovely name for such charismatic plants. In Panama, where I studied *Voyria,* the blue-flowered mycoheterotroph, I started calling them 'mycohets' for short, though I admit it's not much of an improvement.

Monotropa and *Voyria* aren't alone in living this way. About 10 per cent of plant species share the habit. Like lichens and mycorrhizal relationships, mycoheterotrophy is an evolutionary refrain, and has arisen independently in at least forty-six separate plant lineages. Some mycohets, like *Monotropa* and *Voyria,* never photosynthesise. Others behave like mycohets when they're young but become donors when they get older and start to photosynthesise, an approach that Katie Field calls 'take now, pay later'. As Read pointed out to me, *all* 25,000 species of orchid – 'the largest and arguably the most successful plant family on the surface of the Earth' – are mycohets at some stage in their development, whether they take now and pay later, or take now and continue to take later. That mycohets have repeatedly learned to hack the web for their benefit suggests it isn't such a hard trick to pull off. Indeed, for Read and a number of others, mycohets don't exist in an isolated category of their own. They are just the extreme pole of a symbiotic continuum; permanent takers that have lost their ability to pay later. Orchids that take now and pay later fall somewhere further towards the centre of the spectrum, as do Simard's fir seedlings.[12]

Mycohets are striking. Conspicuous, contrarian, they stand out from the ambient vegetation. With no reason to be green or to have leaves, they are free for evolution to carry them off in new aesthetic directions. There is a species of *Voyria* that is entirely yellow. The snow plant (*Sarcodes sanguinea*) is a brilliant red, 'like a bright glowing pillar of fire', wrote the American naturalist John Muir in 1912. It is 'more admired by tourists than any other [plant] in California … Its colour could appeal to one's blood.' (Muir mused on the 'thousand invisible cords' that strung Nature together, but he did not observe that this was literally the case for the snow plant.) It was *Voyria's* dust seeds that had so startled me when I found them germinating into fleshy bundles under a microscope. Marc-André Selosse, a professor at the Muséum National d'Histoire Naturelle in Paris, told me that it was the sight of a bright white mycohet orchid when he was fifteen that seeded his lifelong fascination with symbiosis. The orchid was a reminder of how inseparable plant and fungal lives were. 'The memory of this plant has been with me for my whole career so far,' he reflected fondly.[13]

I find mycohets interesting because of what they indicate about fungal life underground. Amid the riot of plant life in the jungle, *Voyria* were a sign of functioning shared fungal networks; it is by hacking wood wide webs that mycohets are able to live. Without having to perform fiddly experiments, *Voyria* allowed me to gauge whether meaningful quantities of carbon were being transferred between plants. I had the idea when talking to friends in Oregon who hunted for matsutake mushrooms. Matsutake are the fruiting bodies of a mycorrhizal fungus and are sometimes picked before they have poked their way through the forest floor. There's often a clue about where to start looking. Matsutake associate with a mycohet cousin of *Monotropa* that has a red and white striped stem, known as 'candy canes' (*Allotropa virgata*). Candy canes only associate with matsutake, and their presence is as sure an indicator of a thriving matsutake fungus as a matsutake mushroom itself. Candy canes, like many mycohets, serve as periscopes into the mycorrhizal underground.[14]

Given their allure one might expect that mycohets have been understood to indicate something over the years. If candy canes are literal indicators, used by matsutake hunters to locate underground networks of matsutake fungi, *Monotropa* served as a conceptual indicator to biologists. Lichens were the gateway organism for symbiosis in general; *Monotropa* was the gateway organism for shared mycorrhizal networks. Its peculiar appearance implied that material might pass between plants via shared fungal connections in sufficiently large quantities to support a whole way of life.

In all physical systems, energy moves 'downhill', from where there is more to where there is less. Heat travels from the hot sun into cold space. A truffle's scent drifts from areas of high concentration to lower concentration. Neither has to be actively transported. As long as there is an energetic slope, energy will move from the 'source' (at the top) to the 'sink' (at the bottom). What matters most is how steep the slope is between the two.

In many cases, the transfer of resources through mycorrhizal networks occurs downhill, from larger plants to smaller plants. Larger plants tend to have more resources, more developed root systems and more access to light. Relative to smaller plants growing in the shade with less developed root systems, these plants are sources. Smaller plants are sinks. Orchids that take now and pay later start off as sinks, and switch to become sources when they get older. Mycohets like *Monotropa* and *Voyria* remain sinks forever.[15]

Size isn't everything. Source–sink dynamics can switch, depending on the activity of the linked plants. When Simard shaded her fir seedlings – reducing their ability to photosyn-thesise and thereby making them stronger carbon sinks – they received more carbon from their birch 'donors'. In another instance, researchers observed phosphorus to pass from the roots of dying plants to those of nearby healthy plants that shared a fungal network. The dying plants were sources of nutrients, and the living plants were sinks.[16]

In another study of birch and Douglas fir in Canadian forests, the direction of carbon transfer switched *twice* in the course of a single growing season. In the spring, when the fir – an evergreen – was photosynthesising and the leafless birch was just bursting its buds, the birch behaved as a sink, and carbon flowed into it out of the fir. In the summer, when the birch was in full leaf, and the fir found itself in the shaded understory, the direction of carbon flow changed, moving downhill out of the birch, and into the fir. In the autumn, when the birch started to drop its leaves, the trees switched roles again, and carbon moved downhill from the fir into the birch. Resources passed from areas of abundance to areas of scarcity.[17]

These behaviours present a puzzle. At its most basic, the problem is this: why would plants give resources to a fungus that goes on to give them to a neighbouring plant – a potential competitor? At first glance it looks like altruism. Evolutionary theory doesn't cope well with altruism because altruistic behaviour benefits the receiver at the cost of the donor. If a plant donor assists a competitor at a cost to itself, its genes are less likely to make it into the next generation. If the altruist's genes don't make it into the next generation, the altruistic behaviour will soon be weeded out.[18]

There are a number of ways around this impasse. One relies on the idea that the 'costs' to donor plants aren't actually costs. Many plants have plenty of access to light. For such plants, carbon is not a limited resource. If a plant's surplus carbon passes into a mycorrhizal network where it is enjoyed by many as a 'public good', the charge of altruism can be avoided because no-one – whether donor or receiver – has incurred a cost. Another possibility is that *both* sender and receiver plants benefit, but at different times. An orchid might 'take now', but if it 'pays later', then no-one has incurred a cost overall. A birch may benefit when it receives carbon from a fir in the spring, but the fir will surely benefit from the carbon it receives from the birch during the high summer, when it finds itself in the shaded understory.[19]

There are other considerations. In evolutionary terms, it might be beneficial for a plant to assist a close relative to pass on its genes even at a cost to itself – a phenomenon known as 'kin selection'. Some studies have investigated this possibility by comparing the amount of carbon that passes between pairs of Douglas fir seedlings that are siblings, and pairs of seedlings that are unrelated. As one would expect, carbon moved downhill, from a larger donor plant to a smaller receiver plant. But in some cases, more carbon passed between siblings than between strangers: siblings appeared to share more fungal connections than did strangers, providing more pathways for carbon to move between them.[20]

The swiftest route through the puzzle is to switch perspective. You'll notice that in all these stories about shared mycorrhizal networks, plants have been the protagonists. Fungi have featured inasmuch as they connect plants and serve as a conduit between them. They become little more than a system of plumbing which plants can use to pump material between one another.

This is plant-centrism in action.

Plant-centric perspectives can distort. Paying more attention to animals than plants contributes to humans' plant-blindness. Paying more attention to plants than fungi makes us fungus-blind. 'I think many people elaborate about these networks more than they should,' Marc-André Selosse told me. 'Some people talk about trees benefiting from social care or retirement, describe young trees living in nurseries, and say that life is easy and cheap for trees living in a group. I don't much like these views because they portray the fungus as a pipeline. This is not the case. The fungus is a living organism with its own interests. It is an active part of the system. Maybe it is because plants are easier to investigate than fungi that many people take a very plant-centric view of the network.'

I agree. Surely we stumble into plant-centrism because the relevance of plants to our lives is more obvious. We can touch and taste them. Mycorrhizal fungi are evasive. The language of the Wood Wide Web doesn't help. It is a metaphor that tugs us into plant-centrism by implying that plants are equivalent to the webpages, or nodes, in the network, and fungi are the hyperlinks joining the nodes to one another. In the language of the hardware that comprises the Internet, plants are the routers and fungi are the cables.

In fact, fungi are far from being passive cables. As we've seen, mycelial networks can solve complex spatial problems, and have evolved a finely tuned ability to transport substances around themselves. Although material tends to move through fungal networks downhill, from source to sink, transport rarely occurs by passive diffusion alone: it is far too slow. The rivers of cellular fluid that flow within fungal hyphae allow rapid transport and, although these flows are ultimately governed by source–sink dynamics, fungi can direct the flow by growing, thickening and pruning back parts of the network – or indeed, fusing with another network entirely. Without the ability to regulate flow within their networks, much of fungal life – including the intricately choreographed growth of mushrooms – would be impossible.

Fungi are able to manage transport through their networks in other ways. As Toby Kiers' studies suggest, fungi have some degree of control over their trading patterns – whether they 'reward' more co-operative plant partners, 'hoard' minerals within their tissues or move resources around themselves to optimise the 'exchange rate' they obtain. In Kiers' study on resource inequality, phosphorus moved down a gradient from areas of abundance to areas of scarcity, but it did so much faster than passive diffusion would allow – probably transported using fungal microtubule 'motors'. These active transport systems allow fungi to shuttle material around their networks in any direction – even in both directions at once – regardless of the gradient between source and sink.[21]

The Wood Wide Web is a problematic metaphor for other reasons. The idea that there is a single kind of Wood Wide Web at all is misleading. Fungi make entangled webs whether or not they link plants together. Shared mycorrhizal networks are just a special case – fungal networks in which plants find themselves entangled. Ecosystems are riddled with webs of non-mycorrhizal fungal mycelium that stitch organisms into relation. The decomposer fungi that Lynne Boddy studies, for example, range through ecosystems over large distances and link decaying leaves with fallen twigs, large rotting stumps with decomposing roots, as do the world record-breaking networks of honey fungus that stretch for kilometres. These fungi make up wood wide webs of a different sort: webs based around consuming plants rather than sustaining them.

Every link in a wood wide web is a fungus with a life of its own. It's a small point that makes a big difference. Everything changes when we see fungi as active participants. Writing the fungus into the story encourages us to adopt a more fungal point of view. And a fungal point of view is helpful in asking whose interests are being served by shared mycorrhizal networks. Who stands to benefit?

A mycorrhizal fungus that can keep its various plants alive is at an advantage: a diverse portfolio of plant partners insures it against the death of one of them. If a fungus depends on several orchids, and one of them won't be able to supply it with carbon until it grows larger, the fungus will benefit by supporting the young orchid while it grows – to let it 'take now' – provided it will 'pay later'. Adopting a myco-centric perspective helps to avoid the problem of altruism. It also positions fungi front and centre: brokers of entanglement able to mediate the interactions between plants according to their own fungal needs.

Regardless of whether we take a myco-centric or a plant-centric perspective, there are many situations where sharing a mycor-rhizal network provides clear benefits to the plants involved: overall, plants that share a network with others grow more quickly and survive better than neighbouring plants that are excluded from the common network. These findings have fuelled visions of wood wide webs as places of caring, sharing and mutual aid through which plants can free themselves from the rigid hierarchies of competition for resources. These renditions are not unlike the starry-eyed fantasies of the Internet, proclaimed in the fervour of the 1990s to be an escape route from the rigid power structures of the twentieth century and an entrance into a digital utopia.[22]

Ecosystems, like human societies, are rarely so one-dimensional. Some researchers, such as David Read, feel that utopian visions of the soil are a shameless projection of human values onto a non-human system; others, such as Toby Kiers, argue that they ignore the many ways in which collaboration is always an alloy of competition *and* co-operation. The main problem for the myco-utopia is that, like the Internet, shared mycorrhizal networks aren't always beneficial. Wood wide webs are complex amplifiers of plant, fungal and bacterial interactions.

Most of the studies that have found that plants benefit from their involvement with shared mycorrhizal networks have taken place in temperate climates with trees that form relationships with a particular type of mycorrhizal fungus – the 'ectomycorrhizal' fungi. Other types of mycorrhizal fungus can behave differently. In some cases, it appears to make little difference to a plant whether it has its own private fungal network, or whether it shares a fungal network with other plants – although in these situations the fungus still benefits from forming a shared network by gaining access to a larger number of plant partners. In some cases, belonging to a shared network can bring outright disadvantages to plants. Fungi are in control of the supply of minerals they obtain from the soil and can preferentially trade these nutrients with their larger plant partners, which are both more abundant sources of carbon and stronger sinks for soil minerals. These asymmetries can magnify the competitive advantage of larger plants over smaller plants that share the network. In these situations, smaller plants start to benefit only when their connections to the network are severed, or when the bigger plants that share the network – and which have been extracting a disproportionately large quantity of nutrients – are cut back.[23]

Shared mycorrhizal networks can have yet more ambiguous consequences. A number of plant species produce chemicals that stunt or kill plants growing nearby. Under normal conditions, the passage of these chemicals through the soil is slow, and they don't always reach toxic concentrations. Mycorrhizal networks can help to overcome these limitations, in some cases providing a 'fungal fast lane' or 'superhighway' for plants that broadcast poisonous deterrents. In one experiment, a toxic compound released from the fallen leaves of walnut trees was able to travel through mycorrhizal networks and accumulate around the roots of tomato plants, reducing their growth.[24]

Wood wide webs, in other words, are about far more than the movement of resources – whether energy-rich carbon compounds, nutrients or water. Besides poisons, hormones that regulate plant growth and development can pass through shared mycorrhizal networks. In many species of fungus, DNA-containing nuclei and other genetic elements such as viruses or RNA are free to travel through the mycelium, suggesting that genetic material might pass between plants via a fungal channel – although these possibilities have barely been explored.[25]

One of the most surprising properties of wood wide webs is the way they enfold organisms other than plants. Fungal networks provide highways for bacteria to migrate around the obstacle course of the soil. In some cases, predatory bacteria use mycelial networks to pursue and hunt their prey. Some bacteria make their lives within fungal hyphae themselves, and enhance fungal growth, stimulate their metabolisms, produce key vitamins and even influence fungal relationships with their plant partners. One species of mycorrhizal fungus, the thick-footed morel (*Morchella crassipes*), actually farms the bacteria that live within its networks: the fungus 'plants' bacterial populations, then cultivates, harvests and consumes them. There is a division of labour across the network, with some parts of the fungus responsible for food production and some for consumption.[26]

There are yet more extravagant possibilities. Plants emit all sorts of chemicals. When broad bean plants are attacked by aphids, for example, they release plumes of volatile compounds that drift out from the wound and attract parasitic wasps that prey on the aphids. These 'infochemicals' – so-called because they convey information about a plant's condition – are one of the ways plants communicate, both between different parts of their own bodies, and with other organisms.

Might infochemicals pass between plants underground via shared fungal networks? It is a question that came to preoccupy Lucy Gilbert and David Johnson, then working at the University of Aberdeen in Scotland. To find out, they set up a deft experiment. Broad bean plants were either allowed to connect to a shared mycorrhizal network or prevented from doing so using a fine nylon mesh. The mesh allowed water and chemicals to pass, but prevented direct contact between the fungi connected to different plants. Once the plants had grown, aphids were allowed to attack the leaves of one of the plants in the network. Plastic bags placed over the plants prevented the transmission of infochemicals through the air.[27]

Gilbert and Johnson found clear confirmation of their hypothesis. Plants that were connected to the aphid-infested plant via a shared fungal network ramped up their production of volatile defence compounds, even though they had not encountered the aphids themselves. The plumes of volatile compounds produced by the plants were large enough to attract the parasitic wasps, suggesting that information passing between the plants through the fungal channel could make a difference in a real-world setting. Gilbert described this to me as a 'completely new' finding. It revealed a previously unknown role for shared mycorrhizal networks. Not only could a donor plant influence a receiver, but its influence could leak out beyond the receiver in the form of volatile chemicals. A shared mycorrhizal network influenced not only the relationship between two plants, but also the relationship between two plants, their aphid pests and their wasp allies.[28]

Since 2013, it has become clear that Gilbert and Johnson's finding isn't an anomaly. A similar phenomenon has been

Piedmont white truffles, *Tuber magnatum*
COURTESY OF THE AUTHOR

Kika, the truffle-hunting Lagotto Romagnolo
COURTESY OF THE AUTHOR

TRUFFLE-HUNTING.— TRAINED HOGS ROOTING FOR THE VALUABLE ESCULENT.

An illustration from around 1890 captioned, "Truffle-hunting—Trained hogs rooting for the valuable esculent." The hogs are wearing muzzles to prevent them from eating the truffles they unearth. SAMANTHA VUIGNIER/CORBIS VIA GETTY IMAGES

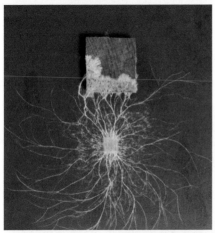
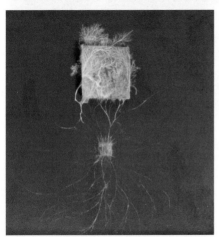

Foraging behavior of the wood-rotting fungus *Phanerochaete velutina*. The three images depict a single fungus growing over a period of 48 days. The mycelium starts in an exploratory mode, proliferating in all directions. When the fungus discovers something to eat, it reinforces the links that connect it with the food and prunes back the links that don't lead anywhere.

COURTESY OF YU FUKASAWA

Mycelium of a wood-rotting fungus exploring and consuming a log.
COURTESY OF ALISON POULIOT

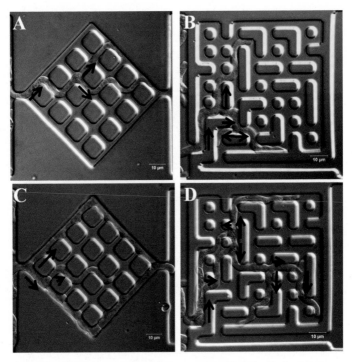

The bread mold *Neurospora crassa*, solving microscopic labyrinths. Black arrows mark the direction of growth of the fungus at branch points and at the entrance to the labyrinths. Image reproduced from Held, et al. (2010).

Bioluminescent ghost mushrooms, *Omphalotus nidiformis*. COURTESY OF ALISON POULIOT

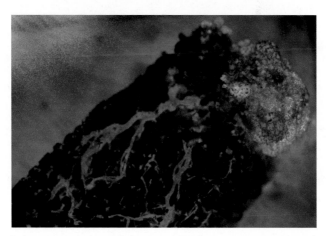

Bioluminescent mycelium of the bitter oyster, *Panellus stipticus*, growing on wood chips. The first submarine, *Turtle*, developed during the American Revolutionary War, used glowing fungi to illuminate its depth gauge. English coal miners in the nineteenth century reported fungus on pit props that cast enough light to see their hands by. COURTESY OF PATRICK HICKEY

Ernst Haeckel's lichens, published in *Art Forms in Nature* (1904).

Beatrix Potter's illustration of a *Cladonia* lichen.

A carpenter ant infected with the "zombie fungus" *Ophiocordyceps lloydii*. Two fungal fruiting bodies sprout from the body of the ant. Sample collected in the Brazilian Amazon. COURTESY OF JOÃO ARAÚJO

A carpenter ant infected with *Ophiocordyceps camponoti-nidulantis*. The fungus is visible as a white furry coating, and the stalk of the fungal fruiting body emerges from the back of the ant's head. Sample collected in the Brazilian Amazon. COURTESY OF JOÃO ARAÚJO

A carpenter ant infected with *Ophiocordyceps camponoti-atricipis*. A fungal fruiting body sprouts from the ant's head. Sample collected in the Brazilian Amazon. COURTESY OF JOÃO ARAÚJO

A carpenter ant infected with *Ophiocordyceps unilateralis*. The white spines belong to a different fungus, a "mycoparasite," that infects *Ophiocordyceps* fungi living on insect bodies. Sample collected in Japan. COURTESY OF JOÃO ARAÚJO

Ophiocordyceps growing around an ant's muscle fibers. Scale bar = 2 micrometers. COURTESY OF COLLEEN MANGOLD

A collection of mushroom stones from Guatemala, photographed in the early 1970s. Around two hundred such stones are thought to survive. These statues suggest that the ceremonial consumption of psilocybin mushrooms dates back at least until the second millenium BCE. GRANT KALIVODA, COURTESY OF CHARLOTTE SCHAARF

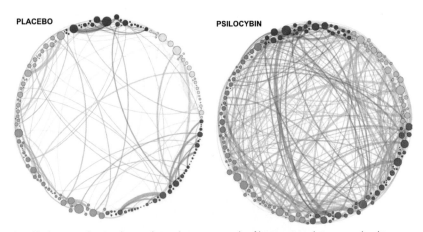

PLACEBO

PSILOCYBIN

Simplified cartoon showing the correlations between networks of brain activity during normal waking consciousness (left), and after an injection of psilocybin (right). Different networks are depicted as small colored circles around the rim of each figure. Following an injection of psilocybin, a tumult of new neuronal pathways arise. The ability of psilocybin to change people's minds seems related to these states of cerebral flux. Image reproduced from Petri, et al. (2014).

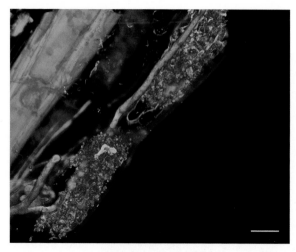

Mycorrhizal fungus living inside a plant root. The fungus is depicted in red, and the plant in blue. The finely branched structures within the plant cells are known as "arbuscules" ("little trees") and are the site of exchange between the plant and fungus. Scale bar = 20 micrometers.
COURTESY OF THE AUTHOR

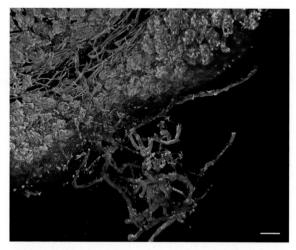

Mycorrhizal fungus growing into a plant root. The fungus is depicted in red, and the edge of the plant's root in blue. The inside of the root is densely inhabited by fungus. Scale bar = 50 micrometers. COURTESY OF THE AUTHOR

The mycoheterotroph *Voyria tenella* growing in a rainforest in Panama. Mycoheterotrophs— "hackers" of the wood wide web—have lost the ability to photosynthesize and draw their nutrients from mycorrhizal fungal networks that lace their way through soil.

The mycoheterotroph *Monotropa uniflora,* or "ghost pipe," growing in Adirondack Park, New York.

The mycoheterotroph *Sarcodes sanguinea,* John Muir's "glowing pillar of fire," growing in the El Dorado National Forest in California.

COURTESY OF TIMOTHY BOOMER

The mycoheterotroph *Allotropa virgata,* or "candycane," growing in Salt Point State Park in California.

COURTESY OF TIMOTHY BOOMER

Intimacies within intimacies. The roots of the mycoheterotroph *Voyria tenella* are densely inhabited by mycorrhizal fungi. In A, fungi are visible as a light-colored ring around the edge of the root. In B, the fungi are depicted in red and the plant material is not shown. Scale bar = 1 millimeter. Image reproduced from Sheldrake, et al. (2017)

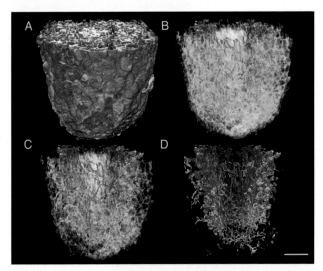

Mycorrhizal fungi living inside a root of the mycoheterotroph *Voyria tenella*. The fungus is depicted in red, and the plant root in grey. A to D show the same section of root with the plant tissue made increasingly transparent. Scale bar = 100 micrometers. Image reproduced from Sheldrake, et al. (2017)

The roots of the mycoheterotroph *Voyria tenella* are poorly adapted to the task of absorbing water and minerals from the soil and have evolved into fungal "farms." Note the mycorrhizal fungal hyphae trailing off the roots. Fragments of soil remain caught in the sticky mycelial web. This is a rare glimpse of the fungal connections that link plant roots with their surroundings. COURTESY OF THE AUTHOR

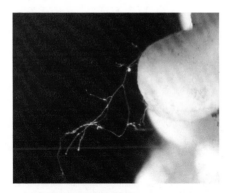

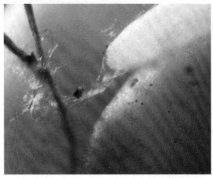

Mycorrhizal fungal mycelium on the roots of the mycoheterotroph *Voyria tenella*. COURTESY OF THE AUTHOR

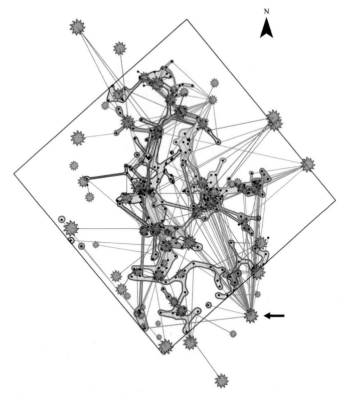

Above: A map of a shared fungal network made by Kevin Beiler. The green shapes are Douglas fir trees, and the straight lines indicate linkages between tree roots and mycorrhizal fungi. The black dots mark the points where Beiler collected samples. Genetically identical fungal networks are outlined in different colors. Networks formed by the mycorrhizal fungus *Rhizopogon vesiculosus* are shaded in blue, and those formed by the fungus *Rhizopogon vinicolor* are shaded in pink. The black border marks the 30 × 30 meter plot, and an arrow points to the most highly connected tree, which was linked to 47 other trees. Image reproduced from Beiler, et al. (2009).

Left: Peter McCoy's experiment with *Pleurotus*. The oyster mushroom is growing on a diet of nothing but used cigarette butts. The outside of a cigarette filter can be seen smeared across the inside of the jar. COURTESY OF PETER McCOY

observed with tomato plants attacked by caterpillars, and between Douglas fir and pine seedlings attacked by budworm. These studies open up exciting new possibilities. Many of the researchers I have talked with share the view that plant communication through fungal networks is one of the most compelling aspects of mycorrhizal behaviour. However, good experiments throw up more questions than they answer. 'What is it that plants are actually responding to, and what is it that the fungus is actually *doing*?' Johnson pondered.[29]

One hypothesis is that infochemicals pass between plants via the shared fungal networks. This seems most likely, given that plants are known to use infochemicals to communicate above ground. Electrical impulses passing along fungal hyphae are another intriguing possibility. As Stefan Olsson and his neuroscientist colleagues found, the mycelium of some fungi – including that of a mycorrhizal fungus – can conduct spikes of electrical activity that are sensitive to stimulation. Plants also use electrical signalling to communicate between different parts of themselves. No-one has investigated whether electrical signals can pass from plant to fungus to plant, though it isn't much of a stretch. However, Gilbert is firm: 'We don't know. That these signals exist at all is such a new finding. We're at the very beginning of a new area of research.' For her, identifying the nature of the signal is a priority. 'Without knowing what the plants are responding to, we can't answer questions about how the signal is controlled, or how it is actually being sent.'[30]

There is so much more to uncover. If information can pass through fungal networks linking small bean plants in pots in a greenhouse, what's going on in natural ecosystems? Compared with the clamour of chemical cues and signals drifting around between plants in the air, how big a role do fungal pathways play? How far can information travel underground through fungal networks? Johnson and Gilbert are conducting experiments where they link several plants in 'daisy chains', to see if information can pass from plant to plant to plant in a relay system. The ecological consequences could be profound, but Johnson is cautious. 'Suddenly to scale up lab findings to whole forests of trees talking and communicating with each other is a bit much,' he told me. 'People are very quick to extrapolate from a pot to an entire ecosystem.'

Precisely what is passing between plants through fungal networks is a thorny question for all researchers investigating wood wide webs. It is a lack of knowledge that leads to some conceptual impasses. For instance, without knowing how information passes between plants, it's impossible to know whether donor plants actively 'send' a warning message, or whether receiver plants simply eavesdrop on their neighbour's stress. In the eavesdropping scenario, there is nothing that we might recognise as deliberate behaviour on the part of the sender. As Toby Kiers explained, 'If a tree gets attacked by an insect then of course it's going to scream in its language: it will produce some sort of chemical to get ready for the attack.' These chemicals could easily spill over from one plant into another through the network. Nothing is actively *sent*. The receiver plant just happens to notice. David Johnson uses the same analogy. If we hear someone screaming, it doesn't mean they are screaming *in order* to warn us of something. Sure, a scream may cause us to change our behaviour, but it doesn't imply any intention on the part of the screamer. 'You're just eavesdropping on their response to a particular situation.'

It may seem like splitting hairs, but a lot turns on which way we read the interaction. Either way, a stimulus moves from one plant to another and allows the receiver to prepare itself for attack. However, if plants do send a message, we would think of it as a signal. If their neighbours are eavesdropping, we would think of it as a cue. How best to interpret the behaviour of shared mycorrhizal networks is a sensitive subject. Some researchers are concerned about how wood wide webs are commonly portrayed. 'Just because we found that plants can respond to a neighbour', Johnson told me, 'doesn't mean that there is some altruistic network in operation.' The idea that plants are talking to each other and warning each other of imminent attack is an anthropomorphic delusion. 'It's very attractive to think that way', he admitted, but it's ultimately 'a load of nonsense'.[31]

The screaming metaphor might not do much to help. It can slide two ways. Humans scream when they're distressed, shocked, excited or in pain. Humans also scream in order to alert other humans to their plight. It's not always easy to disentangle cause and effect, even if one asks the distressed human outright. With plants it's even harder. Perhaps the fraught question of whether plants warn one another of an aphid attack, or just happen to overhear their neighbour's chemical shrieks, is the wrong thing to ask. As Kiers remarked, 'It's the narrative that we tell that needs to be examined. I'd really love to get past the language and try to understand the *phenomenon*.' Once again, it may be more helpful to ask why this behaviour has evolved in the first place. Who stands to benefit?

The receiver bean plant certainly benefits from the warning: by the time aphids arrive, it will have already activated its defences. But why would it benefit the sender bean plant to alert its neighbours? We bump into the altruism problem again. Once more, the swiftest way through the labyrinth is to switch perspectives. Why might it benefit a *fungus* to pass a warning between the multiple plants that it lives with?

If a fungus is connected to several plants and one is attacked by aphids, the fungus will suffer as well as the plant. If a whole clump of plants move into a state of high alert, they will produce a larger plume of wasp-summoning chemicals than a lone plant can. Any fungus that can magnify the chemical beacon will benefit from its ability to do so – of course, the plants benefit too, but without incurring a cost. Similarly, when stress signals pass from a sick plant to a healthy plant, it is the fungus that stands to benefit from keeping the healthy plant alive. 'Imagine that in a forest you've got trees that appear to be giving resources to other trees,' Gilbert explained. 'It seems to me more likely that the fungus notices that tree A is a bit ill at the moment, and tree B isn't, so it shifts some resources over to tree A. If you take a myco-centric point of view it all makes sense.'

Most studies of shared mycorrhizal networks limit themselves to *pairs* of plants. David Read made images of radioactivity passing from the roots of one plant into another. Simard traced radioactive labels from a donor plant into receivers. Only by limiting the scope to a small number of plants could these experiments take place. But wood wide webs potentially sprawl outwards over tens or hundreds of metres, and possibly further. What happens then? Look outside. Trees, shrubs, grass, vines, flowers. Who is connected to whom, and how? What would a map of a wood wide web look like?

Without knowledge about the architecture of shared fungal networks, it is hard to understand what's going on. We know that resources and infochemicals tend to move through networks downhill, from plenty to scarcity, but sources and sinks can't be the whole story. Your heart is a pump that causes blood to flow 'downhill', by creating areas of high pressure and areas of low pressure. Source–sink dynamics can explain why blood circulates, but not why it reaches your organs in the way it does. This has to do with the blood vessels: how thick they are, how branched they are, and the route they take around your body. It is similar for mycorrhizal networks. Material can't pass through the network from source to sink unless there is a network to flow through.

Kevin Beiler, one of Simard's former students, is the lead author of the only two studies that, in the late 2000s, set out to map the spatial structure of a shared mycorrhizal network. Beiler chose a relatively simple ecosystem – a forest in British Columbia made up of Douglas fir trees of different ages. He deployed a technique used to conduct human paternity tests. Within a thirty-by-thirty-metre plot, he identified genetic fingerprints of each individual fungus and tree, which allowed him to work out exactly who associated with whom. This is an unusual level of detail. Many studies have looked at which plant species interact with which fungal species, but few go beyond, and ask which individuals are actually connected to each other.[32]

Beiler's maps are striking. Fungal networks sprawl over tens of metres, but trees are not linked evenly. Young trees have few connections, and older trees have many. The most well-connected tree is linked to forty-seven other trees and would have been linked to 250 others if the plot had been larger than it was. If one uses a finger to jump from tree to tree across the network – which is, of course, a plant-centric thing to do – one doesn't trickle through the forest evenly. One skips across the network through a small number of well-connected older trees. Via these 'hubs', it's possible to get to any other tree in no more than three steps.

In 1999, when Barabási and his colleagues published the first map of the World Wide Web, they discovered a similar pattern. Web pages are linked to other web pages, but not all pages have the same number of links. The great majority of pages have just a few connections. A small number of pages are extremely well connected. The difference between pages with the highest and lowest number of links is huge: around 80 per cent of links on the Web point to 15 per cent of pages. The same goes for many other types of network – from routes of global air travel to neuronal networks in the brain. In each case, well-connected hubs make it possible to traverse the network in a small number of steps. It is in part these properties of a network – known as 'scale-free' properties – that allow diseases, news and fashions to cascade rapidly through populations. It is the same scale-free properties of a shared mycorrhizal network that might allow a young plant to survive in a heavily shaded understory, or infochemicals to ripple out across a stand of trees in a forest. 'A young seedling will quickly become tied up within a complex, interwoven and stable network,' Beiler explained. 'You would expect this to increase its chances of survival and increase the resilience of the forest.' But only up to a point. It is the same scale-free properties that make a wood wide web vulnerable to targeted attacks. Eliminate Google and Amazon and Facebook overnight or shut down the three busiest airports in the world, and you'll cause havoc. Selectively remove large hub trees – as many commercial logging operations do in an effort to extract the most valuable timber – and serious disruption will ensue.[33]

There are no fundamental laws in operation here. Scale-free properties tend to emerge in any network that grows. 'Most networks that arise in the world are the result of some kind of growth process,' Barabási explained. There are more ways for a new node to connect to a well-connected node than to a less well-connected node. Old nodes with lots of links thus end up with even more links. As Beiler put it, 'You can see these mycorrhizal networks as a *contagious* process. You have some founding trees and the network grows from there. Trees with more links to other trees tend to accumulate more links, more quickly.'

Does this mean that the architecture of wood wide webs will be similar in other parts of the world? It's possible, but we haven't mapped enough networks to be sure. Extrapolating from a plant pot to an entire ecosystem brings problems; extrapolating from a thirty-by-thirty-metre plot is no less of an issue. There are many different ways to be a plant and many different ways to be a fungus. Some plants can form relationships with thousands of fungal species; some plants form relationships with fewer than ten and grow into cliquey networks with members of their own species. Some types of fungus have mycelium that easily grafts onto other mycelial networks to form large composite networks; some fungi are more likely to isolate themselves. In Panama, I found that *Voyria* depended on only a single fungal species, but that its specialism was far from limiting: *Voyria's* partner was the most abundant mycorrhizal fungus in the forest and formed relationships with all the common tree species, allowing *Voyria* to connect with the largest possible number of other plants. Other mycohets that grew in the same forest had evolved a different strategy, and associated with a range of fungal species.[34]

Even in the small area of forest that Beiler chose to study – in part for its simplicity – we are missing many pieces of the puzzle. His maps show how trees and fungi were arranged, but we don't know what they were actually *doing*. 'I only looked at one species of tree and two species of fungus – nothing close to the whole community,' he reflected. 'It was just a glimpse; a small window into a vast and open system. Everything I described is a gross underestimate of the actual connectivity in the forest.'

Voyria have lost the ability to form complex root systems. They don't need them; their shared fungal networks are their roots. Where their roots used to be, *Voyria* plants have a cluster of fleshy fingers. Cut them open, and you see hyphae winding and bursting within *Voyria's* cells. Sometimes their roots aren't even buried, and sit like little fists on the surface of the soil. It's easy to pick them up. Their fungal connections snap instantly. It feels strange to cut a plant's life lines with so little effort. *Voyria's* grip on their network is a matter of life and death, yet the physical links are so slight. I often wondered how all the material needed to make an entire plant could traverse such a delicate passage.

Like most research into mycorrhizal networks, asking questions about *Voyria* involved collecting them, thus severing their connections to the web. I've spent days doing it. And days thinking about the irony of cutting the very connections that I was studying. Of course, biologists often destroy the organisms they hope to understand. I was used to this idea, as much as it's possible to get used to it. But severing the connections in a network to study the network felt unusually absurd. The physicists Ilya Prigogine and Isabelle Stengers remarked that attempts to break down complex systems into their components often fail to provide satisfying explanations; we rarely know how to put the pieces back together again. Wood wide webs present a particular challenge. We're still not sure how mycelial networks co-ordinate their own behaviour and stay in touch with themselves, let alone how they manage their interactions with multiple plants in natural soils. However, we do know enough to know that mycelial networks are ongoing happenings rather than *things*. We know that mycelial networks are able to fuse with one another and prune themselves back, redirect flow around themselves, and release – and respond to – plumes of chemicals. We know that mycorrhizal fungi form and re-form their connections with plants, tangling, detangling and retangling. We know, in short, that wood wide webs are dynamic systems in shimmering, unceasing turnover.[35]

Entities that behave in these ways are loosely termed 'complex adaptive systems': complex, because their behaviour is difficult to predict from a knowledge of their constituent parts alone; adaptive, because they self-organise into new forms or behaviours in response to their circumstances. You – like all organisms – are a complex adaptive system. So is the World Wide Web. So are brains, termite colonies, swarming bees, cities and financial markets – to name a few. Within complex adaptive systems, small changes can bring about large effects which can only be observed in the system as a whole. Rarely can a neat arrow be plotted between 'cause' and 'effect'. Stimuli – which may be unremarkable gestures in themselves – swirl into often surprising responses. Financial crashes are a good example of this type of dynamic non-linear process. So are sneezes, and orgasms.[36]

How best to think about shared mycorrhizal networks then? Are we dealing with a super-organism? A metropolis? A living Internet? Nursery schools for trees? Socialism in the soil? Deregulated markets of late capitalism, with fungi jostling on the trading floor of a forest stock exchange? Or maybe it's fungal feudalism, with mycorrhizal overlords presiding over the lives of their plant labourers for their own ultimate benefit. All are problematic. The questions raised by wood wide webs range further than these limited casts of characters allow. However, we do need some imaginative tools. To understand how shared mycorrhizal networks really behave in complex ecosystems – what they are actually doing, rather than what they are capable of doing – perhaps we will have to start to think of them in terms analogous to those that we use to understand other potentially better-studied complex adaptive systems.

Simard draws parallels between shared mycorrhizal networks in forests and neural networks in animal brains. She argues that the field of neuroscience can provide tools to better understand how complex behaviours arise in ecosystems linked by fungal networks. Neuroscience has concerned itself, for longer than mycology has, with the question of how dynamic, self-organising networks can give rise to complex adaptive behaviours. Her point is not that mycorrhizal networks *are* brains. There are countless ways in which the two systems differ. For one, brains are made up of cells belonging to a single organism rather than a multitude of different species. Brains are also anatomically confined and can't range across a landscape in the same way that fungal networks can. Nonetheless, the analogy is seductive. The challenges that face researchers studying wood wide webs and brains are not dissimilar, although neuroscience is several decades and hundreds of billions of dollars ahead. 'Neuroscientists are making slices of brains to map neural networks,' Barabási joked. 'You ecologists need to slice up a forest so you can see exactly where all the roots and fungi are, and who connects to whom.'[37]

Simard observes that there do appear to be some informative – if superficial – points of overlap. Networks of activity in the brain have scale-free properties, with a few well-connected modules that allow information to pass from A to B in a small number of steps. Brains, like fungal networks, reconfigure themselves – or 'adaptively rewire' – in response to new situations. Underused neuronal pathways are pruned back, as are underused patches of mycelium. New connections between neurones – or *synapses* – form and strengthen, as do the connections between fungi and tree roots. Chemicals known as neurotransmitters pass across synapses, allowing information to move from nerve to nerve; similarly, chemical substances pass across mycorrhizal 'synapses' from fungus to plant or plant to fungus, in some cases transmitting information between them. Indeed, the amino acids glutamate and glycine – major signalling molecules in plants, and the most common neurotransmitters in animal brains and spinal cords – are known to pass between plants and fungi at these junctions.[38]

But ultimately, the behaviours of wood wide webs are ambiguous, and our brain analogies – like Internet analogies, or political analogies – are limiting. However these networks co-ordinate themselves, and however it is that cues – or are they signals? – pass between plants via fungal channels, wood wide webs overlap with one another and have soft edges which fray outwards inclusively. Included are the bacteria that migrate from place to place within fungal mycelium. Included are the aphids and also the parasitic wasps lured in for the feast by the volatile compounds produced by broad bean plants. Step back further and humans, too, fall in. Knowingly or not, we have been interacting with mycorrhizal networks for as long as we have interacted with plants.[39]

Are we able to release ourselves from these metaphors, think outside the skull, and learn to talk about wood wide webs without leaning on one of our well-worn human totems? Are we able to let shared mycorrhizal networks be questions, rather than answers known in advance? 'I try just to look at the system and let the lichen be a lichen.' Discussions of wood wide webs often lead me back to the words of Toby Spribille, the researcher who keeps discovering new partners in the lichen symbiosis. Wood wide webs aren't lichens – although to think of them as enormous lichens that we can walk around in brings welcome variety to the range of metaphors currently on offer.

Nonetheless, I wonder if we can learn something from Spribille's patience. Are we able to stand back, look at the system, and let the polyphonic swarms of plants and fungi and bacteria that make up our homes and our worlds be themselves, and quite *unlike* anything else? What would that do to our minds?

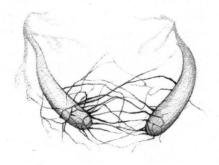

7

Radical Mycology

To use the world well, to be able to stop wasting it and our time in it, we need to re-learn our being in it.

Ursula Le Guin

I lay naked in a mound of decomposing wood chips and was buried up to my neck by the spadeful. It was hot, and the steam smelled of cedar and the fust of old books. I leaned back, sweating under the damp weight, and closed my eyes.

I was in California, visiting one of the only fermentation baths to be found outside Japan. The wood shavings had been moistened and piled into a heap. After two weeks of rotting they had been shovelled into a large wooden tub, and ripened for another week before I arrived. The bath was now cooking, heated by nothing more than the fierce energy of decomposition.

The intense heat made me drowsy, and I thought of the fungi decomposing the wood. How easy it is when one's not being stewed in a heap of rotting wood to take for granted that everything decays. We live and breathe in the space that decomposition leaves behind. I greedily sucked some cold water through a straw and tried to blink the sweat out of my eyes. If we could pause decomposition, starting now, the planet would pile up kilometres deep in bodies. We would think of it as a crisis, but from a fungal point of view it would be an enormous heap of opportunities.

My torpor deepened. It certainly wouldn't be the first time fungi have thrived through a period of dramatic global transformation. Fungi are veteran survivors of ecological disruption. Their ability to cling on – and often flourish – through periods of catastrophic change is one of their defining characteristics. They are inventive, flexible and collaborative. With much of life on Earth threatened by human activity, are there ways we can partner with fungi to help us adapt?

These may sound like the delirious musings of someone buried up to their neck in decomposing wood chips, but a growing number of radical mycologists think exactly this. Many symbioses have formed in times of crisis. The algal partner in a lichen can't make a living on bare rock without striking up a relationship with a fungus. Might it be that we can't adjust to life on a damaged planet without cultivating new fungal relationships?

In the Carboniferous period, 290–360 million years ago, the earliest wood-producing plants spread across the tropics in swampy forests, supported by their mycorrhizal fungal partners. These forests grew and died, pulling huge quantities of carbon dioxide out of the atmosphere. And for tens of millions of years, much of this plant matter didn't decompose. Layers of dead and un-rotted forest built up, storing so much carbon that atmospheric carbon dioxide levels crashed, and the planet entered a period of global cooling. Plants had caused the climate crisis, and plants were hit the hardest by it: huge areas of tropical forest were wiped out in an extinction event known as the Carboniferous rainforest collapse. How had wood become a climate change-inducing pollutant?[1]

From a plant perspective wood was, and remains, a brilliant structural innovation. As plant life boomed, the jostle for light intensified, and plants grew taller to reach it. The taller they became, the greater their need for structural support. Wood was plants' answer to this problem. Today, the wood of some three trillion trees – more than 15 billion of which are cut down every year – accounts for about 60 per cent of the total mass of every living organism on Earth, some 300 gigatonnes of carbon.[2]

Wood is a hybrid material. Cellulose – a feature of all plant cells, whether woody or not – is one of the ingredients, and the most abundant polymer on Earth. Lignin is another ingredient, and the second most abundant. Lignin is what makes wood wood. It is stronger than cellulose, and more complex. Whereas cellulose is made up of orderly chains of glucose molecules, lignin is a haphazard matrix of molecular rings.[3]

To this day, only a small number of organisms have worked out how to decompose lignin. By far the most prolific group are the white rot fungi – so-called because in decomposition, they bleach wood a pale colour. Most enzymes – biological catalysts that living organisms use to conduct chemical reactions – lock onto specific molecular shapes. Faced with lignin, this approach is hopeless: its chemical structure is too irregular. White rot fungi work around the problem using non-specific enzymes that don't depend on shape. These 'peroxidases' release a torrent of highly reactive molecules, known as 'free radicals', which crack open lignin's tightly bonded structure in a process known as 'enzymatic combustion'.[4]

Fungi are prodigious decomposers, but of their many biochemical achievements, one of the most impressive is this ability of white rot fungi to break down the lignin in wood. Based on their ability to release free radicals, the peroxidases produced by white rot fungi perform what is technically known as 'radical chemistry'. 'Radical' has it right. These enzymes have forever changed the way that carbon journeys through its earthly cycles. Today, fungal decomposition – much of it of woody plant matter – is one of the largest sources of carbon emissions, emitting about eighty-five gigatonnes of carbon to the atmosphere every year. In 2018, the combustion of fossil fuels by humans emitted around ten gigatonnes.[5]

How did tens of millions of years' worth of forest go un-rotted over the Carboniferous period? Opinions differ. Some point to climatic factors: tropical forests were stagnant, water-logged places. When trees died, they were submerged in anoxic swamps, where white rot fungi were unable to follow. Others suggest that when lignin first evolved in the early Carboniferous period, white rot fungi weren't yet able to decompose it and required several million more years to upgrade their apparatus of decay.[6]

So what happened to the vast areas of forest that didn't decompose? It's an inconceivably large amount of matter to pile up, kilometres deep.

The answer is coal. Human industrialisation has been pow-ered on these seams of un-rotted plant matter, somehow kept out of fungal reach. (If given the chance, many types of fungi readily decompose coal, and a species known as the 'kerosene' fungus thrives in the fuel tanks of aircraft.) Coal provides a negative of fungal histories: it's a record of fungal absence, of what fungi did not digest. Rarely since then has so much organic material escaped fungal attention.[7]

I lay buried among white rot fungi for twenty minutes, slow-cooked by their radical chemistry. My skin seemed to dissolve into the heat, and I lost track of where my body started and stopped; a complex cuddle, blissful and unbearable in turn. No wonder coal can give off such heat: it is made from wood that hasn't yet been burned. When we burn coal, we physi-cally combust the material that fungi were unable to combust enzymatically. We thermally decompose what fungi were unable to decompose chemically.

It may be rare for wood to escape fungal attention; it is com-mon for fungi to escape ours. In 2009, the mycologist David Hawksworth referred to mycology as 'a neglected megascience'. Animal and plant biology have had their own university depart-ments for generations, but the study of fungi has long been lumped in with plant sciences, and is seldom recognised as a distinct field, even today.[8]

Neglect is a relative term. In China, fungi have been a major source of food and medicines for thousands of years. Today, 75 per cent of the global production of mushrooms – almost 40 million tonnes – occurs in China. In central and eastern Europe, too, fungi have long played important cultural roles. If deaths from mushroom poisoning are any metric of national fungal enthusiasm, compare the one or two deaths a year in the United States with the 200 deaths in Russia and Ukraine in the year 2000.[9]

Nonetheless, for much of the world, Hawksworth's observations hold true. The first State of the World's Fungi report published in 2018 reveals that in the Red List of Threatened Species, compiled by the International Union for the Conservation of Nature (IUCN), only fifty-six species of fungi have had their conservation status evaluated, compared with over 25,000 plants and more than 68,000 animals. Hawksworth proposes several possible solutions for this oversight. One stands out: 'the resources needed to empower "amateur" mycologists' should be increased. His inverted commas speak volumes. Although many fields of science have networks of dedicated and talented amateur practitioners, they are particularly prominent in the field of mycology. Too often, there has been no other outlet for fungal enquiry.[10]

A grassroots scientific movement may sound improbable, but it emerges from a rich tradition. The 'professional' academic study of living organisms only picked up momentum in the nineteenth century. Many major developments in the history of the sciences have been fuelled by amateur enthusiasm and taken place outside dedicated university departments. Today, after a long period of specialisation and professionalisation, there is an explosion of new ways of doing science. 'Citizen science projects', along with 'hackerspaces' and 'makerspaces', have grown increasingly popular since the 1990s, providing opportunities for dedicated non-specialists to carry out research projects. What does one call these practitioners? Do they count as the 'public'? Citizen scientists? Lay experts? Or just amateurs?[11]

Peter McCoy is a hip-hop artist, self-taught mycologist and founder of an organisation called Radical Mycology, which works to develop fungal solutions to the many technological and ecological problems we face. As he explains in his book *Radical Mycology* – a hybrid of fungal manifesto, guidebook and grower's guide – his goal is to create a 'people's mycological movement' versed in 'the cultivation of fungi and the applications of mycology'.

Radical Mycology is part of a larger movement of DIY mycology, which emerged from the psychedelic mushroom-growing scene kickstarted in the 1970s by Terence McKenna and Paul Stamets. The movement took on its modern form as it grew together with hackerspaces, crowdsourced science projects and online forums. Although its centre of gravity remains on the west coast of North America, grassroots mycological organisations are rapidly spreading to other countries and continents. The word 'radical' derives from the Latin *radix*, meaning root. Interpreted literally, the concerns of radical mycology lie with its mycelial base, or its 'grassroots'.[12]

It is for these grassroots fungal enthusiasts that McCoy founded an online mycology school, Mycologos. Knowledge about fungi is often inaccessible and hard to understand. His mission is to reshape human–fungal relations by distributing this information in readily digestible form: 'I envisage teams of Radical Mycologists Without Borders travelling the globe, sharing their skills and discovering new means of working with fungi. Where one Radical Mycologist trains ten, those ten can train a hundred, and from them a thousand – so it is that mycelium spreads.'[13]

In the autumn of 2018, I travelled to a farm in rural Oregon for the biannual Radical Mycology Convergence. There I found over 500 fungal nerds, mushroom growers, artists, budding enthusiasts and social and ecological activists bustling around a farmyard. Wearing a baseball cap, worn sneakers and thick-lensed spectacles, McCoy set the scene in a keynote address: 'Liberation Mycology'.

To grow mushrooms on any kind of scale, growers have to develop a keen nose for material to satisfy voracious fungal appetites. Most mushroom-producing fungi thrive on the mess that humans make. Growing cash crops on waste is a kind of alchemy. Fungi transform a liability with negative worth into a product with value. A win for the waste producer, a win for the cultivator, and a win for the fungi. The inefficiency of many industries is a blessing to mushroom growers. Agriculture is particularly wasteful: palm and coconut oil plantations discard 95 per cent of the total biomass produced. Sugar plantations discard 83 per cent. Urban life isn't much better. In Mexico City, used diapers make up between 5 and 15 per cent by weight of solid waste. Researchers have found that the omnivorous *Pleurotus* mycelium – a white rot fungus which fruits into edible oyster mushrooms – can grow happily on a diet of used diapers. Over the course of two months, diapers introduced to *Pleurotus* lost about 85 per cent of their starting mass when the plastic covering was removed, compared with a mere 5 per cent in fungus-free controls. What's more, the mushrooms produced were healthy and free from human diseases. Similar projects are under way in India. By cultivating *Pleurotus* on agricultural waste – by enzymatically combusting the material – less biomass is thermally combusted, and air quality is improved.[14]

It is no great surprise that the mess humans have made might look like an opportunity from a fungal perspective. Fungi have persisted through Earth's five major extinction events, each of which eliminated between 75 and 95 per cent of species on the planet. Some fungi even thrived during these calamitous episodes. Following the Cretaceous–Tertiary extinction, credited with the dispatch of dinosaurs and the mass destruction of

Oyster mushrooms, *Pleurotus ostreatus,* growing
on agricultural waste

forests across the globe, fungal abundance surged, fuelled by an
abundance of dead woody material to decompose. Radiotrophic
fungi – those able to harvest the energy emitted by radioactive
particles – flourish in the ruins of Chernobyl and are just the
latest players in a longer story of fungi and human nuclear
enterprise. After Hiroshima was destroyed by an atomic bomb,
it is reported that the first living thing to emerge from the
devastation was a matsutake mushroom.[15]

Fungal appetites are diverse, but there are some materials
they won't break down unless they have to. In one of his
workshops, McCoy explained how he had trained *Pleurotus*
mycelium to digest one of the most commonly littered items
in the world: cigarette butts, over 750,000 tonnes of which
are thrown away every year. Unused cigarette butts will break
down, given time, but used cigarette butts are saturated with
toxic residues that impede the process. McCoy had weaned
Pleurotus onto a diet of used butts by gradually phasing out
the alternatives. Over time, the fungus had 'learned' how to
use them as its sole food source. A timelapse video showed
the mycelium seeping steadily upwards through a jam jar filled

with crumpled tar-stained butts. A burly oyster mushroom soon bundled itself up and out of the top.[16]

In fact, it is just as much 'remembering' as 'learning'. A fungus won't produce an enzyme it doesn't need. Enzymes, or even entire metabolic pathways, can lie dormant in fungal genomes for generations. For the *Pleurotus* mycelium to digest the used cigarette butts it might have to dust off an unused metabolic move. Or it might deploy an enzyme normally used for something else and press it into the service of a new cause. Many fungal enzymes, like lignin peroxidases, are not specific. This means that a single enzyme can serve as a multitool, allowing the fungus to metabolise different compounds with similar structures. As it happens, many toxic pollutants – including those in cigarette butts – resemble the byproducts of lignin breakdown. In this sense, to confront *Pleurotus* mycelium with used cigarette butts is to offer it a commonplace challenge.[17]

Much of radical mycology is underwritten by the radical chemistry of white rot fungi. However, it isn't always easy to predict what a given fungal strain will metabolise. McCoy told us about his attempts to grow *Pleurotus* mycelium on dishes spotted with drops of the herbicide glyphosate. Some of the *Pleurotus* strains avoided the drops. Some grew straight through them. Some grew up to the edge of a drop and stopped growing. 'It took those ones a week to work out how to break it down,' McCoy recalled. He likened fungi to jailers with bunches of enzymatic keys that can unlock certain chemical bonds. Some strains might have the right key ready to go. Others might have it buried somewhere inside their genome but choose to avoid the new substance anyway. Others might take a week to riffle through the bunch of keys, trying different ones until they get lucky.

McCoy, like many in the DIY mycology movement, received his first shot of fungal zeal from Stamets. Since his influential work on psilocybin mushrooms in the 1970s, Stamets has grown into an unlikely hybrid between fungal evangelist and tycoon. His TED Talk – 'Six Ways That Mushrooms Can Save the World' – has been viewed millions of times. He runs a multi-million-dollar fungal business, Fungi Perfecti, which does a roaring trade in everything from antiviral throat sprays to fungal dog treats ('Mutt-rooms'). His books on mushroom identification and cultivation – including the definitive *Psilocybin Mushrooms of the World* – continue to provide a crucial reference for countless mycologists, grassroots or otherwise.

As a teenager, Stamets suffered from a debilitating stammer. One day, he took a heroic dose of magic mushrooms and climbed to the top of a tall tree, where he was trapped by a lightning storm. When he came down, his stammer was gone. Stamets was converted. He studied mycology at Evergreen State College as an undergraduate and has dedicated his life to fungal matters ever since. Stamets isn't affiliated with Radical Mycology. However, like McCoy, he is devoted to spreading the fungal message to the widest possible audience. On his website is a letter from a Syrian cultivator who, inspired by Stamets, developed ways to farm oyster mushrooms on agricultural debris. The cultivator taught more than a thousand people how to grow mushrooms in their basements, providing a key foodstuff during six years under siege and bombardment by the Assad regime.

In fact, it is no exaggeration to say that Stamets has done more than anyone else to popularise fungal topics outside university biology departments. However, his relationship with the academic world is not straightforward. From his sensational claims to his speculative theories, Stamets behaves in many ways academic scientists are not supposed to. And yet his maverick approach is undeniably effective. It is a tension that sometimes borders on the absurd. Stamets once described a complaint he received from a university professor he knows. 'Paul, you've created a huge problem. We want to study yeast and these students want to save the world. What do we do?'[18]

*

One of the ways fungi might help save the world is by helping to restore contaminated ecosystems. In 'mycoremediation', as the field is known, fungi become collaborators in environmental clean-up operations.

We have recruited fungi to break things down for millennia. The diverse microbial populations in our guts remind us that in those moments in our evolutionary history when we haven't been able to digest something by ourselves, we've pulled microbes on board. Where this has proved impossible, we've outsourced the process using barrels, jars, compost heaps and industrial fermenters. Human life hinges on many forms of external digestion using fungi: from alcohol to soy sauce, to vaccines, to penicillin, to the citric acid used in all fizzy drinks. This sort of partnering – in which different organisms together sing a metabolic 'song' neither could sing alone – enacts one of the oldest evolutionary maxims. Mycoremediation is just a special case.

And it shows great promise. Fungi have a remarkable appetite for a range of pollutants besides toxic cigarette butts and the herbicide glyphosate. In his book *Mycelium Running*, Stamets writes about a collaboration with a research institute in Washington state, which partnered with the US Department of Defense to develop ways to break down a potent neurotoxin. The chemical – dimethyl methylphosphonate, or DMMP – was one of the deadly components of VX gas, manufactured and deployed in the late 1980s by Saddam Hussein during the Iran–Iraq War. Stamets sent his colleagues twenty-eight different fungal species, which were exposed to the compound in gradually increasing concentrations. After six months, two of the species had 'learned' to consume DMMP as their primary nutrient source. One was *Trametes*, or turkey tail, and the other was *Psilocybe azurescens*, the most potent psilocybin-producing species known, discovered by Stamets several years before and named for the bluish hue on the stems (he later named his son Azureus after the mushroom). Both are white rot fungi.[19]

The mycological literature is filled with hundreds of such examples. Fungi can transform many common pollutants in soil and waterways that endanger lives, whether human or otherwise. They are able to degrade pesticides (such as chlorophenols), synthetic dyes, the explosives TNT and RDX, crude oil, some plastics and a range of human and veterinary drugs not removed by wastewater treatment plants, from antibiotics to synthetic hormones.[20]

In principle, fungi are some of the best-qualified organisms for environmental remediation. Mycelium has been fine-tuned over a billion years of evolution for one primary purpose: to consume. It is appetite in bodily form. For hundreds of millions of years before the plant boom in the Carboniferous, fungi made a living finding ways to decompose the debris that other organisms left behind. They can even boost decomposition by providing mycelial 'highways' that allow bacteria to travel into otherwise inaccessible sites of decay. And yet decomposition is only part of the story. Heavy metals accumulate within fungal tissues, which can then be removed and disposed of safely. The dense meshwork of mycelium can even be used to filter polluted water. 'Mycofiltration' removes infectious diseases such as *E. coli* and can sop up heavy metals like a sponge – a company in Finland uses this approach to reclaim gold from electronic waste.[21]

Despite its promise, however, mycoremediation is no simple fix. Just because a given fungal strain behaves in a certain way in a dish doesn't mean it will do the same thing when introduced to the rumpus of a contaminated ecosystem. Fungi have needs – such as oxygen or additional food sources – that must be taken into account. Moreover, decomposition takes place in stages, achieved by a succession of fungi and bacteria, each able to pick up where the previous ones left off. It is naive to imagine that a lab-trained fungal strain will be able to hustle effectively in a new environment and remediate a site by itself. The challenges faced by mycoremediators are analogous to those faced by brewers – without suitable conditions, yeast will struggle to remediate the sugar in a barrel of grape juice into alcohol – except that the wine barrel is a contaminated ecosystem, and we're inside it.[22]

McCoy advocated a radical approach based on grassroots empiricism. I had been sceptical. The field of mycoremediation, it struck me, needs a big institutional boost. Funky home-grown solutions are all very well, but surely large-scale studies are required. How could the field progress without flagship projects, big grants and institutional attention? I found it hard to imagine that an army of grassroots hobbyists, no matter how dedicated, could be equipped or credible enough to move things forwards.

I soon realised that McCoy was advocating this approach not because of a disregard for institutional research, but because of its scarcity. Many factors contribute. Ecosystems are complex, and there is no single fungal solution that will work in all sites and conditions. To develop scalable off-the-shelf myco-remediation protocols would require large investment, which is uncommon in the remediation sector: on the whole, remediation is undertaken by reluctant companies under pressure to fulfil a legal obligation. Few are interested in solutions seen to be experimental or alternative. Moreover, there is a conventional remediation industry in full swing, which scrapes up polluted soil by the tonne, transports it elsewhere and burns it. Despite the expense and ecological disruption this causes, it is an industry in no hurry to be replaced.

Radical mycologists have little choice but to take matters into their own hands. And since the early 2000s, inspired in part by Stamets' evangelism, a number of projects have been set up to test fungal solutions. One of the older organisations, CoRenewal, has been conducting research into the ability of fungi to detoxify the poisonous byproducts of crude oil extrac-tion left behind by Chevron's twenty-six-year operation in the Ecuadorian Amazon. In an alliance with partners in polluted areas, researchers are investigating the microbial communities and local 'petrophilic' fungal strains found in contaminated soils. It is classic radical mycology – local mycologists learning how to partner with local fungal strains to solve local prob-lems. There are other examples. A grassroots organisation in

California has laid out kilometres of straw-filled tubes filled with *Pleurotus* mycelium in the hope that they will remediate the toxic run-off from houses destroyed in the 2017 wildfires. In 2018, floating booms filled with *Pleurotus* mycelium were installed in a Danish harbour to help mop up fuel spills. Most of these projects have only just begun, others are under way. None has reached maturity.[23]

Will mycoremediation take off? It's too early to say. But it's clear that now, as we fret at the edge of a toxic puddle of our own making, radical mycological solutions based on the ability of certain fungi to decompose wood offer some hope. Our favoured method of accessing the energy in wood has been to burn it. This too is a radical solution. And it is this energy – the fossilised remnants of a wood boom in the Carboniferous – that has helped get us into trouble. Could the radical chemistry of white rot fungi – an evolutionary response to the very same wood boom – now help to pull us through?

For Peter McCoy, Radical Mycology means more than just solving particular problems in particular places. A distributed network of grassroots practitioners is also capable of advancing the state of fungal knowledge as a whole. One way this can happen is through the discovery and isolation of potent fungal strains. Fungi isolated from a contaminated environment may have already learned how to digest a given pollutant and, as locals, be able to remediate a problem *and* thrive. This was the approach used by a team of researchers in Pakistan who in 2017 screened soil from a city landfill site in Islamabad and found a novel fungal strain that could degrade polyurethane plastic.[24]

Crowdsourcing fungal strains may sound implausible, but it has resulted in some major discoveries. The industrial production of the antibiotic penicillin was only possible because of the discovery of a high-yielding strain of *Penicillium* fungus. In 1941, this 'pretty golden mould' was found on a rotting cantaloupe in an Illinois market by Mary Hunt, a laboratory assistant, after the lab put out a call for civilians to submit moulds. Before this point, penicillin had been expensive to produce and remained largely unavailable.²⁵

Finding fungal strains is one thing. Isolating them and testing their activity is more difficult. Hunt may have found the mould, but it had to be taken into the lab to be examined. This was my main doubt about McCoy's approach. How could radical mycologists isolate and grow new strains without access to well-provisioned facilities? Sterile benches pumping clean air, ultra-pure chemicals, expensive machines whirring away in equipment rooms – surely this was all needed to make any kind of real progress?

I wanted to find out more, so I attended one of McCoy's weekend mushroom cultivation courses in Brooklyn, New York. The class was an eclectic mix: artists, educators, community planners, computer programmers, a university lecturer, entrepreneurs and chefs. McCoy stood behind a table piled high with dishes, plastic bags filled with grain and boxes stacked with syringes and scalpels – staple tackle of the modern mushroom cultivator. A large pot of water simmered on the stove, filled with gelatinous wood ear mushrooms which we ladled into mugs during the tea break. This was Radical Mycology at its growing tip. Or rather, at one of its growing tips.

Over the course of the weekend, it became clear that the field of amateur fungal cultivation is in a state of wild proliferation. A well-connected, actively experimenting network of fungal enthusiasts are already accelerating the production of fungal knowledge. Techniques like DNA sequencing remain out of reach for most, but recent advances make it possible to perform operations that would have been impossible for amateurs even ten years ago. Most are ingenious low-tech solutions developed by kitchen-sink magic mushroom growers. Many are improvements and tweaks on methods developed and published by Terence McKenna and Paul Stamets in their grower's guides. Although McCoy's vision of mycological transformation includes community lab spaces, a lot can be done without them.

The most revolutionary innovation emerged in 2009. The founder of the magic mushroom-growing forum mycotopia.net, known only by the handle hippie3, devised a method to grow fungi without fear of contamination. This changed everything. Contamination is the menace of all fungal cultivators. Freshly sterilised material is a biological vacuum; if exposed to the busy world of the open air, life rushes in. Using hippie3's 'injection port' method, amateur mushroom cultivators can ditch the most expensive kit and fiddly procedures. All one needs is a syringe and a modified jam jar. The knowledge spread quickly. In McCoy's view this was one of the most important developments in the history of mycology – 'lab results without the lab' – and has changed the cultivation of mushrooms forever. He grinned and expelled a small libation from the syringe he held. 'A squirt for hippie3.'

I laughed at the thought of teams of mycohackers tinkering around at the edge of problems, just as McCoy's *Pleurotus* mycelium had hovered at the edge of the puddle of glyphosate, experimenting with different enzymes until it found a way through. McCoy was training radical mycologists to cultivate fungi at home, so they could then train fungal strains to make an opportunity out of yet another toxic human oversight. Even with relatively small incentives, the field could advance rapidly. I pictured crowds of enthusiasts gathering to race their home-grown fungal strains through fiendish cocktails of toxic waste, competing for a million-dollar annual award.[26]

So much remains to be seen. Mycology, whether radical or not, is in its infancy. Humans have been cultivating and domesticating plants for more than 12,000 years. But fungi? The earliest records of mushroom cultivation date from around 2,000 years ago in China. Wu San Kwung, who is credited with working out how to grow shiitake mushrooms – another white rot fungus – in China around the year 1000, is commemorated with an annual feast day, and temples throughout the country are dedicated to his achievements. By the late nineteenth century, in the limestone catacombs that riddle the subsurface of Paris, hundreds of mushroom farmers produced more than a thousand tonnes of 'Paris' mushrooms every year. Yet lab-based techniques only arose around a hundred years ago. Many of the techniques that McCoy teaches, including hippie3's injection port method, are only about a decade old.[27]

McCoy's course ended in a flutter of excitement, ideas flying around. 'There's lots of ways to play,' he grinned, a quiet blend of incitement and encouragement. 'There's a lot we just don't know.'

For as long as fungi have existed, they have been bringing about a 'change from the roots'. Humans are latecomers to the story. Over hundreds of millions of years, many organisms have formed radical partnerships with fungi. Many – such as plants' relationships with mycorrhizal fungi – have been blockbuster moments in the history of life, with world-changing consequences. Today there are plenty of non-humans cultivating fungi in sophisticated ways, with radical outcomes. Can these relationships be thought of as ancient precursors to radical mycology?[28]

African *Macrotermes* termites are some of the more striking examples. *Macrotermes,* like most termites, spend much of their lives foraging for wood, although they aren't able to eat it. Instead, the termites cultivate a white rot fungus – *Termitomyces* – that digests it for them. The termites chew wood into a slurry which they regurgitate in fungal gardens, known as the 'fungus comb', by contrast with bees' honeycomb. The fungus uses radical chemistry to decompose the wood. The termites consume the compost that remains. To house the fungus, *Macrotermes* build towering mounds that reach heights of nine metres, some of which are more than 2,000 years old. Societies of *Macrotermes* termites, like those of leafcutter ants, are some of the most complex formed by any insect group.[29]

Macrotermes mounds are giant, externalised guts – prosthetic metabolisms that allow the termites to decompose complex materials they can't break down themselves. Like the fungi they cultivate, *Macrotermes* muddle the concept of individuality. An individual termite can't survive apart from its society. A termite society can't survive separate from the cultures of fungi and other microbes that feed them, and that they feed. The partnership is prolific: a substantial proportion of the wood decomposed in the African tropics passes through *Macrotermes* mounds.[30]

Whereas humans access the energy bound up in lignin by physically burning it, *Macrotermes* help a white rot fungus to burn it chemically. Termites deploy white rot fungi just as a radical mycologist might enlist *Pleurotus* to break down crude oil or cigarette butts. Or in the way that a no less radical mycologist might outsource their metabolism to fungi in the barrels and jars used to ferment wine, miso or cheese. However, there's no question who got there first. *Macrotermes* had been cultivating fungi for over 20 million years by the time the genus *Homo* evolved. And indeed, when it comes to *Termitomyces* fungi, termites' cultivation techniques far outstrip those of humans. *Termitomyces* mushrooms are a delicacy (and can grow to a metre across, making them some of the largest mushrooms in the world). But despite prolonged effort, humans have not found a way to cultivate them. The fungus requires the finely balanced conditions furnished by the termites through a combination of their bacterial symbionts and the architecture of their mounds.

The expertise of termites has not been lost on the humans who live around them. The radical chemistry of white rot fungi – and its astonishing force – has long been wrapped into human lives. Termites are reported to consume between $1.5 and $20 billion a year of property in the United States every year. (As Lisa Margonelli observes in *Underbug*, North American termites are most commonly described as eating 'private' property, as if they had some intentional anarchist or anti-capitalist sentiment.) In 2011 termites found their way into a bank in India and ate ten million rupees in banknotes

Macrotermes termites

– around \$225,000. In a twist on the theme of radical fungal partnerships, one of Paul Stamets' 'six ways that fungi can save the world' involves tweaking the biology of certain disease-causing fungi so they are able to bypass termites' defences, and exterminate their colonies (this is the same fungus – the mould *Metarhizium* – that shows promise in eliminating populations of malarial mosquitoes).[31]

The anthropologist James Fairhead describes how farmers in many parts of west Africa encourage *Macrotermes* termites because of the way they 'wake up' the soil. Earth from inside termite mounds is sometimes eaten by humans or smeared on wounds, and has been found to have a number of benefits: as a mineral supplement, or an antidote to toxins, or an antibiotic – *Macrotermes* cultivate an antibiotic-producing bacterium, *Streptomyces*, within their mounds. The partnership between *Macrotermes* and their fungi has even been weaponised by humans for radical political causes. In the early twentieth century in coastal west Africa, locals secretly released termites in the military outpost of a colonising French army. Driven by the voracious appetites of their fungal partners, the termites destroyed the buildings and chewed up the bureaucrats' papers. The French garrison quickly abandoned their post.[32]

In a number of west African cultures, termites lie above humans in spiritual hierarchies. In some, *Macrotermes* are portrayed as messengers between humans and gods. In others, it was only with the help of a termite assistant that God was able to create the universe in the first place. In these myths, *Macrotermes* are not just portrayed as breaking things down. They are builders on the largest possible scale.[33]

Around the world, the idea that fungi can be used to build things as well as break them down is starting to catch on. A material made from the outer layers of portabello mushrooms shows promise in replacing graphite in lithium batteries. The mycelium of some species makes an effective skin substitute, used by surgeons to help wounds to heal. And in the United States, a company called Ecovative is growing building materials out of mycelium.[34]

I went to visit Ecovative's research and manufacturing facility in an industrial park in upstate New York. Stepping into the lobby, I found myself surrounded by mycelial products. There were boards, bricks, acoustic tiles and moulded packaging for wine bottles. All were light grey with a rough texture and looked like cardboard. Next to a mycelium lampshade and stool was a box filled with white cubes of squishy mycelial foam. Next to this was a piece of fungal leather. I felt as if I'd stumbled into an elaborate prank, the set of a satirical TV show making fun of people with big claims about how fungi can save the world.

Eben Bayer, the young CEO of Ecovative, found me prodding a piece of mycelium. 'Dell ship their servers in packaging like that. We send them about half a million pieces a year.' He gestured to a stool. 'Safe, healthy, sustainably grown furniture.' Its seat was covered with mycelium leather and padded with mycelium foam. If you ordered one it would arrive in mycelial packaging. Whereas mycoremediation is all about decomposing the consequences of our actions, 'mycofabrication' is all about recomposing the types of material we choose to use in the first place. It is the yang to the yin of decomposition.

Like the radical mycologists I had met in Oregon and Brooklyn, Ecovative reroutes agricultural waste streams to feed its fungi. Out of sawdust or corn stalks grew a valuable commodity. It was the familiar fungal win-win-win: for waste producer, cultivator and fungus. However, in the case of Ecovative there were some additional wins. One of Bayer's long-standing ambitions has been to disrupt polluting industries. The packaging materials that Ecovative grows are designed to replace plastics. Their construction materials are designed to replace brick, concrete and particle board. Their leather-like textiles replace animal leather. Hundreds of square feet of mycelial leather can be grown in less than a week on materials that would otherwise be disposed of. At the end of their life, mycelial products can be composted. Ecovative's materials are lightweight, water-resistant and fire-retardant. They are stronger than concrete when subjected to bending forces, and resist compression better than wood framing. They have a better insulation value than expanded polystyrene, and can be grown in a matter of days into an unlimited number of forms (researchers in Australia are working to create a termite-resistant brick by combining *Trametes* mycelium with crushed glass – a product that would avoid the need for Paul Stamets' termite-killing fungi).[35]

The potential of mycelial materials has not gone unnoticed. The designer Stella McCartney is working with fungal leather grown using Ecovative's methods. Ecovative has a close relationship with IKEA, who are developing ways to replace their polystyrene packaging with a mycelial alternative. Researchers at NASA have taken an interest in 'mycotecture' and its potential role in growing structures on the Moon. Ecovative has just received a ten-million-dollar research-and-development contract from DARPA, the Defence Advanced Research Projects Agency, a wing of the US military. DARPA is interested in growing barracks out of mycelium that repair themselves when damaged and decompose when their job is done. Growing housing for soldiers hadn't been part of Bayer's original vision, but these are adaptable techniques. 'We can use these methods to grow relief shelters in disaster zones,' Bayer pointed out. 'Using mycelium, you can grow a lot of housing for a lot of people at really low cost.'[36]

The basic idea is simple. Mycelium weaves itself into a dense fabric. The living mycelium is then dried into a dead material. The final product depends on how the mycelium is encouraged to grow. The bricks and packaging material are formed as mycelium 'runs' through a slurry of damp sawdust packed into moulds. The flexible materials are made from pure mycelium. Tan it, and you get leather. Dry it, and you get a foam that can be used to make anything from insoles for sports shoes to dock floats. Whereas McCoy and Stamets tempt fungi into new metabolic behaviours, Bayer tempts them into new growth forms. Mycelium can always be trusted to pour itself into its environment, whether it be a puddle of neurotoxin or a mould shaped like a lampshade.[37]

Bayer and I pushed our way through a set of doors, and entered a hangar large enough to build an aeroplane in. Wood chips and other raw materials slithered down chutes into mixing drums where they were combined in ratios digitally controlled via banks of computer screens. Twenty-foot-long Archimedes' screws trammelled streams of sawdust through heating and cooling chambers at rates of half a tonne an hour. Towering stacks of plastic moulds were wheeled between growth chambers and ten-metre-high drying racks. The chambers were digitally controlled microclimates – light, humidity, temperature, oxygen and carbon dioxide levels all varied in carefully programmed cycles. It was the industrial human equivalent of a *Macrotermes* termite mound.

Like Ecovative's growth facility, *Macrotermes* mounds are carefully regulated microclimates, built around the requirements of the fungus. By opening and closing tunnels within a system of chimneys and galleries, termites are able to regulate temperature, moisture and levels of oxygen and carbon dioxide. In the middle of the Sahara, termites can create the cool, damp conditions that allow the fungus to thrive.

As in *Macrotermes* mounds, the fungi grown at Ecovative are species of white rot fungi. Most products are grown from *Ganoderma* mycelium, the species that fruits into reishi mushrooms. Some use *Pleurotus*, and others *Trametes*, which fruits into turkey tail mushrooms. It was *Pleurotus* that McCoy had trained to digest glyphosate and cigarette butts. It was *Trametes* that Stamets' collaborators had trained to digest the toxic precursor to VX gas. Just as different fungal strains vary in their willingness to break down toxic nerve agents or glyphosate, different strains vary in how fast they grow, and what sort of material their mycelium will make.[38]

Ecovative holds a patent on its process and grows more than 400 tonnes of furniture and packaging every year, but its business model does not depend on it being the primary producer of mycelial materials. There are people and organisations licensed to use Ecovative's Grow It Yourself (GIY) kits in thirty-one countries, producing everything from furniture to surfboards. Lighting is popular (the MushLume lamp has recently launched). A designer in the Netherlands is making mycelial slippers. The US National Oceanic and Atmospheric Administration replaced the plastic foam in the buoyant rings used to float tsunami detection devices with a mycelial alternative.[39]

One of the more ambitious visions for building with mycelium is Fungal Architectures, or FUNGAR. FUNGAR is an international consortium of scientists and designers who intend to create a building made entirely of fungus, combining mycelial composites with fungal 'computing circuits' that will detect and respond to light levels, temperature and pollution. One of the lead researchers is Andrew Adamatzky, of the Unconventional Computing Lab, the researcher who proposes that mycelial networks can be harnessed to compute information using electrical impulses that pass along their hyphae. Mycelial networks only generate electrical impulses when they are alive, a problem

Adamatzky hopes to overcome by encouraging living mycelium to absorb electrically conductive particles. Once killed and preserved, these mycelial networks will create electrical circuits consisting of mycelial wires, transistors and capacitors – 'a computing network that will fill every cubic millimetre of the building'.[40]

Walking around Ecovative's production facility, it's hard to escape the feeling that a handful of white rot fungal species are doing very well out of this arrangement. Sure, they are killed before the materials are used. But only after their appetites have been fulfilled. And after they have been introduced, for yet another time, into hundreds of pounds of freshly pasteurised sawdust. Like McCoy and the radical mycologists, who literally – and figuratively – spread spores around the world, Ecovative serves as a global dispersal system for a number of fungal species. The fungi are at once a 'technology' and partners with humans in a new type of relationship.

It's too early to say where the relationships being forged at Ecovative will end up. Confronted with the problem of how to access the energy in plant matter, *Macrotermes* termites have been cultivating huge quantities of white rot fungi in purpose-built production facilities for 30 million years. *Macrotermes* and *Termitomyces* have lived with one another for so long that neither can survive without the other. Whether or not myco-fabrication will draw humans into a co-dependent symbiosis remains to be seen, but already it's clear that once more, a global crisis is turning into a suite of fungal opportunities. Yet again, human waste streams are being reimagined in terms of fungal appetites. Some trends go viral. I started to reflect on what it would mean to go fungal.

If anyone knows about going fungal, it's Paul Stamets. I have often wondered whether he has been infected with a fungus that fills him with mycological zeal – and an irrepressible urge to persuade humans that fungi are keen to partner with us in new and peculiar ways. I went to visit him at his home on the west coast of Canada. The house is balanced on a granite bluff, looking out to sea. The roof is suspended on beams that look like mushroom gills. A *Star Trek* fan since the age of twelve, Stamets christened his new house Starship Agarikon – agarikon is another name for *Laricifomes officinalis*, a medicinal wood-rotting fungus that grows in the forests of the Pacific North-west.

I've known Stamets since I was a teenager, and he has done a lot to inspire my own interest in fungi. Every time I see him I'm met with a flurry of electrifying fungal newsflashes. Within minutes his mycological patter picks up speed, and he leaps between bulletins almost faster than he can talk, a ceaseless torrent of fungal enthusiasm. In his world, fungal solutions run amok. Give him an insoluble problem and he'll toss you a new way it can be decomposed, poisoned or healed by a fungus. Much of the time, he wears a hat made from amadou – a felt-like material produced from the fruiting body of the tinder fungus, or *Fomes fomentarius*, another white rot fungus. It carries fitting associations. Amadou has been used by humans as a firestarter for thousands of years – it was carried by the 'Iceman', the 5,000-year-old corpse preserved in glacial ice. As a tool of – thermal – combustion, it is one of the most ancient examples of human radical mycology currently known.

Not long before I arrived, Stamets had been contacted by the creative team behind the TV series *Star Trek: Discovery*, who wanted to know more about his work. He had agreed to brief them on the ways in which fungi could be used to save worlds. Sure enough, *Star Trek: Discovery*, which premiered the next year, was laced with mycological themes. A new character was introduced – a brilliant astromycologist called Lieutenant Paul Stamets – who uses fungi to develop powerful technologies which can be deployed to save humanity in a fight against a series of terminal threats. The *Star Trek* team have taken plenty of licence, though they hardly needed to. By tapping into intergalactic mycelial networks – 'an infinite number of roads, leading everywhere' – (the fictional) Stamets and his team work out how to travel in the 'mycelial plane' faster than the speed of light. Following his first mycelial immersion, Stamets comes to, dazed and transformed. 'I've spent my whole life trying to grasp the essence of mycelium. And now I do. I saw the network. An entire universe of possibilities I never dreamed existed.'

One of the problems (the real) Stamets hoped to address by collaborating with the *Star Trek* team is the neglected state of mycology. Art imitates life and life imitates art. Fictional astromycologist heroes might be able to shape the non-fictional future of fungal knowledge by inspiring a generation of young people to get excited by fungi. For (the real) Stamets, a surge of interest in fungi could fuel the development of mycological technologies that might 'help save the planet that's in jeopardy'.

When I showed up at Starship Agarikon, I found Stamets sitting on the deck fiddling around with a mason jar and a blue plastic dish. It was the prototype for a bee feeder he had invented. The jar dribbled sugar water laced with fungal extracts

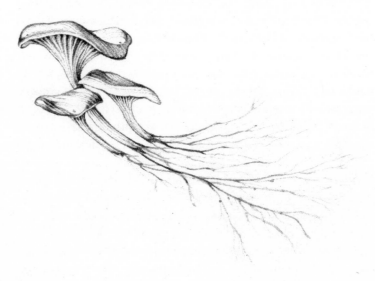

Oyster mushroom, *Pleurotus ostreatus*

into the dish, and bees crawled through a chute to get to it. It was his latest venture; a seventh way mushrooms could help save the world. Even by Stamets' standards, this project was a big headline. His latest study, co-authored with entomologists at the Washington State University bee lab, had been accepted by the prestigious journal *Nature Scientific Reports*. He and his team had shown that extracts of certain white rot fungi could be used to reduce bee mortality dramatically.[41]

About a third of global agricultural output depends on pollination from animals, particularly honeybees, and the precipitous decline in bee populations is one of the many pressing threats to humanity. A number of factors contribute to the syndrome known as colony collapse disorder. Widespread use of insecticides is one. Habitat loss is another. The most insidious problem, however, is the varroa mite, appropriately named *Varroa destructor*. Varroa mites are parasites that suck fluid from bees' bodies and are vectors for a range of deadly viruses.[42]

Wood-rotting fungi are a rich source of antiviral compounds, many of which have long been used as medicines, particularly in China. After 9/11, Stamets collaborated with the US National Institute of Health and Department of Defense in Project BioShield, a search for compounds that could be used to fight viral storms unleashed by biological terrorists. Of the thousands of compounds tested, some of Stamets' extracts from wood-rotting fungi had the strongest activity against a number of deadly viruses, including smallpox, herpes and flu. He had been producing these extracts for human consumption for several years – it is largely these products that have made Fungi Perfecti into a multi-million-dollar business. But the idea of using them to treat bees was a more recent brainwave.[43]

The effects of the fungal extracts on the bees' viral infections were unambiguous. Adding a one per cent extract of amadou (or *Fomes*) and reishi (*Ganoderma*, the species used to grow materials at Ecovative) to bees' sugar water reduced deformed wing virus eighty-fold. *Fomes* extracts reduced levels of Lake Sinai virus nearly ninety-fold, and *Ganoderma* extracts reduced it 45,000-fold. Steve Sheppard, a professor of entomology at Washington State University and one of Stamets' collaborators on the study, observed that he had not encountered any other substance that could extend the life of bees to this extent.[44]

Stamets told me how he had come up with the idea. He was day-dreaming. All of a sudden, separate lines of thought came together and hit him 'like a lightning bolt'. If fungal extracts had antiviral properties, then maybe they'd help reduce the viral load of bees – and yes, in fact, he remembered that in the late 1980s he had watched bees from his hives visiting a pile of rotting wood chips in his garden, moving the chips aside to feed on the mycelium underneath. 'Oh my God': Stamets woke up. 'I think I know how to save the bees.' It was a big moment, even for someone who has spent decades dreaming up fungal solutions to obstinate problems.

It is easy to see why *Star Trek* borrowed Stamets. His narrative style is straight out of an American blockbuster movie. Many of his accounts feature fungal heroes, poised to save the planet from almost certain doom. *Viral storms of unprecedented proportions threaten global food security. Critical pollinators struggle on under grave threat from the virus-bearing parasites, poised to inflict global famine. The future of the world hangs in the balance. But wait. Is that …? Yes! Once again, fungi come to the rescue with the help of their human accomplice, Stamets.*

Will antiviral compounds produced by wood-rotting fungi really save the bees? Stamets' findings are promising, but it's too early to say whether the fungal extracts will translate into fewer collapsed colonies in the long term. Viruses are just one of many problems that bees face. Whether the fungal antivirals perform equally well in other countries and contexts isn't known. More important, to save bee populations, Stamets' solution has to be widely adopted, a feat he hopes to accomplish by recruiting the efforts of millions of 'citizen' scientists.

I travelled down to the Olympic Peninsula in Washington state to visit Stamets' production facility. Headquarters is a cluster of large, hangar-like sheds, surrounded by woods, several miles off the beaten track. This was where Stamets grew and extracted the fungi used in the study. It was where production was soon to be 'ramped up' to bring a product to market for widespread use. In the few months after the bee study had been published, he had received tens of thousands of requests for the BeeMushroomed Feeder. Unable to keep up with demand, Stamets plans to open-source the 3D-printed design in the hope that others will start to manufacture them.

I met one of Stamets' directors of operations who had agreed to show me around. There was a strict dress code: no shoes, a lab coat and a hairnet – beard nets were also provided. We kitted up and passed through a special set of double doors designed to reduce the inflow of contaminant-filled air from outside.

We entered the fruiting rooms, which were warm and damp, the air thick and cloying. There were ranks of shelves lined with clear plastic growing bags stitched solid with mycelium sporting all sorts of startling protrusions, from woody reishi mushrooms with their shiny chestnut scalps to lion's mane, tumbling out of the bags like delicate cream-coloured corals. In the reishi fruiting room, the air was so thick with spores I could taste their soft, damp bitterness. After just a couple of minutes my hands were dusted a cappuccino brown.

Once again, humans were going out of their way to divert tonnes of food into fungal networks. Once again, a global crisis was turning into a set of fungal opportunities. Like the challenge faced by *Pleurotus* mycelium paused at the edge of a puddle of toxic waste, radical mycological solutions are less about inventing than remembering. Somewhere in the

Pleurotus genome there is probably an enzyme that will do the job. Perhaps it has done the job before. Perhaps it hasn't, but can be repurposed to serve a new cause. Similarly, somewhere in the history of life there may be a fungal ability or relationship that can inspire a new, old solution to one of our many dire problems. I thought of the bee story. Stamets's eureka moment happened when he remembered something he had seen decades earlier – bees appearing to medicate themselves using fungi. Stamets didn't discover the idea of curing bees using fungi. Bees did, we presume, during a biochemical squabble with viruses in a damp corner of their shared history. Somewhere deep in the psycho-spiritual compost heap of his dream world, Stamets metabolised an old radical mycological solution into a new one.

I walked into the growing rooms, packed with shelving units three metres tall. This was the fungus comb. Thousands of bags charged with soft blocks of furry mycelium filled the space. Some were white, some off-yellow, some a pale orange. If the fans filtering the air had stopped, I felt I might have heard the crackling of millions of miles of mycelium running through its food. Upon harvest, the bags of mycelium were extracted in large barrels full of alcohol to produce the cure for the bees. Like so many radical mycological solutions, it is still uncertain; the first tender steps towards the possibility of mutually assured survival, symbiosis in its earliest infancy.

8

Making Sense of Fungi

It matters which stories tell stories, which concepts think
concepts ... which systems systematise systems.

Donna Haraway

The fungi that share the most intimate history with humans are
yeasts. Yeasts live on our skin, in our lungs and in our gastro-
intestinal tract, and line our orifices. Our bodies have evolved
to regulate these populations, and have been doing so for long
stretches of our evolutionary history. For thousands of years,
too, human cultures have evolved sophisticated ways to regulate yeast populations outside the confines of our bodies, in
barrels and jars.[1] Today, yeasts are some of the most widely
used model organisms in cell biology and genetics: they are the
simplest envelopes of eukaryotic life, and many human genes
have equivalents in yeasts. In 1996, *Saccharomyces cerevisiae*,
the species of yeast used in brewing and baking, became the first
eukaryotic organism to have its genome sequenced. Since 2010,
more than a quarter of Nobel Prizes for Physiology or Medicine
have been awarded for work on yeast. Yet it was only in the
nineteenth century that yeasts were discovered to be microscopic organisms.[2]

Exactly when humans first started to work with yeast remains
an open question. The first unambiguous evidence dates from

around 9,000 years ago in China, but microscopic starch grains have been unearthed on stone tools in Kenya that date from 100,000 years ago. The shape of the starch grains suggests that the tools had been used to process the African wine palm, *Hyphaene petersiana*, which is still used to make liquor. Given that any sugary liquid left for longer than a day will start to ferment by itself, it is probable that humans have been brewing for far longer.[3]

Yeasts oversee a transformation of sugar into alcohol; the anthropologist Claude Lévi-Strauss argued that they also oversaw one of the most dramatic cultural transformations in human history: the passage of humans from hunter-gatherers to agriculturalists. He considered mead – a drink made from fermented honey – to be the first alcoholic beverage, and imagined the transition from 'natural' fermentation to cultural 'brewing' using the example of a hollow tree. The alcohol was part of nature if the honey had fermented 'by itself', and part of culture if humans had placed the honey to ferment in an artificially hollowed-out trunk. (It is an interesting distinction; by extension, *Macrotermes* termites and leafcutter ants made the transition from nature to culture tens of millions of years before humans.)[4]

Lévi-Strauss may or may not be correct about mead. However, the yeast resembling modern brewing yeast arose around the same time as domesticated goats and sheep. The origins of agriculture around 12,000 years ago – the so-called Neolithic Transition – can be understood, at least in part, as a cultural response to yeast. It was either for bread or for beer that humans started to give up their nomadic lifestyles and settle into sedentary societies (the beer-before-bread hypothesis has steadily gained traction among scholars since the 1980s). And whether in bread or in beer, yeasts were the primary beneficiaries of humans' earliest agricultural efforts. In the preparation of either, humans feed yeast before they feed themselves. The cultural developments associated with agriculture – from fields

Brewer's yeast, *Saccharomyces cerevisiae*

of crops to cities, accumulation of wealth, grain stores, new diseases – form part of our shared history with yeast. In many ways, you might argue, yeasts have domesticated us.[5]

My own relationship with yeast underwent a transformation at university. One of my neighbours had a boyfriend who visited regularly. Without fail, soon after he arrived, large plastic mixing bowls filled with fluid and covered with cling film would appear on the kitchen windowsills. It was wine, he told me. He had learned how to make alcohol from a friend who had spent time in prison in French Guyana. I was fascinated, and soon had a collection of mixing bowls of my own. It turned out to be remarkably simple. Yeast does almost all the work. It likes to be warm, but not too hot, and reproduces most happily in the dark. Fermentation starts when you add the yeast to a warm sugary solution. In the absence of oxygen, yeast converts sugar into alcohol and releases carbon dioxide. Fermentation stops when the yeast runs out of sugar or dies of alcohol poisoning.

I filled a mixing bowl with apple juice, sprinkled in a couple of teaspoons of dried baker's yeast, and left it by the heater in my bedroom. I watched as swathes of froth appeared and the plastic covering puffed out into a bubble. Now and then a small jet of gas would escape, carrying increasingly alcoholic fumes. After three weeks I could contain my curiosity no longer and took the bowl along to a party, where it disappeared in a matter of minutes. The brew was drinkable, if a little sweet, and judging by its effects, had an alcohol percentage around that of a strong beer.

Things escalated quickly. After a couple of years I had several large brewing containers, including a fifty-litre saucepan, and had started to brew drinks from recipes I found documented in historical texts. There were spiced meads, from *The Closet of Sir Kenelm Digby* published in 1669, and medieval 'gruit' ales made with bog myrtle that I harvested from a nearby marsh. Soon followed hawthorn wines, nettle beers, and a medicinal ale recorded in the seventeenth century by Dr William Butler, the physician to James I, and said to be a remedy for everything from 'London plague' to measles and 'diverse other diseases'. My room was lined with barrels of bubbling liquids and my wardrobe filled up with bottles.[6]

I brewed the same fruit with cultures of yeast gathered from different places. Some were rich and savoury. Some were cloudy and delicious. Others tasted of socks or turpentine. There was a fine line between foul and fragrant, but it didn't matter. Brewing granted me access to the invisible worlds of these fungi, and I delighted in being able to taste the difference between yeasts gathered from apple skins, or collected in plates of sugar water placed overnight on shelves in old libraries.

The transformational power of yeast has long been personified as a divine energy, a spirit, or a god. How could it escape this treatment? Alcohol and inebriation are some of the oldest magics. An invisible force conjures wine from fruit, beer from grain, mead from nectar. These liquids alter our minds and have been enfolded within human cultures in many ways: from ritual feasting and statecraft to a means to pay for labour. For just as long, they have been responsible for dissolving our senses, for wildness and ecstasy. Yeasts are both makers and breakers of human social orders.

Ancient Sumerians – who left written beer recipes dating back 5,000 years – worshipped a goddess of fermentation, Ninkasi. In the Egyptian *Book of the Dead*, prayers are addressed to 'givers of bread and beer'. Among the Ch'orti' people in South America, the onset of fermentation was understood as 'the birth of the good spirit'. The ancient Greeks had Dionysus: god of wine, wine-making, madness, drunkenness and domesticated fruit in general – a personification of the power of alcohol both to forge and corrode human cultural categories.[7]

Today, yeasts have become biotechnological tools engineered to produce drugs from insulin to vaccines. Bolt Threads, a company working with Ecovative to produce mycelial leather, has genetically engineered yeasts to produce spider silk. Researchers are working to modify yeast metabolisms to allow them to make sugar from woody plant material for use in biofuels. One team is working on 'Sc2.0', a human-designed synthetic yeast built from the bottom up – an artificial life form that engineers will be able to program to produce any number of compounds. In all these instances, yeast with its transformational power blurs the line between nature and culture, between an organism that self-organises and a machine that is built.[8]

In my experimentations, I learned that the brewer's art involves a subtle negotiation with the cultures of yeast. Fermentation is domesticated decomposition – rot rehoused. If successful, the brew would end up on the right side of the line. But as is so often the case with fungi, nothing is for certain. By attending to cleanliness and temperature and ingredients – all important constraints on the possible routes fermentation could take – I could lure the fermentation in promising directions, but there could be no coercion. For this reason the outcome was always surprising.

Many of the historical brews were fun to drink. The meads brought on laughter. The gruit ales made people talkative.

Dr Butler's ale induced a peculiar golden heaviness. Some were bottled havoc. Whatever their effect, I was fascinated by the process of brewing a historical text into being. Old brewing recipes are records of how yeasts have etched themselves into human lives and minds over the last few hundred years. In all the pages of these books, yeasts are a silent companion, an invisible participant in human culture. Ultimately, these recipes were stories that made sense of how substances decomposed. They reminded me that it matters what stories we use to make sense of the world. The story you hear about grain determines whether you end up with bread or beer. The story you hear about milk determines whether you end up with yoghurt or cheese. The story you hear about apples determines whether you end up with sauce or cider.

Yeasts are microscopic, which makes it easy for a thick narrative sediment to build up around their lives. Fungi that grow mushrooms have generally been understood in simpler terms. Mushrooms, it has long been known, might be delicious, but might also poison you, cure you, feed you or give you visions. For hundreds of years, east Asian poets have written rhapsodic verse about mushrooms and their flavours. 'Oh matsutake: / The excitement before finding them', enthused Yamaguchi Sodo in seventeenth-century Japan. European authors, on the whole, have been more dubious. Albertus Magnus, in his thirteenth-century herbal *De Vegetabilibus*, warned that mushrooms 'of a moist humour' might 'stop up in the head the mental passages of the creatures [that eat them] and bring on insanity'. John Gerard, writing in 1597, warned his readers to stay clear entirely:

> Few mushrooms are good to be eaten and most of them do suffocate and strangle the eater. Therefore, I give my advice unto those that love such strange and new fangled meates to beware of licking honey amongst thorns lest the sweetness

of the one do countervaile the sharpness and pricking of
the other.

But humans have never stayed clear.[9]

In 1957, Gordon Wasson – who first popularised 'magic
mushrooms' in the 1957 article in *Life* magazine – and his wife
Valentina developed a binary system by which all cultures could
be categorised: 'mycophilic' (fungus-loving cultures) as opposed
to 'mycophobic' (fungus-fearing cultures). Present-day cultural
attitudes to mushrooms were, the Wassons speculated, a 'latter-
day echo' of ancient psychedelic mushroom cults. Mycophilic
cultures were descendants of those who had worshipped mushrooms. Mycophobic cultures were descendants of those who had
considered their power diabolical. Mycophilic attitudes might
lead Yamaguchi Sodo to write poems in praise of matsutake,
or urge Terence McKenna to proselytise about the benefits of
taking large doses of psilocybin mushrooms. Mycophobic attitudes might fuel a moral panic that leads to their illegalisation,
or prompt Albertus Magnus and John Gerard to issue stark
warnings about the dangers of these 'new fangled meates'. Both
positions recognise the power of mushrooms to affect people's
lives. Both make sense of this power in different ways.[10]

We shoehorn organisms into questionable categories all the
time. It's one of the ways we make sense of them. In the nineteenth century, bacteria and fungi were classified as plants.[11]
Today both are recognised as belonging to their own distinct
kingdoms of life, although it wasn't until the mid 1960s that they
won their independence. For much of recorded human history
there has been little consensus over what fungi actually are.[12]

Theophrastus, a student of Aristotle, wrote about truffles –
but could only say what they were not: he described them as
having no root, no stem, no branch, no bud, no leaf, no flower,
no fruit; nor bark, nor pith, nor fibres, nor veins. In the opinion
of other classical writers, mushrooms were spontaneously generated by lightning strikes. For others, they were outgrowths of the
Earth, or 'excrescences'. Carl Linnaeus, the eighteenth-century

Swedish botanist who devised the modern taxonomic system, wrote in 1751 that 'The order of Fungi is still Chaos, a scandal of art, no botanist knowing what is a Species and what is a Variety'.[13]

To this day, fungi slip around the systems of classification we build for them. The Linnaean system of taxonomy was designed for animals and plants, and doesn't easily cope with fungi, lichens or bacteria. A single species of fungus can grow into forms that bear no resemblance to each other whatsoever. Many species have no distinctive characteristics that can be used to define their identity. Advances in gene sequencing make it possible to order fungi into groups that share an evolutionary history, rather than groups based on physical traits. However, deciding where one species starts and another stops based on genetic data presents as many problems as it solves. Within the mycelium of a single fungal 'individual' there can exist multiple genomes. Within the DNA extracted from a single pinch of dust, there might be tens of thousands of unique genetic signatures, but no way to assign them to known fungal groups. In 2013, in a paper called 'Against the naming of fungi', the mycologist Nicholas Money went so far as to suggest that the concept of fungal species should be abandoned altogether.[14]

Systems of classification are just one of the ways people make sense of the world. Out-and-out value judgements are another. Charles Darwin's grand-daughter, Gwen Raverat, described the disgust evoked in her aunt Etty – Darwin's daughter – by the stinkhorn mushroom, *Phallus impudicus*. Stinkhorns are notorious for their phallic form. They also produce a pungent-smelling slime that attracts the flies that help disperse their spores. In 1952, Raverat reminisced:[15]

> In our native woods there grows a kind of toadstool called in the vernacular The Stinkhorn (though in Latin it bears a grosser name). The name is justified for the fungus can be hunted by scent alone, and this was Aunt Etty's great invention. Armed with a basket and a pointed stick, and wearing a special hunting

cloak and gloves, she would sniff her way through the wood, pausing here and there, her nostrils twitching when she caught a whiff of her prey. Then with a deadly pounce she would fall upon her victim and poke his putrid carcass into her basket. At the end of the day's sport the catch was brought back and burnt in the deepest secrecy on the drawing room fire with the door locked – because of the morals of the maids.

Crusade or fetish? Mycophobe or closet mycophile? It isn't always easy to tell the difference. For someone repulsed by the stinkhorns, Aunt Etty spent more time than most seeking them out. In her 'sport', she would no doubt have performed better than any number of flies in spreading the stinkhorns' spores. The noisome odour, presumably irresistible to the flies, proved irresistible to Aunt Etty as well – although her attraction was refracted through distaste. Motivated by her horror, she enveloped stinkhorns within a Victorian morality and became a passionate recruit to a fungal cause.

The ways in which we try to make sense of fungi often tell us as much about ourselves as the fungi we try to understand. The yellow staining mushroom (*Agaricus xanthodermus*) is described in most field guides as poisonous. A keen mushroom hunter with a large mycological library once told me about an old guidebook he owned, in which the same mushroom was described as 'delicious, when fried', although the author did add as an afterthought that the mushroom 'may cause a light coma in those of weak constitution'. How you make sense of the yellow staining mushroom depends on your physiological make-up. Although poisonous to most people, some are able to eat it without ill effect. How it is described will depend on the physiology of the person doing the describing.[16]

This kind of bias is particularly apparent in discussions of symbiotic relationships, which have been understood in human terms since the word was first coined in the late nineteenth century.

The analogies used to make sense of lichens and mycorrhizal fungi say it all. Master and slave, cheater and cheated, humans and domesticated organisms, men and women, the diplomatic relationships between nations ... The metaphors change over time, but attempts to dress up more-than-human relationships in human categories continue to the present day.

As the historian Jan Sapp explained to me, the concept of symbiosis behaves like a prism through which our own social values are often dispersed. Sapp is fast-talking and has a sharp eye for ironic detail. The history of symbiosis is his speciality. He has spent decades with biologists, in labs, conferences, symposiums and jungles, as they've wrestled with questions about how organisms interact with one another. A good friend of Lynn Margulis and Joshua Lederberg, he has watched from the front row as the modern science of microbiology has 'gone big'. The politics of symbiosis have always been fraught. Is nature fundamentally competitive or co-operative? A lot turns on this question. For many, it changes the way we understand ourselves. It isn't surprising that these issues remain a conceptual and ideological tinderbox.[17]

The dominant narrative in the United States and western Europe since the development of evolutionary theory in the late nineteenth century was one of conflict and competition, and it mirrored views of social progress within an industrial capitalist system. Examples of organisms co-operating with one another to their mutual benefit 'remained close to the margins of polite biological society', in Sapp's words. Mutualistic relationships, such as those that give rise to lichens, or plants' relationships with mycorrhizal fungi, were curious exceptions to the rule – where they were acknowledged to exist at all.[18]

Opposition to this view didn't fall neatly down East–West lines. Nonetheless, the ideas of mutual aid and co-operation in evolution were more prominent in Russia than in western European evolutionary circles. The strongest riposte to the dog-eat-dog vision of 'Nature red in tooth and claw' came from the Russian anarchist Peter Kropotkin in his bestselling book *Mutual Aid: A Factor of Evolution*, in 1902. In it, he stresses that 'sociability' was as much a part of nature as the struggle for existence. Based on his interpretation of nature, he advocated a clear message: 'Don't compete!' 'Practise mutual aid! That is the surest means of giving to each and to all the greatest safety, the best guarantee of existence and progress, bodily, intellectual, and moral.'[19]

For much of the twentieth century, discussion of symbiotic interactions remained loaded with political charge. Sapp points out that the Cold War prompted biologists to take more seriously the question of co-existence in the world at large. The first international conference on symbiosis was held in London in 1963, six months after the Cuban Missile Crisis had brought the world to the brink of nuclear war. This was no accident. The editors of the conference proceedings commented that 'the pressing problems of co-existence in world affairs may have influenced the Committee in their choice of subject for this year's Symposium'.[20]

It is well established in the sciences that metaphors can help to generate new ways of thinking. The biochemist Joseph Needham described a working analogy as a 'net of co-ordinates' that could be used to arrange an otherwise formless mass of information, much as a sculptor might use a wire frame to provide support for wet clay. The evolutionary biologist Richard Lewontin pointed out that it is impossible to 'do the work of science' without using metaphors, given that almost 'the entire body of modern science is an attempt to explain phenomena that cannot be experienced directly by human beings'. Metaphors and analogies, in turn, come laced with human stories and values, meaning that no discussion of scientific ideas – this one included – can be free of cultural bias.[21]

Today, the study of shared mycorrhizal networks is one of the fields most commonly beset with political baggage. Some portray these systems as a form of socialism by which the wealth of the forest can be redistributed. Others take inspiration from mammalian family structures and parental care, with young trees nourished by their fungal connections to older and larger 'mother trees'. Some describe networks in terms of 'biological markets', in which plants and fungi are portrayed as rational economic individuals trading on the floor of an ecological stock exchange, engaging in 'sanctions', 'strategic trading investments' and 'market gains'.[22]

The Wood Wide Web is a no less anthropomorphic term. Not only are humans the only organisms to build machines, but the internet and World Wide Web are some of the most overtly politicised technologies that exist today. Using machine metaphors to understand other organisms can be as problematic as borrowing concepts from human social lives. In reality, organisms grow; machines are built. Organisms continually remake themselves; machines are maintained by humans. Organisms self-organise; machines are organised by humans. Machine metaphors are sets of stories and tools that have helped make countless discoveries of life-changing importance. But they aren't scientific facts, and can lead us into trouble when prioritised over all other types of story. If we understand organisms to be machines, we'll be more likely to treat them as such.[23]

Only with hindsight can we see which metaphors are most helpful. Today it would be absurd to try to bundle all fungi into categories of 'agents of disease' or 'parasites', as was common in the late nineteenth century. Yet before lichens had led Albert Frank to coin the word 'symbiosis', there was no other way to describe relationships between different types of organism. In recent years the narratives surrounding symbiotic relationships have become more nuanced. Toby Spribille – the researcher who found that lichens consist of more than two players – makes the case that lichens have to be understood as systems. Lichens don't seem to be the product of a fixed partnership, as had long been thought. Rather, they arise out of an array of possible relationships between a number of different players. For Spribille, the relationships that underpin lichens have become a question, rather than an answer known in advance.

Similarly, plants and mycorrhizal fungi are no longer thought of as behaving either mutualistically or parasitically. Even in the relationship between a single mycorrhizal fungus and a single plant, give-and-take is fluid. Instead of a rigid dichotomy, researchers describe a mutualism-to-parasitism continuum. Shared mycorrhizal networks can facilitate co-operation and also competition. Nutrients can move through the soil via fungal connections, but so can poisons. The narrative possibilities are richer. We have to shift perspectives and find comfort in – or just endure – uncertainty.

Nonetheless, some still relish politicising the debate. One biologist in particular, Sapp relayed with amusement, 'calls me the biological left, and himself the biological right'. They had been discussing the idea of biological individuals. In Sapp's view, the developments in microbial sciences had made it hard to define the boundaries of an individual organism. For his detractor, who had positioned himself on the biological right, neat individuals had to exist. Modern capitalist thought is founded on the idea of rational individuals acting in their own interest. Without individuals, everything comes crashing down. From his perspective, Sapp's argument belied a fondness for collectives and an underlying socialist tendency. Sapp laughed. 'Some people just like to make artificial dichotomies.'[24]

In *Braiding Sweetgrass*, the biologist Robin Wall Kimmerer writes about a word in the Native American language Potawatomi, *puhpowee*. *Puhpowee* translates as 'the force which causes mushrooms to push up from the earth overnight'. Kimmerer recalls that she later learned that '*puhpowee* is used not only for mushrooms, but also for certain other shafts that rise mysteriously in the night'. Is it anthropomorphic to describe a mushroom's emergence in the same language used to describe human male sexual arousal? Or is it mycomorphic to describe human male sexual arousal in the same language used to describe a mushroom's growth? Which way does the arrow point? If you say that a plant 'learns', 'decides', 'communicates' or 'remembers', are you humanising the plant, or vegetalising a set of human concepts? The human concept might take on new meanings when applied to a plant, just as plant concepts might take on new meanings when applied to a human: blossom, bloom, robust, root, sappy, radical ...[25]

Natasha Myers, the anthropologist who introduced the word 'involution' to describe the tendency for organisms to associate with one another, points out that Charles Darwin seemed quite ready to vegetalise himself, to practise 'phytomorphism'. Writing about orchid flowers in 1862, Darwin observed that 'The position of the antennae in this *Catasetum* may be compared with that of a man with his left arm raised and bent so that his hand stands in front of his chest, and with his right arm crossing his body lower down so that the fingers project just beyond his left side.'[26] Is Darwin humanising the flower, or is he being vegetalised by the flower? He is describing plant features in human terms, a sure sign of anthropomorphism. But he is also reimagining the male body – including his own – in floral form, suggesting that he is open to exploring the flower's anatomy on its own terms.

This is an old story. It is hard to make sense of something without a little part of that something rubbing off on you. Sometimes it is intentional. Radical Mycology, for example, is an organisation without a well-defined shape. This is no accident. Its founder, Peter McCoy, points out that fungi have the power to change the way we think and imagine. Trees crop up in everything from our depictions of genealogy and relationship (whether human, biological or linguistic families), to the tree-like data structures in computer science, to 'dendrites' in nervous systems (from *dendron*, the Greek for tree). Why shouldn't mycelium? Radical Mycology organises itself using decentralised mycelial logics. Regional networks loosely associate with the larger movement. Periodically, the Radical Mycology network coalesces into a fruiting body, such as the Radical Mycology Convergence I attended in Oregon.

How different would our societies and institutions look if we thought of fungi, rather than animals or plants, as 'typical' life forms?[27]

Sometimes we imitate the world without conscious effort. Dog owners often look like their dogs; biologists often come to behave like their subject matter. Since the term 'symbiosis' was first coined by Frank in the late nineteenth century, researchers studying the relationships between organisms have been coaxed into forming unusual interdisciplinary collaborations. As Jan Sapp pointed out to me, it was an unwillingness to make daring leaps across institutional boundaries that contributed to the neglect of symbiotic relationships for much of the twentieth century. As the sciences became increasingly professionalised, disciplinary chasms separated geneticists from embryologists, botanists from zoologists, microbiologists from physiologists.

Symbiotic interactions reach across species boundaries; studies of symbiotic interactions must reach across disciplinary boundaries. This is no less the case today. 'Sharing resources for mutual benefit: crosstalk between disciplines deepens the understanding of mycorrhizal symbioses …' So began a write-up of the international conference on mycorrhizal biology in 2018. The study of mycorrhizal fungi requires that an academic symbiosis form between mycologists and botanists. The study of the bacteria that live in fungal hyphae requires symbiotic interactions between mycologists and bacteriologists.[28]

I never behave more like a fungus than when I'm investigating them, and quickly enter into academic mutualisms based on an exchange of favours and data. In Panama I acted like the growing front of mycorrhizal mycelium, up to my elbows in red mud for days on end. I anxiously ferried large coolers of samples to other countries through customs, X-ray checks and sniffer dogs. I peered down microscopes in Germany, pored over fungal lipid profiles in Sweden and extracted and sequenced fungal DNA in England. I sent gigabytes of data coughed out of a machine in Cambridge to be processed in Sweden, and then on to collaborators in the United States and Belgium. If my movements had left a trail behind them, they would have traced a complex network, complete with the bidirectional movement of information and resources. Like plants, my collaborators in Sweden and Germany got access to a greater volume of soil by associating with me. They were unable to travel to the tropics themselves, so I extended their reach. In return, like a fungus, I got access to funds and techniques that would have otherwise remained out of my reach. My collaborators in Panama benefited from the grants and technical expertise of my colleagues in England. Similarly, my colleagues in England benefited from the grants and expertise of my Panamanian collaborators. To study a flexible network, I had to assemble a flexible network. It is a recurring theme: look at the network, and it starts to look back at you.

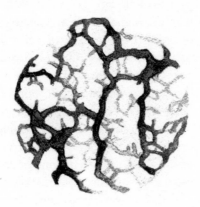

'Drunkenness', the French theorist Gilles Deleuze writes, is 'a triumphant eruption of the plant in us.' It is no less the triumphant eruption of the fungus in us. Can intoxication help us rediscover parts of ourselves in the fungal world? Might there be ways to make sense of fungi by loosening our grip on our humanness, or finding in our humanness a bit of something else, something fungal? This something else might be a scrap or two left over from a time when we were more closely related to fungi. Or perhaps something that we have learned in our long and entangled history with these extraordinary creatures.[29]

About ten million years ago, the enzyme our bodies use to detoxify alcohol, known as alcohol dehydrogenase, or ADH4, underwent a single mutation that left it forty times more efficient. The mutation occurred in the last common ancestor we shared with gorillas, chimpanzees and bonobos. Without a modified ADH4, even small quantities of alcohol are poisonous. With a modified ADH4, alcohol can be consumed safely and used by our bodies as a source of energy. Long before our ancestors became human, and long before we evolved stories to make cultural and spiritual sense of alcohol and the cultures of yeast that produce it, we evolved the enzymes to make metabolic sense of them.[30]

Why would the ability to metabolise alcohol arise so many millions of years before humans developed technologies of fermentation? Researchers point out that ADH4 upgraded at a time when our primate ancestors were spending less time in trees and adapting to life on the ground. The ability to metabolise alcohol, they speculate, played a crucial role in the ability of primates to make a living on the forest floor by opening up a new dietary niche: overripe, fermented fruit that had fallen from trees.

The ADH4 mutation provides support for the 'drunken monkey hypothesis', proposed by the biologist Robert Dudley to explain the origins of humans' fondness for alcohol. In this view, humans are tempted by alcohol because our primate ancestors were. The scent of alcohol produced by yeasts was a reliable way to find ripe fruit as it rotted on the ground. Both our human attraction to alcohol, and the entire ecology of gods and goddesses that oversee fermentation and intoxication, are remnants of a much more ancient fascination.[31]

Primates aren't the only animals attracted by alcohol. Malaysian treeshrews – small mammals with feathery tails – climb into the flower buds of the bertam palm and drink fermented nectar in quantities that, scaled to body weight, would intoxicate a human. The plume of alcoholic vapours produced by yeasts attracts the treeshrews to the palm flowers. Bertam palms depend on treeshrews to pollinate them, and their flower buds have evolved into specialised fermentation vessels – structures that harbour communities of yeast and encourage such rapid fermentation that their nectar froths and bubbles. Treeshrews for their part have evolved a remarkable ability to detoxify the alcohol, and appear not to suffer from any negative effects of inebriation.[32]

The mutation in ADH4 helped our primate ancestors to extract energy from alcohol. In a twist on the drunken monkey hypothesis, humans continue to look for ways to extract energy from alcohol, although we burn it as biofuel in combustion engines rather than as metabolic fuel in our bodies. Billions of gallons of ethanol biofuels are produced every year from corn in the United States and sugarcane in Brazil. In the States, an area larger than the size of England is used to grow corn which is processed and fed to yeast. The rate of conversion of grassland to biofuel crops is comparable to the rate of deforestation, by percentage of land cover, in Brazil, Malaysia and Indonesia. The ecological consequences of the biofuel boom are far-reaching. Vast government subsidies are required; conversion of grassland into farmland releases large amounts of carbon into the atmosphere; huge quantities of fertiliser run off into streams and rivers and are responsible for the dead zone in the Gulf of Mexico. Once again, yeasts and the ambiguous power of the alcohol they produce are participating in human agricultural transformation.[33]

Inspired by the drunken monkey hypothesis, I resolved to ferment some overripe fruit. This would be a way to consummate a narrative, to let it modify my perceptions of the world, to make decisions under the influence of it, to be intoxicated by it. Drunkenness may be the eruption of the fungus in us; this would be the eruption of a fungal story. How often stories change our perceptions, and how often we don't notice.

The idea occurred to me while on a tour of the Cambridge Botanical Gardens given by their charismatic director. In his company, clouds of stories emanated from even the most unremarkable shrub. One plant, a large apple tree near the entrance, stood out. It grew, we were told, from a cutting taken from a 400-year-old apple tree in the garden of Isaac Newton's family home, Woolsthorpe Manor. It was the only apple tree that grew there, and was old enough to have been around when Newton formulated his theory of universal gravitation. If any tree had dropped an apple that inspired Newton, it was this one.

Having been grown from a cutting, the tree in front of us was, the director reminded us, a clone of the famous tree. This made it, at least genetically, the same tree that had done the deed. Or rather, that *would have* done the deed if the deed had happened. Given that the apple story had no basis in firm fact, we were quickly assured, it was unlikely that an apple had been involved in the theory of gravitation at all. Nevertheless, this was far and away the most likely candidate for the tree that *didn't* drop the apple that inspired the theory of gravitation.

This wasn't the only clone. The director informed us that there were two more: one on the site of Newton's alchemical laboratory, at the front of Trinity College, the other outside the maths faculty. (It later transpired that there are even more – in the president's garden at the Massachusetts Institute of Technology, among other places.) The myth was strong enough to make three separate academic committees – known for their caution and indecision above all else – decide to plant the trees in auspicious places around the town. All the while, the official position remained unchanged: the story of Newton's apple was apocryphal and had no basis in firm fact.

As botanical theatre goes, it doesn't get much better. A plant's involvement in one of the most significant theoretical breakthroughs in the history of Western thought was being affirmed and denied *at the same time*. Out of this ambiguity grew actual trees, with actual apples, that fell to the ground and rotted into a pungent alcoholic mess.

The story of Newton's apple is apocryphal because Newton himself left no written account of it. However, there are several versions of the story recorded by Newton's contemporaries. The most detailed account was written by William Stukeley, a young fellow of the Royal Society and antiquarian best known today for his works on Britain's stone circles. In 1726, Stukeley recalled, he and Newton ate together in London:[34]

> After dinner, the weather being warm, we went into the garden & drank thea under the shade of some apple tree; only he & myself ... Amid other discourse, he told me, he was just in the same situation, as when formerly the notion of gravitation came into his mind. Why shd that apple always descend perpendicularly to the ground, thought he to himself; occasion'd by the fall of an apple, as he sat in contemplative mood. Why shd it not go sideways, or upwards? But constantly to the Earth's centre? Assuredly the reason is, that the Earth draws it. There must be a drawing power in matter.

The modern story of Newton's apple is a story about a story about what Newton said. This is what made the trees so narratively rich. It was impossible to verify the story either way. In response to this quandary, the academics acted as if it were both true and false. The story shuttled in and out of legend. The trees were saddled with an impossible narrative, an example of the way non-human organisms stretch the seams of our categories to breaking point. Whether an apple had actually inspired Newton to derive his theory of gravitation had long ceased to matter. The trees grew; the story thrived.

Politely, I asked the director if I could pick some apples from the tree. It hadn't occurred to me that this might be a problem. We had been told that the apples – a rare variety called the

'Flower of Kent' – were famously unpleasant to eat. It was something to do with their particular combination of sourness and bitterness, the director explained, a combination that some likened to Newton's character in his later life. I was surprised to be met with a hard no, and asked why. 'The apples have to be seen by the tourists to fall from the tree,' the director confessed apologetically, 'to add verisimilitude to the myth.'

Who was kidding whom? How could so many respectable people become so intoxicated by a story, so comforted by it, constrained by it, enraptured with it, blinded by it? Then again, how could they not be? Stories are told to modify our perceptions of the world, so it's rare that they *don't* do all these things to us. But it is uncommon to find a situation where the absurdity is so apparent, where a plant is made to clown for us quite so explicitly. I picked up one of the already fallen, decomposing apples, smelled the alcohol, and decided that this would be my rotting fruit.

The trouble was that I had no way to press apples into juice. I looked online and read about an apple problem afflicting communities in a suburb of Cambridge. Residents' apple trees overhanging the road were dropping their fruit into the street. Local youngsters used them as missiles. Windows had been broken and cars dented. In an inspired political gambit, a residents' association had provided a community apple press to manage the problem and reduce waste. It appeared to have worked. Community violence was pressed into juice. Juice was fermented into cider. Cider was drunk into community spirit. The principle was sound. A human crisis was being decomposed by a fungus. In yet another way, humans were organising themselves to divert waste into waiting fungal appetites. In turn, fungal metabolisms were acting back on human lives and culture. Beer, penicillin, psilocybin, LSD, biofuels ... how many times had this happened before?

I contacted the custodian of the press to ask for a turn. It was in high demand, and had to be transferred directly from borrower to borrower. I was put in touch with a local vicar, who pulled up in a battered Volvo a couple of days later, the elegant device in tow. There were vicious-looking toothed cogs to mulch apples into a pulp, a large screw to apply pressure, and a spout for the juice to run out.

I harvested Newton's apples by night with a friend and large camping backpacks. We left some apples on the tree for the sake of the myth, but I'm sorry to say we made off with most of them. I later discovered that we were 'scrumping' – a dialect word of West Country provenance, originally used to describe the collection of windfalls, and later the taking of fruit without permission. The difference was that in the West Country apples entailed cider, and cider had value: landowners used to include a daily pintage of cider as part of their labourers' wages, one of the many ways in which yeasts' metabolisms fed back into the agricultural systems made to house them. Under the Newton tree, however, apples meant mess and a gardener's nuisance. The press was working its magic. Waste was pressed into juice, and juice fermented into cider. A win-win.

Pressing the apples was hard work. Two or three people would steady the press, while one cranked the handle. As the apples were mulched, two people would wash and chop. It grew into a production line. The room was filled with the sharp, musty smell of crushed apples. There were apples everywhere, in various states. Our hair had pulp in it and our clothes were sodden. The carpets were sticky and damp and the walls were stained. By the end of the day there were thirty litres of juice.

When fermenting cider, you're faced with a choice. Either you add an established yeast culture that comes in a packet, or you add nothing and let the resident yeast on the apple skins take on the task. Different varieties of apple have their own indigenous yeast cultures on their skins, each fermenting at its own pace, preferentially preserving and transforming different elements of the fruit's flavour. Like all fermentations, it is a fine line. If rogue yeasts or bacteria become established, the juice goes rotten. A cider made with a single cultivated strain from a packet would be less likely to veer into a rot, but wouldn't represent the apples' own cultures of yeast. There was no question that the wild yeasts would have to handle this job. Newton's apples came ready dusted in Newton's yeast. I would have no way to know exactly which yeast strains would end up running the brew, but so it had been for most of human history.

The juice fermented in about two weeks, resulting in a cloudy, pungent liquid, which I bottled. After a few days, once it had settled, I poured myself a glass. To my amazement it was delicious. The bitterness and sourness of the apples had transformed. The taste was floral and delicate, dry with a gentle fizz. Drunk in larger quantities, it elicited elation and light euphoria. I didn't experience the muddiness of emotion I had felt after drinking some ciders. Nor did I feel clumsy, although yeast had most certainly made a nonsense of me. I was intoxicated with a story, comforted by it, constrained by it, dissolved in it, made senseless by it, weighed down by it. I called the cider 'Gravity', and lay heavy and reeling under the influence of yeast's prodigious metabolism.

Epilogue

This Compost

Our hands imbibe like roots,
so I place them on what is beautiful in this world.

Saint Francis of Assisi

As a child I loved the autumn. Leaves fell from a large chestnut tree and gathered into drifts in the garden. I raked them into a pile and tended it carefully, adding fresh armfuls as the weeks went by. Before long, the piles grew large enough to fill several bathtubs. Again and again, I'd leap into the leaves from the low branches of the tree. Once inside, I'd wriggle until I was entirely submerged and lie buried in the rustle, lost in the curious smells.

My father encouraged me to immerse myself in the world head first. He used to carry me around on his shoulders and bury my face in flowers as if I were a bee. We must have pollinated countless flowers as we shuffled from plant to plant, my cheeks smeared with yellows and oranges, my face scrunched into new shapes to better fit inside the pavilions that the petals made, both of us delighted with the colours and smells and mess.

My leaf piles were both places to hide and worlds to explore. But as months went by, the piles shrank. It became harder to submerge myself. I investigated, reaching down into the deepest regions of the heap, pulling out damp handfuls of what looked less and less like leaves, and more and more like soil. Worms started to appear. Were they carrying the soil up into the pile, or the leaves down into the soil? I was never sure. My sense was that the pile of leaves was sinking, but if it was sinking, what was it sinking into? How deep was the soil? What kept the world afloat on this solid sea?

I asked my father. He gave me an answer. I replied with another 'Why?' No matter how many times I asked why he always had an answer. These games of 'Why' would go on until I exhausted myself. It was in one of these binges that I first learned about decomposition. I struggled to imagine the invisible creatures that ate all the leaves, and how such small beings could have such enormous appetites. I struggled to imagine how they could devour my leaf piles as I lay submerged in them. Why couldn't I see it happening? If their hunger was so fierce, surely I would be able to catch them in the act if I buried myself in the heap of leaves and lay there quietly enough? They always eluded me.

My father proposed an experiment. We cut the top off a clear plastic bottle. Into the bottle we placed alternating layers of soil, sand, dead leaves, and finally a handful of earthworms. Over the next days I watched the worms wind their way between the layers. They mixed and stirred. Nothing stayed still. Sand crept into soil and leaves crept into sand. The hard edges of the layers dissolved into each other. The worms might be visible, my father explained, but there are many more creatures that behave like this that you can't see. Tiny worms. And creatures smaller than tiny worms. And creatures still smaller that don't look like worms but are able to mix and stir and dissolve one thing into another just like these worms can. Composers make pieces of music. These were decomposers, who unmake pieces of life. Nothing could happen without them.

This was such a useful idea. It was as if I'd been shown how to reverse, how to think backwards. Now there were arrows that pointed in both directions at once. Composers make; decomposers unmake. And unless decomposers unmake, there isn't anything that the composers can make *with*. It was a thought that changed the way I understood the world. And from this thought, from my fascination with the creatures that decompose, grew my interest in fungi.

It is out of this compost heap of questions and fascinations that this book has composed itself. There have been so many questions, so few answers – and this feels exciting. Ambiguity isn't as itchy as it was; it's easier for me to resist the temptation to remedy uncertainty with certainty. In my conversations with researchers and enthusiasts I've found myself acting as an unwitting go-between, answering questions about what people are doing in different, far-flung fields of mycological enquiry, sometimes carrying a few grains of sand into the soil, sometimes a few clumps of soil into the sand. There is more pollen on my face than when I began. New whys have fallen on top of old whys. There is a bigger pile to leap into, and it smells just as mysterious as it did at the start. But there is more damp, more space to bury myself, and more to explore.

Fungi might make mushrooms, but first they must unmake something else. Now that this book is made, I can hand it over to fungi to unmake. I'll dampen a copy and seed it with *Pleurotus* mycelium. When it has eaten its way through the words and pages and endpapers and sprouted oyster mushrooms from the covers, I'll eat them. From another copy I will remove the pages, mash them up and using a weak acid break the cellulose of the paper into sugars. To the sugar solution I'll add a yeast. Once it's fermented into a beer, I'll drink it and close the circuit.

Fungi make worlds; they also unmake them. There are lots of ways to catch them in the act: when you cook mushroom soup, or just eat it; when you go out gathering mushrooms or buy them; when you ferment alcohol, plant a plant, or just bury your hands in the soil. And whether you let a fungus into your mind, or marvel at the way that it might enter the mind of another; whether you're cured by a fungus, or watch it cure someone else; whether you build your home from fungi, or start growing mushrooms in your home, fungi will catch *you* in the act. If you're alive, they already have.

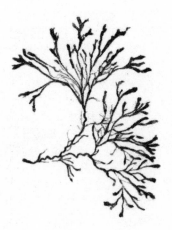

Acknowledgements

Without the guidance, teaching and patient assistance of many experts, scholars, researchers and enthusiasts this book would have been inconceivable. In particular I would like to thank: Ralph Abraham, Andrew Adamatzky, Phil Ayres, Eben Bayer, Kevin Beiler, Luis Beltran, Michael Beug, Martin Bidartondo, Lynne Boddy, Ulf Büntgen, Duncan Cameron, Keith Clay, Yves Couder, Bryn Dentinger, Julie Deslippe, Katie Field, Emmanuel Fort, Mark Fricker, Maria Giovanna Galliani, Lucy Gilbert, Rufino Gonzales, Trevor Goward, Christian Gronau, Omar Hernandez, Allen Herre, David Hibbett, Stephan Imhof, David Johnson, Toby Kiers, Callum Kingwell, Albert László-Barábasi, Natuschka Lee, Charles Lefevre, Egbert Leigh, David Luke, Scott Mangan, Michael Marder, Peter McCoy, Dennis McKenna, Pål Axel Olsson, Stefan Olsson, Magnus Rath, Alan Rayner, David Read, Dan Revillini, Marcus Roper, Jan Sapp, Carolina Sarmiento, Justin Schaffer, Jason Scott, Marc-André Selosse, Jason Slot, Sameh Soliman, Toby Spribille, Paul Stamets, Michael Stusser, Anna Tsing, Raskal Turbeville, Ben Turner, Milton Wainwright, Håkan Wallander, Joe Wright and Camilo Zalamea.

My agent Jessica Woollard and my editors Will Hammond at Bodley Head and Hilary Redmon at Random House have provided a steady stream of encouragement, clear vision and wise counsel for which I'm immensely grateful. At Bodley Head/

Vintage I've been lucky to work with Graham Coster, Suzanne Dean, Sophie Painter and Joe Pickering, and at Random House I have had an excellent team in Karla Eoff, Lucas Heinrich, Tim O'Brian, Simon Sullivan, Molly Turpin and Ada Yonenaka. Collin Elder experimented with ink made from the shaggy ink cap mushroom and produced a beautiful set of fungal illustrations. For their help with various pieces of translation I thank Xavier Buxton, Julia Hart, Anna Westermeier, Simi Freund and Pete Riley. Pam Smart provided valuable help with transcription, and Chris Morris from 'Spores for Thought' collected spore prints. Christian Ziegler joined me in the forest in Panama and was able to photograph the strange magic of mycoheterotrophic plants.

I am enormously grateful to those who read parts or all of the book at various stages of its growth: Leo Amiel, Angelika Cawdor, Nadia Chaney, Monique Charlesworth, Libby Davy, Tom Evans, Charles Foster, Simi Freund, Stephan Harding, Ian Henderson, Johnny Lifschutz, Robert Macfarlane, Barnaby Martin, Uta Paszkowski, Jeremy Prynne, Jill Purce, Pete Riley, Erin Robinsong, Nicholas Rosenstock, Will Sapp, Emma Sayer, Rupert Sheldrake, Cosmo Sheldrake, Sara Sjölund, Teddy St Aubyn, Erik Verbruggen and Flora Wallace. I could not have done without their insight and sensitivity.

For many kinds of humour, care, and inspiration along the way I thank: David Abram, Mileece Abson, Matthew Barley, Fawn Baron, Finn Beames, Gerry Brady, Dean Broderick, Caroline Casey, Udavi Cruz-Márquez, Mike de Danann Datura, Lindy Dufferin, Andréa de Keijzer, Zac Embree, Amanda Feilding, Johnny Flynn, Tom Fortes Mayer, Viktor Frankel, Dana Frederick, Charlie Gilmour, Lucy Hinton, Rick Ingrasci, James Keay, Oliver Kelhammer, Erica Kohn, Natalie Lawrence, Sam Lee, Andy Letcher, Jane Longman, Luis Eduardo Luna, Robert Macfarlane, Vahakn Matossian, Sean Matteson, Evan McGown, Zayn Mohammed, Mark Morey, Misha Mullov-Abbado, Viktoria Mullova, Charlie Murphy, Dan Nicholson, Richard Perl, Sara Perl Egendorf, John Preston, Jeremy Prynne, Anthony

Ramsay, Vilma Ramsay, Steve Rooke, Gryphon Rower Upjohn, Matt Segall, Rupinder Sidhu, Wayne Silby, Paulo Roberto Silva e Souza, Joel Solomon, Anne Stillman, Peggy Taylor, Robert Temple, Jeremy Thres, Mark Vonesch, Flora Wallace, Andrew Weil, Khari Wendell-McClelland, Kate Whitley, Heather Wolf and Jon Young. I am indebted to many wonderful teachers and mentors who have helped me over the years, in particular: Patricia Fara, William Foster, Howard Griffiths, David Hanke, Nick Jardine, Mike Majerus, Oliver Rackham, Fergus Read, Simon Schaffer, Ed Tanner and Louis Vause.

I am grateful for the support of several institutions: Clare College, Cambridge, and the Cambridge Department of Plant Sciences and Department of History and Philosophy of Science, where I spent several exciting years; the Smithsonian Tropical Research Institute, for its support while I lived in Panama and for its ongoing care of the Barro Colorado Nature Monument; and Hollyhock, British Columbia, for providing me with a beautiful place to work through the winter.

Countless hours of music have helped to me to think and feel my way through this book. Of particular importance have been the sounds of: the Aka people, Johann Sebastian Bach, William Byrd, Miles Davis, João Gilberto, Billie Holiday, Charles Mingus, Thelonius Monk, Moondog, Bud Powell, Thomas Tallis, Fats Waller and Teddy Wilson. The two places that have most guided the emergence of this book are Hampstead Heath and Cortes Island. To these places, and to all those who inhabit and protect them, I owe more than I can say. Above all, for their inspiration, love, wit, wisdom, generosity and endless patience I am grateful to Erin Robinsong, Cosmo Sheldrake, and my parents Jill Purce and Rupert Sheldrake.

The author and publisher gratefully aknowledge the copyright holders for granting their permission to reproduce excerpts from the following material: 'Heaven is Jealous' from *A Year With Hafiz: Daily Contemplations*, translation © Daniel Ladinsky 2011; 'Like Roots' from *Love Poems From God: Twelve Sacred Voices From the East and West*, translation © Daniel Ladinsky 2002; 'Fayan Wenyi' from 'The Book of Silences' from *Selected Poems* © Robert Bringhurst 2009; 'Green Grass', words and music by Kathleen Brennan & Thomas Waits © 2004 Jalma Music. Universal Music Publishing MGB Limited. All rights reserved. International copyright secured. Used by permission of Hal Leonard Europe Limited; 'A New Year Greeting' © The Estate of W. H. Auden.

Grateful aknowledgement is made for permission to reproduce the image on page 58, redrawn by Collin Elder from an original image © Symbolae.

Notes

Introduction: What Is It Like To Be A Fungus?

Epigraph: Hafiz (1315–190), in Ladinsky, *A Year with Hafiz* (2010).

1   Ferguson et al., 2003. There are numerous other reports of enormous networks of *Armillaria*. A study published by Anderson et al. (2018) investigated a mycelial network in Michigan with an estimated age of 2,500 years and a weight of at least 400 tonnes, sprawling over an area of 75 hectares. The researchers found that the fungus had an extremely low rate of genetic mutation, suggesting that it has a way of protecting itself against damage to its DNA. How exactly the fungus is able to maintain such a stable genome is not known, but this probably helps to account for its ability to live to such a great age. Apart from *Armillaria*, some of the largest organisms are clonal sea grasses (Arnaud-Haond et al., 2012).

2   Moore et al. (2011), ch. 2.7; Honegger et al. (2018). The fossilised remains of *Prototaxites* have been found in North America, Europe, Africa, Asia and Australia. Biologists have puzzled over what *Prototaxites* was since the mid nineteenth century. It was first thought to be a rotted tree. Shortly afterwards, it was promoted to the status of giant marine alga, despite overwhelming evidence that it grew on land. In 2001, after decades of debate, it was argued that *Prototaxites* was in fact the fruiting bodies of a fungus. It is a persuasive argument: *Prototaxites* was formed from thickly woven filaments that look more like fungal hyphae than anything else. Analysis of the carbon isotopes indicates that it survived by consuming its surroundings rather than by photosynthesis. More recently, Selosse (2002) has argued that it is more plausible that *Prototaxites* were giant lichen-like structures, made up of a union of fungi with photosynthetic algae. He argues that *Prototaxites* were too large to support themselves by decomposing plants. If *Prototaxites* were partly photosynthetic, they would have been able to supplement their diet of dead plants with energy from photosynthesis. They would have both the means and the incentive to grow into structures taller than anything else around. What's more, *Prototaxites* contained tough polymers found in algae of the time, suggesting that algal cells lived interwoven with the fungal hyphae. The lichen hypothesis also helps to explain why they went extinct. After 40 million years of global dominance, *Prototaxites* mysteriously died out just as plants were evolving into trees and shrubs. This

observation fits with *Prototaxites* being lichen-like organisms, because more plants mean less light.

3 For a broad discussion of fungal diversity and distribution see Peay (2016); for marine fungi see Bass et al. (2007); for fungal endophytes see Mejía et al. (2014), Arnold et al. (2003), Rodriguez et al. (2009). For an account of specialist fungi found in distilleries, where they thrive on the alcohol fumes that evaporate from whiskey barrels as they age, see Alpert (2011).

4 For rock-digesting fungi see Burford et al. (2003) and Quirk et al. (2014); for plastics and TNT see Peay et al. (2016), Harms et al. (2011), Stamets (2011), Khan et al. (2017); for radiation-resistant fungi see Tkavc et al. (2018); for radiographic fungi see Dadachova and Casadevall (2008) and Casadevall et al. (2017).

5 For spore ejection see Money (1998), Money (2016) and Dressaire et al. (2016). For spore mass and influence on weather see Fröhlich-Nowoisky et al. (2009). For a review of the many colourful solutions fungi have evolved in response to the problems of spore dispersal see Roper et al. (2010) and Roper and Seminara (2017).

6 For flow see Roper and Seminara (2017); for electrical impulses see Harold et al. (1985), Olsson and Hansson (1995). Yeasts make up about 1 per cent of the fungal kingdom and multiply by 'budding' or splitting in two. Some yeasts can form hyphal structures under certain conditions (Sudbery et al., 2004).

7 For accounts of fungi pushing through asphalt and lifting paving stones see Moore (2013b), ch. 3.

8 Leafcutter ants don't just feed and house their fungi: they medicate them as well. Leafcutter ants' fungal gardens are monocultures, consisting of a single type of fungus. Like human monocultures, the fungi are vulnerable. Particularly threatening is a type of specialist parasitic fungus that can destroy a fungal garden. Leafcutters harbour bacteria in elaborate chambers in their cuticles, fed by specialised glands. Each nest cultures its own specific strain of bacteria, recognised and favoured by the ants above other strains, even closely related ones. These domesticated bacteria produce antibiotics which powerfully inhibit the parasitic fungus and boost the growth of the cultivated one. Without these fungi, leafcutter ant colonies wouldn't be able to grow to such a large size. See Currie et al. (1999), Currie et al. (2006) and Zhang et al. (2007).

9 For Roman god Robigus see Money (2007), ch. 6, and Kavaler (1967), ch. 1. For fungal superbugs see Fisher et al. (2012, 2018), Casadevall et al. (2019), Engelthaler et al. (2019); for fungal disease of amphibians see Yong (2019); for banana disease see Maxman (2019). Among animals, diseases caused by bacteria pose more of a threat than those caused by fungi. By contrast, among plants, diseases caused by fungi pose a greater threat than those caused by bacteria. It is a pattern that holds through sickness and through health: animal microbiomes tend to be dominated by bacteria, while plant microbiomes tend to be dominated by fungi. This is not to say that animals don't suffer from fungal diseases at all. Casadevall (2012) hypothesises that the rise of mammals and decline of reptiles following the extinction event that wiped out the dinosaurs – the Cretaceans – Tertiary (K–T) extinction – was due to the ability of mammals to fight fungal diseases. Compared with reptiles, mammals have a number of handicaps: it is energetically costly to be warm-blooded, and even more so to produce milk and deliver intensive parental care. But it may be that mammals' elevated body temperatures were exactly what made it possible to replace reptiles as the dominant land-dwelling animals. Mammals' elevated body temperatures help to deter the growth of fungal pathogens that are hypothesised to have proliferated in the

'global compost heap' that followed the widespread dieback of forests during the K–T extinction. To this day, mammals are more resistant to common fungal diseases than reptiles or amphibians.

10 For study of Neanderthals see Weyrich et al. (2017); for Iceman see Peintner et al. (1998). How the Iceman used the birch polypore (*Fomitopsis betulina*) can't be known for certain, but they are bitter and indigestibly corky, so clearly not 'nutritional' in any conventional sense. The Iceman's careful preparation of these fungi – which were mounted like keyrings on leather thongs – indicates a well-developed knowledge of their value and application.

11 For mould cures see Wainwright (1989 a and b). Human remains from archaeological sites in Egypt, Sudan and Jordan dating from around the year 400 CE have been found to have high levels of the antibiotic tetracycline in their bones, indicating a long-term sustained intake, most likely in a therapeutic context. Tetracycline is produced by a bacterium, not a fungus, but its likely source was mouldy grains, likely used to make a medicinal beer (Bassett et al., 1980, Nelson et al., 2010). The journey from Fleming's first observation to penicillin's emergence onto the world stage was not a straightforward one and required a great deal of human effort: experiments, industrial know-how, investment, political support. For a start, it was difficult for Fleming to persuade anyone to take an interest in his discovery. In the words of Milton Wainwright, a microbiologist and historian of science, Fleming was eccentric, a 'messer-abouter'. 'He had a reputation for being a nutter and doing daft things, like creating pictures of the Queen on a petri dish using different bacteria cultures.' Dramatic proof of penicillin's therapeutic value didn't come until twelve years after Fleming's first observations. In the 1930s, a research group in Oxford developed a method to extract and purify penicillin, and in 1940 conducted trials that demonstrated its astonishing ability to fight infections. Nonetheless, penicillin remained difficult to produce. In the absence of a widely available product, instructions on how to grow the mould were published in the medical press. Crude 'kitchen sink' extracts, together with chopped mycelium on surgical gauze – 'mycelial pads' – were used by some doctors to treat infections, treatments observed to be remarkably effective (Wainwright, 1989 a and b). It was in the United States that penicillin production was industrialised. This was partly due to the well-developed American methods to cultivate fungi in industrial fermenters, and partly due to the discovery of higher-yielding strains of *Penicillium* mould; strains that were further enhanced by rounds of mutation. The industrialisation of penicillin led to a massive effort to search for new antibiotics, and thousands of fungi and bacteria were screened.

12 For drugs see Linnakoski et al. (2018), Aly et al. (2011), Gond et al. (2014). For psilocybin see Carhart-Harris et al. (2016a), Griffiths et al. (2016), Ross et al. (2016). For vaccines and citric acid see the *State of the World's Fungi* (2018). For market in edible and medicinal mushrooms see www.knowledge-sourcing.com/report/global-edible-mushrooms-market [accessed 29 October 2019]. In 1993, a study published in the journal *Science* reported that paclitaxel (sold under the brand name Taxol) was produced by an endophytic fungus isolated from the bark of the Pacific yew (Stierle et al., 1993). It has subsequently emerged that paclitaxel is produced far more widely by fungi than by plants – by around two hundred endophytic fungi, spread across several fungal families (Kusari et al. 2014). A potent antifungal, it plays an important defensive role: fungi that are able to produce paclitaxel are able to deter other fungi. It acts against fungi in the same way it acts against cancer, by interrupting cell division. Paclitaxel-producing fungi are immune to its effects, as

are other fungal endophytes of yew (Soliman et al., 2015). A number of other fungal anti-cancer drugs have made their way into mainstream pharmaceutical practice. Lentinan, a polysaccharide from the shiitake mushroom, has been found to stimulate the immune system's ability to fight cancers, and is medically approved in Japan for the treatment of gastric and breast cancers (Rogers, 2012). PSK, a compound isolated from turkey tail mushrooms, extends the survival time of patients suffering from a range of cancers and is used alongside conventional cancer treatments in China and Japan (Powell, 2014).

13 For fungal melanins see Cordero (2017).

14 For estimates of the number of fungal species see Hawksworth (2001) and Hawksworth and Lücking (2017).

15 Among neuroscientists, the involvement of our expectations in perception is known as top-down influence, or sometimes Bayesian inference (after Thomas Bayes, a mathematician who made a founding contribution to the mathematics of probability, or 'the doctrine of chances'). See Gilbert and Sigman (2007) and Mazzucato et al. (2019).

16 Adamatzky (2016); Latty and Beekman (2011); Nakagaki et al. (2000); Bonifaci et al. (2012); Tero et al. (2010), Oettmeier et al. (2017). In *Advances in Physarum Machines*, researchers detail many surprising properties of slime moulds. Some use slime moulds to make decision gates and oscillators, some simulate historical human migrations and model possible future patterns of human migrations on the moon. Mathematical models inspired by slime moulds include a non-quantum implementation of Shor's factorisation, calculation of shortest paths, and the design of supply chain networks (Adamatzky 2016). Oettmeier et al. (2017) note that Hirohito, the emperor of Japan between 1926 and 1989, was fascinated by slime moulds, and in 1935 published a book on the subject. Slime moulds have been a high-prestige subject of research in Japan ever since.

17 The system of classification devised by Carl Linnaeus and published in his *Systema Naturae* in 1735, a modified version of which is used today, extended this hierarchy to human races. At the top of the human league tables were Europeans: 'Very smart, inventive. Covered by tight clothing. Ruled by law.' Americans followed: 'Ruled by custom'. Then Asians – 'Ruled by opinion' – then Africans: 'sluggish, lazy … [c]rafty, slow, careless. Covered by grease. Ruled by caprice.' (Kendi, 2017). The way hierarchical classification systems order different species can be seen, by extension, as species racism.

18 For different microbial communities in different parts of the body see Costello et al. (2009) and Ross et al. (2018). For comparison with stars in the galaxy see Yong (2016), ch. 1. W. H. Auden, in his 'New Year Greeting', offers up the ecosystems of his body to his microbial inhabitants:

> For creatures your size I offer
> a free choice of habitat,
> so settle yourselves in the zone
> that suits you best, in the pools
> of my pores or the tropical
> forests of arm-pit and crotch,
> in the deserts of my fore-arms,
> or the cool woods of my scalp.

19 For organ transplants and human cell cultures see Ball (2019). For estimate of the size of our microbiome see Bordenstein & Theis. (2015). For viruses within

viruses see Stough et al. (2019). For a general introduction to the microbiome see Yong (2016), and a special issue of *Nature* on the human microbiome (May 2019): www.nature.com/collections/fiabfcjbfj [accessed 29 October 2019].

20 In a sense, all biologists are now ecologists – but disciplinary ecologists have a head start, and their methods are starting to seep into new fields: a number of biologists are starting to call for the application of ecological methods to historically non-ecological fields of biology. See Gilbert and Lynch (2019), and Venner et al. (2009). There are a number of examples of the knock-on effects of the microbes that live within fungi. A study published by Márquez et al. (2007) in the journal *Science* in 2007 described 'a virus in a fungus in a plant'. The plant – a tropical grass – grows naturally in high soil temperatures. But without a fungal associate that grows in its leaves, the grass can't survive at high temperatures. When grown alone, without the plant, the fungus fares little better and is unable to survive. However, it turns out not to be the fungus that confers the ability to survive high temperatures after all. Rather, it is a virus that lives within the fungus that confers heat tolerance. When grown without the virus, neither fungus nor plant can survive high temperatures. The microbiome of the fungus, in other words, determines the role that the fungus plays in the microbiome of the plant. The outcome is clear: life or death. One of the most dramatic examples of microbes that live within microbes comes from the notorious rice blast fungus (*Rhizopus microsporus*). The key toxins produced by *Rhizopus* are actually produced by a bacterium living within its hyphae. In a dramatic indication of how entwined the fates of fungi and their bacterial associates can be, *Rhizopus* requires the bacterium not only to cause the disease, but also to produce spores. Experimentally 'curing' *Rhizopus* of its bacterial residents impedes the fungus's ability to reproduce. The bacterium is responsible for the most important features of *Rhizopus's* lifestyle, from its diet to its sexual habits. See: Araldi-Brondolo et al. (2017), Mondo et al. (2017), Deveau et al. (2018).

21 For remark about loss of self-identity see Relman (2008). The questions of whether human beings are singular or plural is not a new one. In nineteenth-century physiology, the bodies of multicellular organisms were thought of as being made up of a community of cells, with each cell an individual in its own right, by analogy to the individual human member of a nation state. These questions are complicated by developments in the microbial sciences, because the multitude of cells in your body aren't strictly related to each other, as, for example, an average liver cell would be related to an average kidney cell. See Ball (2019), ch. 1.

Chapter 1: A Lure

Epigraph: Prince, 'Illusion, Coma, Pimp & Circumstance' on the album *Musicology* (2004).

1 Psychoactive 'truffles' sold in Amsterdam are not, as their name suggests, fruiting bodies. They are storage organs known as 'sclerotia', which are called truffles because of their superficial resemblance.

2 For trillion odours see Bushdid et al. (2014); for olfactory navigation see Jacobs et al. (2015); for olfactory flashbacks and general discussion of human olfactory abilities see McGann (2017). Some humans are classed as 'supersmellers' or hyperosmic individuals. A study published by Trivedi et al. (2019) reported

that a 'super smeller' was able to detect Parkinson's disease using their sense of smell alone.

3 For discussion of the smell of different chemical bonds see Burr (2012), ch. 2.

4 These receptors belong to a large family called G-protein coupled receptors, or GPCRs. For study of human olfactory sensitivity see Sarrafchi et al. (2013), who report that humans can detect some odours at concentrations of 0.001 parts per trillion.

5 For *turmas de tierra* see Ott (2002). Truffles, according to Aristotle, were 'a fruit consecrated to Aphrodite'. They are reputed to have been used as aphrodisiacs by Napoleon and the Marquis de Sade, and George Sand described them as the 'black magic apple of love'. The French gastronome Jean Anthelme Brillat-Savarin documented that 'truffles are conducive to erotic pleasure'. In the 1820s, he set out to investigate this commonly held belief, and embarked on a series of consultations with ladies ('all the replies I received were ironical or evasive'), and men ('who by their profession are invested with special trust'). He concluded that 'the truffle is not a true aphrodisiac but in certain circumstances it can make women more affectionate and men more attentive' (Hall et al., 2007, p. 33).

6 For Laurent Rambaud see Chrisafis (2010). The reporter Ryan Jacobs documents the foul play that occurs all the way along truffle supply lines. Some poisoners use meatballs laced with strychnine, others poison pools of water in the forest so that dogs with muzzles can still be poisoned, some deploy meat spiked with shards of glass, others use rat poison or antifreeze. Based on vets' reports, hundreds of poisoned dogs receive treatment each truffle season. The authorities have taken to using poison-sniffing dogs to patrol certain woods (Jacobs, 2019, pp. 130–4). In 2003, the *Guardian* reported that Michel Tournayre, a French truffle expert, had his truffle dog stolen. Tournayre suspected that the thieves had not sold the dog but rather were using her to steal truffles from other people's land (Hall et al., 2007, p. 209). What better ways to steal truffles than with a stolen dog?

7 For elk with bloodied noses see Tsing (2015), ch. 'Interlude. Smelling'; for fly-pollinated orchids see Policha et al. (2016); for orchid bees collecting complex aromatic compounds see Vetter and Roberts (2007); for similarity with compounds produced by fungi see de Jong et al. (1994). Orchid bees secrete a fatty substance which they apply to the scented object. Once the scent has been absorbed, they scrape the fat back up again and store it in pockets on their hind legs. This approach is identical in principle to enfleurage, the method used by humans for hundreds of years to capture fragrances like jasmine that are too delicate to extract using heat (Eltz et al., 2007).

8 Naef (2011).

9 For de Bordeu see Corbin (1986), p. 35.

10 For record-breaking truffle see: news.bbc.co.uk/1/hi/world/europe/7123414.stm [accessed 29 October 2019].

11 For discussion of the role of truffle microbiome in odour production see Vahdatzadeh et al. (2015). When I was out with Daniele and Paride I noticed that a truffle excavated from the silty soil near a river smelled quite different from one found in the more clay-heavy soil further up the valley. These differences are unlikely to make much difference to a hungry shrew. But a white truffle found in Alba will sell for four times as much as a white truffle found near Bologna (although the fact that some truffle dealers regularly pass off Bolognese truffles as being from Alba would suggest that not everyone is able to tell the difference).

Regional differences in truffles' volatile profiles have been confirmed in formal studies (Vita et al., 2015).

12 For original report that truffles produce androstenol see Claus et al. (1981); for follow-up study nine years later see Talou et al. (1990).

13 The number of volatiles produced by a single species of truffle has steadily increased over the years as the sensitivity of detection methods has improved. These methods are still less sensitive than the human nose, and the number of truffle volatiles is likely to increase yet further in the future. For white truffle volatiles see Pennazza et al. (2013) and Vita et al. (2015); for other species see Splivallo et al. (2011). There are a number of reasons why it is risky to pin all of truffles' allure on a single compound. In the study by Talou et al. (1990), a small sample of animals was used and only a single species of truffle was tested, at a single shallow depth, at a single site. Different subsets of the profile of volatile compounds might be more prominent at different depths or in different places. Moreover, in the wild, a range of animals are attracted to truffles, from wild pigs to voles to insects. It might be that different elements of the cocktail of volatile compounds that truffles produce attract different animals. It may be that androstenol acts on animals in more subtle ways. It might not be effective on its own, as tested in the study, but only in conjunction with other compounds. Alternatively, it may be less important in finding the truffles, and more important in the animals' experience of eating them. For more on poisonous truffles see Hall et al. (2007). Besides *Gautieria*, the truffle species *Choiromyces meandriformis* is reported to smell 'overpowering and nauseous' and is considered toxic in Italy (although it is popular in northern Europe). *Balsamia vulgaris* is another species considered to be mildly toxic, although dogs appear to enjoy its aroma of 'rancid fat'.

14 For truffle export and packaging see Hall et al. (2007), p. 219, 227.

15 In areas of exploring mycelium, hyphae usually grow away from other hyphae without ever touching. In more mature parts of the mycelium, hyphal inclinations pivot. Growing tips instead become attracted to each other and start to 'home' (Hickey et al., 2002). How hyphae attract and repel each other remains poorly understood. Work on the model organism, the bread mould *Neurospora crassa*, is starting to provide some clues. Each hyphal tip takes it in turn to release a pheromone that attracts and 'excites' the other. Through this back-and-forthing – 'as if throwing a ball', write the authors of one study – hyphae are able to entrain and home in on each other by falling into rhythm. It is this oscillation – a chemical rally – that allows them to lure the other without stimulating themselves. When they are serving, they aren't able to detect the pheromone. When the other serves, they are stimulated (Read et al., 2009; Goryachev et al., 2012).

16 For discussion of mating types of *Schizophyllum commune* see McCoy (2016), p. 13; for fusion between sexually incompatible hyphae see Saupe (2000), and Moore et al. (2011), ch. 7.5. The ability of hyphae to fuse with each other is determined by their 'vegetative compatibility'. Once hyphal fusion has taken place, a separate system of mating types determines which nuclei can undergo sexual recombination. These two systems are regulated differently, although sexual recombination cannot take place unless hyphae have fused with each other and shared genetic material. The outcome of vegetative fusions between different mycelial networks can be complex and unpredictable (Rayner et al., 1995, Roper et al., 2013).

17 For details of truffle sex see Selosse et al. (2017), Rubini et al. (2007), and Taschen et al. (2016); for examples of intersexuality in the animal world see

Roughgarden (2013). If truffle cultivators really want to understand how to crack truffle cultivation, they must understand truffle sex. The problem is that they don't. Truffle fungi have never been caught in the act of fertilisation. Perhaps this isn't so surprising given their inaccessible lifestyles. More peculiar is that no-one has ever found a paternal hypha. Despite searching, researchers have only found maternal hyphae growing on tree roots and in soil, whether '+' or '-'. Paternal truffles seem to be short-lived, and vanish after fertilisation: 'birth, then a drop of sex, then nothing' (Dance, 2018).

18 The hyphae of some types of mycorrhizal fungi can withdraw themselves back into their spores and re-sprout at a later date (Wipf et al., 2019).

19 For fungal influence on plant roots see Ditengou et al. (2015), Li et al. (2016), Splivallo et al. (2009), Schenkel et al. (2018), Moisan et al. (2019).

20 For discussion of the evolution of communication in mycorrhizal symbioses, including suspension of immune response, see Martin et al. (2017); for discussion of plant–fungus signalling and its genetic basis see Bonfante (2018); for plant–fungus communication in other types of mycorrhizal association see Lanfranco et al. (2018). The chemical propositions released by fungi are nuanced and have a wide dynamic range. The volatiles used to communicate with a plant might also be used to communicate with the surrounding bacterial populations (Li et al., 2016, Deveau et al., 2018). Fungi use volatile compounds to deter rival fungi; plants use volatile compounds to deter unwanted fungi (Li et al., 2016, Quintana-Rodriguez et al., 2018). The same volatile can have different effects on plants depending on its concentration. The plant hormones produced by some truffles to manipulate their hosts' physiology can kill plants at higher concentrations and may serve as competitive weapons to deter plants that might compete with their own plant partners (Splivallo et al., 2007, 2011). Some species of truffle fungus are parasitised by other fungi, probably attracted by their chemical announcements. The truffle parasite, *Tolypocladium capitata*, is a cousin of the *Ophiocordyceps* fungi that parasitise insects and is known to parasitise certain species of truffle such as the deer truffle, *Elaphomyces* (Rayner et al., 1995, for photos see mushroaming.com/cordyceps-blog [accessed 29 October 2019]).

21 For the first report of fruiting *Tuber melanosporum* in the British Isles – thought to be due to climate change – see Thomas & Büntgen (2017). The 'modern' method used to cultivate *Tuber melanosporum* wasn't developed until 1969, and resulted in the first batch of artificially inoculated truffles in 1974. Seedling roots are incubated with the mycelium of *Tuber melanosporum* and planted out when the roots are thoroughly inhabited by the fungus. After several years, given the right conditions, the fungus will start to produce truffles. There is an increasingly large area of land in truffle cultivation (over 40,000 hectares worldwide), and Périgord truffle orchards are successfully fruiting in countries from the United States to New Zealand (Büntgen et al., 2015). Charles Lefevre explained that even if he wrote his method down point by point, it would be difficult for someone else to replicate. There is so much intuitive knowledge that is hard to communicate and keep track of. The tiniest details – from the vagaries of the season to the conditions in the nursery – make a huge difference. Secrecy is part of the problem. Truffle cultivators spend much of their time in a fog of uncertainty, picking their way around jealously tended 'proprietary insight'. 'It is a tradition with old roots in mushroom-picking,' Ulf Büntgen told me. 'Many people go out to pick mushrooms in the woods, but they never tell you anything. If you ask someone how their day was and they say, "Oh,

I found a huge crop!" they probably found nothing. It is an attitude that persists over generations and makes research very slow.' Undeterred, Lefevre still grows a number of trees with the mycelium of the elusive *Tuber magnatum* every year in the hope that something, somehow, might just prompt them to fruit. Armed with the same optimism, he continues to experiment in pairing European truffle species with American trees (*Tuber magnatum* turns out to develop a healthy – though fruitless – partnership with aspen). Other cultivators isolate bacteria from truffles in the hope that they will encourage the growth of *Tuber* mycelium (some groups of bacteria do seem to be helpful). Did many people buy his *Tuber magnatum* trees for their truffle orchards, I wondered? 'Not many,' he replied, 'but we sell the trees in the spirit that if no one tries, no one will succeed.'

22 For discussion of chemical eavesdropping see Hsueh et al. (2013).

23 Nordbring-Hertz (2004); Nordbring-Hertz et al. (2011)

24 Nordbring-Hertz (2004).

25 Today, the field of biology most inflamed by debates about anthropomorphism is the study of plants and the ways they sense and respond to their environment. In 2007, thirty-six prominent plant scientists signed a letter that dismissed the nascent field of 'plant neurobiology' (Alpi et al., 2007). Those who put forward the term argued that plants have electrical and chemical signalling systems equivalent to those found in humans and other animals. The thirty-six authors of the letter argued that these were 'superficial analogies and questionable extrapolations'. A spirited debate ensued (Trewavas, 2007). From an anthropological perspective, these controversies are fascinating. Natasha Myers, an anthropologist at York University in Canada, interviewed a number of plant scientists about how they understood plants to behave (Myers, 2014). She describes the troubled politics of anthropomorphism, and the different ways in which researchers dealt with the issue.

26 Kimmerer (2013), ch. 'Learning the Grammar of Animacy'.

27 'Its relationship with its host trees is very poorly understood,' Charles Lefevre explained: 'even in places where truffle productivity is high, the proportion of tree roots colonised by the fungus is often extremely low. This means that productivity can't be explained in terms of the amount of energy that the fungus receives from the host tree.'

28 For smells and their likenesses see Burr (2012), ch. 2. The anthropologist Anna Tsing writes that in the Edo period in Japan (1603–1868) the smell of matsutake mushrooms became a popular subject for poetry. Trips to pick matsutake grew into the autumn equivalent of cherry-blossom parties in the spring, and references to 'the autumn aroma' or the 'aroma of the mushroom' became familiar poetic moods.

Chapter 2: Living Labyrinths

Epigraph: Cixous (1991).

1 For fungal navigation of mazes see Hanson et al. (2006), Held et al. (2009, 2010, 2011, 2019). For excellent videos see supplementary information of Held et al. (2011) – www.sciencedirect.com/science/article/pii/S1878614611000249 [accessed 29 October 2019] and Held et al. (2019) – www.pnas.org/content/116/27/13543/tab-figures-data [accessed 29 October 2019].

2   For marine fungi see Hyde et al. (1998), Sergeeva & Kopytina (2014) and Peay (2016); for fungi in dust see Tanney et al. (2017); for estimate of the length of fungal hyphae in soils see Ritz and Young (2004).

3   This is a commonly reported phenomenon. See Boddy et al. (2009) and Fukusawa et al. (2019)

4   Fukusawa et al. (2019). Did the new block of wood cause changes in chemical concentrations or gene expression across the network? Or did the mycelium rapidly redistribute itself within the original block of wood, making regrowth in one direction more likely? Lynne Boddy and her colleagues aren't sure. The researchers who challenge fungi with microscopic mazes have observed that structures within fungal growing tips behave like internal gyroscopes, and provide hyphae with a directional memory that allows them to recover the original direction of growth after being diverted around an obstacle (Held et al., 2019). However, it is unlikely that this mechanism is responsible for the effect Boddy and her colleagues observed because all the hyphae – including their tips – were removed from the original block of wood before it was placed in the fresh dish.

5   Fungal hyphae are unlike cells in animal or plant bodies, which (usually) have clear boundaries. In fact, strictly speaking, hyphae shouldn't be described as cells at all. Many fungi have hyphae with divisions along their length, known as 'septa', but these can be opened or closed. When open, hyphal contents can flow between the 'cells', and the mycelial networks are referred to as being in a 'supracellular' state (Read, 2018). One mycelial network can fuse with many others to make sprawling 'guilds', in which the contents of one network may be shared with others. Where does one cell start and stop? Where does one network start and stop? These questions are often unanswerable. For a recent study on swarms see Bain and Bartolo (2019), and commentary by Ouellette (2019). This study treats swarms as entities in themselves, rather than a collection of individual agents behaving according to local rules. By treating the swarm as a pattern of fluid flow, its behaviour can be more effectively modelled. It's possible that these top-down 'hydrodynamic' models could be used to model the growth of hyphal tips more effectively than swarm models based on local rules of interaction.

6   For slime moulds see Tero et al. (2010), Watanabe et al. (2011) and Adamatzky (2016); for fungi see Asenova et al. (2016) and Held et al. (2019).

7   For discussion of mycelial trade-offs see Bebber et al. (2007).

8   For discussion of natural selection of links in mycelial networks see Bebber et al. (2007).

9   For discussion of the role of fungal bioluminescence and insect spore dispersal see Oliveira et al. (2015); for foxfire and the *Turtle* see www.cia.gov/library/publications/intelligence-history/intelligence/intelltech.html [accessed 29 October 2019], and Diamant (2004), p. 27. In a guidebook to fungi published in 1875, Mordecai Cooke wrote that bioluminescent fungi were commonly found on the timber props used in coal mines. Miners 'are well acquainted with phosphorescent fungi, and the men state that sufficient light is given "to see their hands by". The specimens of *Polyporus* were so luminous that they could be seen in the dark at a distance of twenty yards.'

10 Stefan Olsson's videos are available online at doi.org/10.6084/m9.figshare.c.4560923.v1 [accessed 29 October 2019].

11 A study published by Oliveira et al. (2015) found that the bioluminescent mycelium of *Neonothopanus gardneri* was regulated by a temperature-regulated circadian clock. The authors hypothesise that by increasing bioluminescence

at night, fungi are better able to attract insects that disperse their spores. The phenomena that Stefan Olsson observed can't be explained on the basis of a circadian rhythm because it took place only once over the course of several weeks.

12 For hyphal diameter see Fricker et al. (2017). The ecologist R. H. Whittaker observed that animal evolution is a story of 'change and extinction', whereas fungal evolution is one of 'conservatism and continuity'. The great diversity of animal body plans in the fossil record illustrates the many ways animals have found to ingest features of their worlds. The same can't be said about fungi. Mycelial fungi have had much longer to evolve than many organisms, but ancient fossilised fungi are remarkably similar to those alive today. It appears that there are only so many ways to make a life as a network. See Whittaker (1969).

13 For mycelial nets that catch falling leaves see Hedger (1990).

14 For measurement of the pressure exerted by the rice-blast pathogen see Howard et al. (1991); for 8-tonne school bus figure and for general discussion of invasive fungal growth see Money (2004a). To exert such high pressures, the penetrative hyphae must glue themselves to the plant to prevent themselves from pushing away from the surface. They do this by producing an adhesive that can resist pressures of over 10 megapascals (MPa) – superglue can resist pressures of 15–25 MPa, though probably not on the waxy surface of a plant leaf (Roper and Seminara, 2017).

15 The cellular 'bladders' are known as 'vesicles'. Fungal tip growth is managed by a cellular structure or 'organelle' called the *Spitzenkörper*, or tip body. Unlike most organelles, the *Spitzenkörper* does not have a clearly defined boundary. It is not a singular structure like a nucleus, although it appears to move as one. The *Spitzenkörper* is thought of as a 'vesicle supply centre', receiving and sorting vesicles from inside the hyphae, and distributing them to the hyphal tip. The *Spitzenkörper* pilots both itself and its hypha. Hyphal branching is triggered when *Spitzenkörpers* divide. When growth stops, the *Spitzenkörper* disappears. If one changes the position of the *Spitzenkörper* within the growing tip, one can steer the hypha in a different direction. What the *Spitzenkörper* makes, it can also unmake, dissolving hyphal walls to allow fusion between different parts of a mycelial network. For an introduction to the *Spitzenkörper* and '600 vesicles per second' see Moore (2013a), ch. 2; for further discussion of the *Spitzenkörper* see Steinberg (2007); for observation that the hyphae of some species can extend in real time see Roper and Seminara (2017).

16 The French philosopher Henri Bergson described the passage of time in terms reminiscent of a fungal hypha: 'Duration is the continuous progress of the past which gnaws into the future and which swells as it advances' (Bergson 1911, p. 7). For the biologist J. B. S. Haldane, life was not populated with things, but with stabilised processes. Haldane went as far as to deem 'the conception of a "thing", or material unit' to be 'useless' in biological thinking (Dupré and Nicholson 2018). For a general introduction to processual biology see Dupré and Nicholson (2018); for William Bateson quote see Bateson (1928), p. 209.

17 For stinkhorns growing through asphalt see Niksic et al. (2004); for Mordecai Cooke see Moore (2013b), ch. 3. Tip growth occurs in other organisms besides fungi, but it is an exception not the rule. Animal neurones grow by elongating at the tip, as do some types of plant cell, like pollen tubes. But neither can prolong themselves indefinitely, as fungal hyphae can under the right conditions (Riquelme 2012).

18 Frank Dugan describes the 'herb wives' or 'wise women' of Reformation Europe as 'midwives' to the field of modern mycology (Dugan, 2011). Many lines

of evidence suggest that women were primary holders of fungal lore. Such women were the source of much of the information about mushrooms that were formally described by male scholars of the time, including Carolus Clusius (1526–1609) and Francis van Sterbeeck (1630–93). A number of paintings, from *The Mushroom Seller* (Felice Boselli, 1650–1732), to *Women Gathering Mushrooms* (Camille Pissarro, 1830–1903), to *The Mushroom Gatherers* (Felix Schlesinger, 1833–1910), portray women working with mushrooms. Numerous European travellers' accounts from the nineteenth and twentieth century describe women selling or gathering mushrooms.

19 For discussion and broad definition of polyphony see Bringhurst (2009), ch. 2: 'Singing with the frogs: the theory and practice of literary polyphony'.

20 For estimates of flow rates through cords and rhizomorphs see Fricker et al. (2017). It is generally thought that fungi use chemicals to regulate their development, but little is known about these growth-regulating substances (Moore et al., 2011, ch. 12.5, Moore, 2005). How can such well-defined forms arise from a uniform mass of hyphal strands? An animal's finger is an elaborate form. But it is made up of an elaborate combination of different sorts of cell, with its blood cells, bone cells, nerve cells and all the rest. Mushrooms are elaborate forms too, but they are sculpted tufts of one type of cell: hyphae. How fungi make mushrooms has long proved a mystery. In 1921, the Russian developmental biologist Alexander Gurwitsch puzzled over the development of mushrooms. A mushroom's stalk, the ring around its stalk and its cap are all made of hyphae, tousled like 'shaggy uncombed hair'. This is what baffled him. Building a mushroom from nothing but hyphae is like trying to build a face from nothing but muscle cells. For Gurwitsch, the way hyphae grew together to make complex forms was one of the central riddles in all of developmental biology. An animal's organisation is specified at the earliest point in their development. Animal form arises from highly organised parts; regularity gives rise to further regularity. But the form of mushrooms arises from *less* organised parts. A regular form arises from an irregular material (von Bertalanffy, 1933, pp. 112–17). Inspired in part by mushroom growth, Gurwitsch hypothesised that the development of organisms was guided by fields. Iron filings can be rearranged using a magnetic field. In an analogous way, Gurwitsch advanced, the arrangement of cells and tissues within an organism could be shaped by form-giving biological fields. Gurwitsch's field theory of development has been picked up by a number of contemporary biologists. Michael Levin, a researcher at Tufts University in Boston, describes how all cells are bathed in a 'rich field of information', whether made up of physical, chemical or electrical cues. These fields of information help to explain the way in which complex forms can arise (Levin, 2011 and 2012). A study published in 2004 built a mathematical model that simulated fungal mycelial growth – a 'cyberfungus' (Meskkauskas et al., 2004, Money, 2004b, Moore, 2005). In the model, each hyphal tip is able to influence the behaviour of other hyphal tips. The study reports that mushroom-like forms can emerge when all the hyphal tips follow exactly the same rules of growth. These findings imply that mushrooms' forms can emerge from the 'crowd behaviour' of hyphae without the need for the sort of top-down developmental co-ordination found in animals and plants. But for this to work, tens of thousands of hyphal tips must obey the same sets of rules at the same time, and switch to different sets of rules at the same time – a modern reframing of Gurwitsch's riddle. The researchers who created the cyberfungus hypothesise that developmental changes might be co-ordinated using a cellular

'clock', but no such mechanism has yet been identified, and the means by which living fungi co-ordinate their development remains a mystery.

21 For microtubule motors see Fricker et al. (2017); for *Serpula* in Haddon Hall see Moore (2013b), ch. 3; for a discussion of the role of flow in fungal development see Alberti (2015) and Fricker et al. (2017). Flow rates in fungal hyphae range from 3–70 micrometres per second, sometimes over one hundred times faster than passive diffusion alone (Abadeh and Lew, 2013). Alan Rayner is keen on the river analogy because rivers are 'systems that both shape and are shaped by their landscape'. A river flows between its banks. In the process, it shapes the banks that it flows within. Rayner understands hyphae to be blunt-ended rivers that flow within banks that they build for themselves. As in any flow system, pressure is everything. Hyphae absorb water from their surroundings. The inward flow of water increases the pressure in the network. But pressure itself doesn't lead to flow. For material to flow through mycelium, hyphae have to make space for it to flow into. This is hyphal growth. Hyphal contents flow towards hyphal growing tips. Water flows through a mycelial network towards a rapidly inflating mushroom. If one reverses the pressure gradients, one reverses the flow (Roper et al., 2013). Hyphae appear to be able to regulate flow in more precise ways, however. A study published in 2019 traced the movement of nutrients and signalling compounds through hyphae in real time. In certain large hyphae, the flow of cellular fluid changed direction every few hours, allowing signalling compounds and nutrients to flow along the network in both directions. For around three hours, flow occurred in one direction. For the next three hours, flow occurred in the other direction. How hyphae are able to control the flow of material inside them isn't known, but by rhythmically changing the direction of cellular flow, substances are distributed more efficiently through the network. The authors speculate that co-ordinated opening and closing of hyphal pores are a 'major factor' in the co-ordination of bidirectional flow along transport hyphae (Schmieder et al., 2019, see also commentary by Roper and Dressaire, 2019). 'Contractile vacuoles' are another way in which fungi might direct flow through themselves. These are tubes within hyphae along which waves of contraction are able to pass, and which have been reported to play a part in transport through mycelial networks (Shepherd et al., 1993, Rees et al., 1994, Allaway and Ashford, 2001, Ashford and Allaway, 2002).

22 Roper et al. (2013); Hickey et al. (2016); Roper and Dressaire (2019). Videos available on YouTube: 'Nuclear dynamics in a fungal chimera', www.youtube. com/watch?v=_FSuUQP_BBc [accessed 29 October 2019]; 'Nuclear traffic in a filamentous fungus', www.youtube.com/watch?v=AtXKcro5030 [accessed 29 October 2019].

23 Cerdá-Olmeda (2001); Ensminger (2001), ch. 9.

24 For 'most intelligent ...' see Cerdá-Olmeda (2001); for avoidance response see Johnson and Gamow (1971) and Cohen et al. (1975).

25 Many aspects of mycelial life are influenced by light, from mushroom development to relationship-building with other organisms – the dreaded rice blast fungus only infects its plant hosts at night (Deng et al., 2015). For light-sensing in fungi see Purschwitz et al. (2006), Rodriguez-Romero et al. (2010), Corrochano and Galland (2016); for sensing surface topography see Hoch et al. (1987) and Brand and Gow (2009); for sensitivity to gravity see Moore (1996), Moore et al. (1996), Kern (1999), Bahn et al. (2007), and Galland (2014).

26 Darwin and Darwin (1880), p. 573. For arguments in favour of 'root-brains' see Trewavas (2016) and Calvo Garzón and Keijzer (2011); for arguments against

brain analogies see Taiz et al. (2019); for an introduction to the 'plant intelligence' debate see Pollan (2013).

27 For behaviour of hyphal tips see Held et al. (2019).

28 For fairy rings see Gregory (1982).

29 Some researchers have reported sudden hyphal contractions, or twitches, which
might be used to transmit information. But they aren't regular enough to be
useful on a moment-to-moment basis. See McKerracher and Heath (1986 a
and b), Jackson and Heath (1992), Reynaga-Peña and Bartnicki-García (2005).
Some propose that information can be transmitted across mycelial networks by
changing the patterns of flow within the network, in some cases changing the
direction of flow in rhythmic oscillations (Schmieder et al., 2019, Roper and
Dressaire, 2019). This is a promising line of research, and it may be helpful
to think of mycelial networks as a type of 'liquid computer', many versions
of which have been built and deployed in systems from fighter jets to nuclear
reactor control systems (Adamatzky, 2019). However, changes in mycelial flow
are still too slow to explain many phenomena. The regular pulses of metabolic
activity that pass across mycelial networks are a plausible way for mycelial
networks to co-ordinate their behaviour, but are also too slow to explain many
phenomena (Tlalka et al., 2003, 2007, Fricker et al., 2007a and b, Fricker et
al., 2008). The poster organism for network living is the puzzle-solving slime
mould. Although they're not fungi, slime moulds have evolved ways to co-
ordinate their sprawling, shape-shifting bodies and provide a helpful model for
thinking about the challenges and opportunities faced by mycelial fungi. They
grow more quickly than fungal mycelium, which makes them easier to study.
Slime moulds communicate between different parts of themselves using rhythmic pulses that ripple down the branches of their networks in rolling waves
of contraction. Branches that have found food produce a signalling molecule
that increases the strength of contraction. Stronger contractions cause a greater
volume of cellular contents to flow along that branch of the network. For a
given contraction, more material will pass along a shorter route than a longer
route. The more material passes along a route, the more it is strengthened. It
is a feedback loop that allows the organism to redirect itself along 'successful'
routes at the expense of less 'successful' ones. Pulses from the different parts
of its network combine, interfere and reinforce one another. In this way, slime
moulds can integrate information from their various branches and solve complex
routing problems without needing a special place to do so (Zhu et al., 2013,
Alim et al., 2017, Alim, 2018).

30 One researcher observed in the mid-1980s that 'fungal electrobiology is about as
far as one can get from the present mainstream of biological research' (Harold
et al. 1985). Nevertheless, fungi have since been found to respond to electrical
stimulation in potentially surprising ways. Treating mycelium with bursts of
electrical current can substantially increase mushroom crops (Takaki et al., 2014).
Crops of the highly prized matsutake mushroom – a mycorrhizal species that has
so far resisted cultivation – can be nearly doubled by jolting the ground around
its partner trees with a 50 kilovolt pulse of electricity. Researchers conducted the
study following reports from matsutake pickers that bumper crops of mushrooms
could be found in the area around a lightning strike several days after it hit
(Islam and Ohga, 2012). For action potentials in plants see Brunet and Arendt
(2015); for early reports of action potentials in fungi see Slayman et al. (1976);
for general discussion of fungal electrophysiology see Gow and Morris (2009);
for 'cable bacteria' see Pfeffer et al. (2012); for action potential-like waves of

activity in bacterial colonies see Prindle et al. (2015), Liu et al. (2017), Martinez-Corral et al. (2019), and summary in Popkin (2017).

31 Olsson measured the speed of travel by timing the gap between stimulation and measuring a response. This estimated speed thus includes the time taken for the fungus to sense the stimulus, for the stimulus to travel from A to B, and for the response to register with the microelectrodes. The actual speed of travel of the impulse could thus be considerably faster than this estimate. The fastest rate of bulk flow measured in fungal mycelium is around 180 millimetres per hour (Whiteside et al., 2019). The action potential-like impulses that Stefan Olsson measured travelled at 1,800 millimetres per hour.

32 Olson and Hansson (1985), Olsson (2009). For Olsson's recording of the change in action potential-like firing rate see doi.org/10.6084/m9.figshare.c.4560923.v1 [accessed 29 October 2019].

33 In a paper titled 'The brain: a concept in flux', Oné Pagán (2019) points out that there is no generally accepted definition of a brain. He argues that it makes more sense to define brains in terms of what they *do*, rather than based on specific details of their anatomy. For regulation of pores in fungal networks see Jedd and Pieuchot (2012) and Lai et al. (2012).

34 Adamatzky (2018 a and b).

35 For examples of network computing see van Delft et al. (2018) and Adamatzky (2016).

36 Adamatzky (2018 a and b).

37 As I discuss in 'Radical Mycology', Andrew Adamatzky is part of an interdisciplinary collaboration called Fungal Architectures (FUNGAR), which hopes to incorporate fungal computing circuits into architectural structures.

38 I asked Olsson why no-one had followed up his studies from the 1990s. 'When I presented the work at conferences people were really, really interested,' Olsson said, 'but they thought it was weird.' All the researchers I have asked about his study are fascinated and want to know more. The study has since been cited many times. But he was unable to get funding for further work into the subject. It was considered too likely to come to nothing – 'too risky', in technical parlance.

39 For 'archaic myth' see Pollan (2013); for ancient cellular processes underlying brain behaviour see Manicka and Levin (2019). The 'moving hypothesis' posits that brains evolved as a cause and a consequence of the need for animals to move around. Organisms that don't move around aren't faced with the same type of challenge, and have evolved different types of network to deal with the problems they face (Solé et al., 2019).

40 For 'minimal cognition' see Calvo Garzón and Keijzer (2011); for 'biologically embodied cognition' see Keijzer (2017); for plant cognition see Trewavas (2016); for 'basal' cognition and degrees of cognition see Manicka and Levin (2019); for discussion of microbial intelligence see Westerhoff et al. (2014); for discussion of different types of 'brain' see Solé et al. (2019).

41 For 'network neuroscience' see Bassett and Sporns (2017) and Barbey (2018). Scientific advances that make it possible to grow cultures of human brain tissue in a dish – known as brain 'organoids' – complicate our understanding of intelligence yet further. The philosophical and ethical questions raised by these techniques – and the absence of clear answers – are a reminder of how the limits of our own biological selves remain far from clear. In 2018, several leading neuroscientists and bioethicists published an article in *Nature* in which they raised some of these questions (Farahany et al., 2018). Over the coming decades, advances in brain

tissue culturing will make it possible to grow artificial 'mini-brains' which more closely mimic the functioning of human brains. The authors write that

> as brain surrogates become larger and more sophisticated, the possibility of them having capabilities akin to human sentience might become less remote. Such capacities could include being able to feel (to some degree) pleasure, pain or distress; being able to store and retrieve memories; or perhaps even having some perception of agency or awareness of self.

> Some are concerned that brain organoids might one day outsmart us (Thierry 2019).

42 For flatworm experiment see Shomrat and Levin (2013); for nervous systems of octopuses see Hague et al. (2013) and Godfrey-Smith (2017), ch. 3.
43 Bengtson et al. (2017), Donoghue and Antcliffe (2010). With studied caution, Bengtson and colleagues point out that their specimens might not be *actual* fungi but might belong to a separate lineage of organisms resembling modern fungi in every observable way. One can understand their hesitance. The authors point out that if these mycelial fossils were true fungi, they would 'overturn' our current understanding of where and how fungi first evolved. Fungi do not fossilise well, and exactly when fungi first branched off the tree of life is disputed. DNA-based methods – using the so-called 'molecular clock' – suggest that the earliest fungi diverged around a billion years ago. In 2019, researchers reported fossilised mycelium found in Arctic shale that dates from around a billion years ago (Loron et al., 2019, Ledford, 2019). Prior to this finding, the earliest undisputed fungal fossils date from around 450 million years ago (Taylor et al., 2007). The earliest fossilised gilled mushroom dates from around 120 million years ago (Heads et al., 2017).
44 For Barbara McClintock see Keller (1984).
45 *Ibid.*
46 von Humboldt (1849), vol. 1, p. 20.

Chapter 3: *The Intimacy of Strangers*

Epigraph: Rich (1994).

1 BIOMEX is one of several astrobiological projects. For BIOMEX see de Vera et al. (2019); for EXPOSE facility see Rabbow et al. (2009).
2 For 'limits and limitations' quote see Sancho et al. (2008); for a review of organisms sent into space, including lichens, see Cottin et al. (2017); for lichens as models for astrobiological research see Meeßen et al. (2017) and de la Torre Noetzel et al. (2018).
3 Wulf (2015), ch. 22.
4 For discussion of Schwendener and the 'dual hypothesis' see Sapp (1994), ch. 1.
5 For 'A useful and invigorating' see Sapp (1994), ch. 1; for 'sensational romance' see Ainsworth (1976), ch. 4. Some of Beatrix Potter's biographers have suggested that she was a proponent of Schwendener's dual hypothesis, and it is possible she changed her mind over the course of her life. Nonetheless, in 1897, in a letter to Charles MacIntosh, a rural postman and amateur naturalist, she appeared to take a clear stance on the question:

You see we do not believe in Schwendener's theory, and the older books say that the lichens pass gradually into hepaticas, through the foliaceous species. I should like very much to grow the spore of one of those large flat lichens, and also the spore of a real hepatica in order to compare the two ways of sprouting. The names do not matter as I can dry them. If you could get me any more spores of the lichen and the hepatica when the weather changes I should be very much obliged. (Kroken 2007).

6 The tree is one of the founding images in modern theories of evolution, and famously the only illustration in Darwin's *On the Origin of Species*. Darwin was by no means the first to deploy the image. For centuries, the branching form of trees has provided a framework for human thought in fields from theology to mathematics. Perhaps most familiar are genealogical trees, which have their roots in the Old Testament ('the Tree of Jesse').

7 For debate about Schwendener's portrayal of lichens see Sapp (1994), ch. 1 and Honegger (2000); for Albert Frank and 'symbiosis' see Sapp (1994), ch. 1, Honegger (2000) and Sapp (2004). Albert Frank first used the word 'symbiotismus' (which translates literally as 'symbiotism').

8 Ancestors of green sea slugs – *Elysia viridis* – ingested algae which continued to live within their tissues. Green sea slugs obtain their energy from sunlight, as a plant would. For new symbiotic discoveries see Honegger (2000); for 'animal lichens' see Sapp (1994), ch. 1; for 'microlichens' see Sapp (2016).

9 For Huxley quote see Sapp (1994), p. 21.

10 For 8 per cent estimate see Ahmadjian (1995); for a greater area than tropical forests see Moore (2013a), ch. 1; for 'hung in hashtags' see Hillman (2018); for the diversity of lichen habitats, including erratics and lichens that live on insects, see Seaward (2008); for interview with Kerry Knudsen see //aeon.co/videos/how-lsd-helped-a-scientist-find-beauty-in-a-peculiar-and-overlooked-form-of-life [accessed 29 October 2019].

11 For 'every monument' quote see twitter.com/GlamFuzz [accessed 29 October 2019]; for Mount Rushmore see Perrottet (2006); for Easter Island heads see www.theguardian.com/world/2019/mar/01/easter-island-statues-leprosy [accessed 29 October 2019].

12 For lichens' approach to weathering see Chen et al. (2000), Seaward (2008) and Porada et al. (2014); for lichens and soil formation see Burford et al. (2003).

13 For history of panspermia and related ideas see Temple (2007) and Steele et al. (2018).

14 In response to Lederberg's concerns about interplanetary infection, NASA developed ways to sterilise spacecraft before departure from Earth. These have not been entirely successful: there is a thriving volunteer population of bacteria and fungi aboard the International Space Station (Novikova et al., 2006). When the Apollo 11 mission returned from the first trip to the moon in 1969, the astronauts were isolated in stringent quarantine in a converted Airstream trailer for three weeks (Scharf, 2016).

15 It had been known that bacteria are capable of acquiring DNA from their surroundings since the work of Frederick Griffith in the 1920s, later confirmed by Oswald Avery and his colleagues in the early 1940s. What Lederberg showed was that bacteria could actively exchange genetic material with each other – a process known as 'conjugation'. For discussion of Lederberg's findings see Lederberg (1952), Sapp (2009), ch. 10 and Gontier (2015a). Viral DNA has had a profound influence on the history of animal life: it is thought that viral genes

played key roles in the evolution of placental mammals from their egg-laying ancestors (Gontier, 2015a, Sapp 2016).

16 Bacterial DNA is found in the genomes of animals (for a general discussion see Yong, 2016, ch. 8). Bacterial and fungal DNA is found in plant and algal genomes (Pennisi, 2019b). Fungal DNA is found in lichen-forming algae (Beck et al., 2015). Horizontal gene transfer is pervasive in fungi (Gluck-Thaler and Slot, 2015, Richards et al., 2011, Milner et al., 2019). At least 8 per cent of the human genome started off in viruses (Horie et al., 2010).

17 For foreign DNA 'short-circuiting' evolution on Earth see Lederberg and Cowie (1958).

18 For hostile conditions in space see de la Torre Noetzel et al. (2018).

19 Sancho et al. (2008).

20 Even at 18 kilograys of gamma irradiation, samples of the lichen *Circinaria gyrosa* only suffered a 70 per cent reduction in photosynthetic activity. At 24 kilograys, photosynthetic activity was reduced by 95 per cent, but wasn't eliminated entirely (Meeßen et al., 2017). To put these results in context, one of the most radiotolerant organisms ever documented, an archaea isolated from deep sea hydrothermal vents (appropriately named *Thermococcus gammatolerans*), can withstand levels of gamma irradiation up to 30 kilograys (Jolivet et al., 2003). For a summary of lichen space studies see Cottin et al. (2017), Sancho et al. (2008), and Brandt et al. (2015); for effects of high-dose irradiation on lichens see Meeßen et al. (2017), Brandt et al. (2017) and de la Torre et al. (2017); for tardigrades in space see Jönsson et al. (2008).

21 Some disciplines are routinely 'informed' by lichens. Lichens are so sensitive to some forms of industrial pollution that they are used as reliable indicators of air quality – 'lichen deserts' extend downwind of urban areas and can be used to map the zone affected by industrial pollution. In some cases, lichens serve as indicators in a more literal sense. They are used by geologists to determine the age of rock formations (a discipline known as lichenometry). And litmus, the pH-sensitive dye used to make the 'indicator paper' found in all school science departments, comes from a lichen.

22 Recent work by Thijs Ettema and his group at Uppsala University suggests that eukaryotes arose within archaea. The exact sequence of events remains much debated (Eme et al., 2017). Bacteria have long been thought of as having no internal cellular structures, known as 'organelles'. This view is changing. Many bacteria appear to have organelle-like structures that perform specialised functions. For a discussion see Cepelewicz (2019).

23 Margulis (1999); for 'intimacy of strangers' see Mazur (2009).

24 For 'fusion and merger' see Margulis (1996); for origins of endosymbiosis see Sapp (1994), ch. 4 and ch. 11; for Stanier quote see Sapp (1994), p. 179; for 'serial endosymbiosis theory' see Sapp (1994), p. 174; for bacteria within bacteria within insects see Bublitz et al. (2019); for Margulis's original paper (under the name Sagan) see Sagan (1967).

25 For 'quite analogous' quote see Sagan (1967); for 'remarkable examples' quote see Margulis (1981), p. 167. For de Bary, in 1879, the most significant implication of symbiosis was that it could result in evolutionary novelty (Sapp 1994, p. 9). 'Symbiogenesis' – 'becoming by living together' – was the term given to the process by which symbiosis could give rise to new species by its earliest Russian proponents, Konstantin Mereschkowsky (1855–1921) and Boris Mikhaylovich

Kozo-Polyansky (1890–1957) (Sapp 1994, pp. 47–8). Kozo-Polyansky included several references to lichens in his work.

> One should not think that lichens are just a simple sum of certain algae and fungi. Rather, they have many specific features found neither in algae nor in fungi … everywhere – in its chemistry, its shape, its structure, its life, its distribution – the composite lichen exhibits new features not characteristic of its separated components (Kozo-Polyansky, trans. 2010, pp. 55–6).

26 For Dawkins and Dennett quotes, among others, see Margulis (1996).

27 'The evolutionary "tree of life" seems like the wrong metaphor,' the geneticist Richard Lewontin remarked. 'Perhaps we should think of it as an elaborate bit of macramé' (Lewontin 2001). It's not entirely fair on trees. The branches of some species can fuse with each other. It is a process known as 'inosculation', from the Latin *osculare*, which means 'to kiss'. But look at the tree nearest to you. The chances are that it forks more than it fuses. The branches of most trees are not like fungal hyphae, which meld with each other as part of their daily practice. Whether the tree is an appropriate metaphor for evolution has been debated for decades. Darwin himself worried about whether the 'coral of life' would make a better image, though he decided in the end that it would make things 'excessively complicated' (Gontier 2015b). In 2009, in one of the most acrimonious upwellings of the tree question, *New Scientist* published an issue that proclaimed on its cover that 'Darwin was wrong'. '*Uprooting Darwin's tree*', shrieked the editorial. Predictably, it inflamed a furious response (Gontier 2015b). Amid the storm of reaction a letter sent by Daniel Dennett stands out: '*What on earth were you thinking when you produced a garish cover proclaiming that "Darwin was wrong"* …?' You can understand why Dennett was cross. Darwin wasn't wrong. It is just that he came up with his theory of evolution before DNA, genes, symbiotic mergers and horizontal gene transfer were known to exist. Our understanding of the history of life has been transformed by these discoveries. But Darwin's central thesis that evolution proceeds by natural selection is not contested – though the extent to which it is the primary driving force in evolution is debated (O'Malley, 2015). Symbiosis and horizontal gene transfer provide new ways that novelty can be generated; they are new *co-authors* of evolution. But natural selection remains the editor. Nonetheless, in the light of symbiotic mergers and horizontal gene transfer, many biologists have begun to reimagine the tree of life as a reticulate meshwork formed as lineages branch, fuse and entangle one another: a 'network', or a 'web', a 'net', a 'rhizome' or a 'cobweb' (Gontier 2015b, Sapp 2009, ch. 21). The lines on these diagrams knot and melt into each other, connecting different species, kingdoms and even domains of life. Links loop in and out of the world of viruses, genetic entities not even considered to be alive. If anyone wanted a new poster organism for evolution they needn't look far. This is a vision of life that resembles fungal mycelium more than anything else.

28 In some lichens, specialised dispersal structures called 'soredia' form, which consist of fungal and algal cells. In some cases, a newly germinated lichen fungus might team up with a photobiont that doesn't quite satisfy its needs, and survive as a small 'photosynthetic smudge' known as a 'prethallus' until the real thing comes along (Goward, 2009a). Some lichens can disassemble and reassemble without producing spores. If certain lichens are placed in a petri dish with the right kind of nutrients, the partners disentangle and creep apart. Once separated, they can

re-form their relationship (though usually imperfectly). In this sense, lichens are reversible. At least in some cases, the honey can be stirred out of the porridge. However, to date only in the case of a single lichen – *Endocarpon pusillum* – have the partners been separated from each other, grown apart, and then recombined to form all the stages of the lichen including functional spores – known as a 'spore-to-spore' resynthesis (Ahmadjian and Heikkilä, 1970).

29　The symbiotic nature of lichens presents some interesting technical problems. Lichens have long been small nightmares for taxonomists. As the situation stands, lichens are referred to by the name of the fungal partner. For example, the lichen that arises through the interaction of the fungus *Xanthoria parietina* and the alga *Trebouxia irregularis* is known as *Xanthoria parietina*. Similarly, the combination of the fungus *Xanthoria parietina* and the alga *Trebouxia arboricola* is known as *Xanthoria parietina*. Lichen names are a synecdoche, in that they refer to a whole by reference to a part (Spribille, 2018). The current system implies that the fungal component of the lichen *is the lichen*. But this isn't true. Lichens emerge out of a negotiation between several partners. 'To see lichens as fungi', Trevor Goward bemoans, 'is to miss seeing lichens altogether' (Goward 2009a). It is as if chemists called any compound that contained carbon – from diamond to methane to methamphetamine – *carbon*. You'd be forced to admit that they might be missing something. This is more than semantic grumbling. To name something is to acknowledge that it exists. When any new species is found, it is 'described' and given a name. And lichens do have names, plenty of them. Lichenologists aren't taxonomically ascetic. It's just that the only names they can give glance off the phenomenon they aim to describe. It is a structural issue. Biology is built around a taxonomic system that has no way to recognise the symbiotic status of lichens. They are literally unnameable.

30　Sancho et al. (2008).

31　de la Torre Noetzel et al. (2018).

32　For unique lichen compounds and human uses see Shukla et al. (2010) and *State of the World's Fungi* (2018); for metabolic legacies of lichen relationships see Lutzoni et al. (2001).

33　For report from the Deep Carbon Observatory see Watts (2018).

34　For lichens in deserts see Lalley and Viles (2005) and *State of the World's Fungi* (2018); for lichens within rocks see de los Ríos et al. (2005) and Burford et al. (2003); for Antarctic Dry Valleys see Sancho et al. (2008); for liquid nitrogen see Oukarroum et al. (2017); for lichen longevity see Goward (1995).

35　Sancho et al. (2008).

36　For shock of ejection see Sancho et al. (2008) and Cockell (2008). In a number of studies, bacteria have proved to be more resistant to high temperatures and shock pressures than lichens. For re-entry see Sancho et al. (2008).

37　Sancho et al. (2008); Lee et al. (2017).

38　For origins of lichens see Lutzoni et al. (2018) and Honegger et al. (2012). There is a lot of debate about the identity of ancient lichen-like fossils and their relationship to extant lineages. Marine lichen-like organisms have been found dating from 600 million years ago (Yuan et al., 2005), and some argue that these marine lichens played a role in the movement of lichens' ancestors onto the land (Lipnicki, 2015). For multiple evolution of lichens and re-lichenisation see Goward (2009a); for de-lichenisation see Goward (2010); for optional lichenisation see Selosse et al. (2018).

39　Hom and Murray (2014).

40 For 'the song, not the singer' see Doolittle and Booth (2017).

41 *Hydropunctaria maura* used to be known as *Verrucaria maura* (or 'warty midnight'). For a long-term study of the arrival of lichens on a newly born island see the case of Surtsey: www.anbg.gov.au/lichen/case-studies/surtsey.html [accessed 29 October 2019].

42 For 'wholes' and 'collections of parts' see Goward (2009b).

43 Spribille et al. (2016).

44 For discussion of diversity of fungi within lichens see Arnold et al. (2009); for additional partners in wolf lichens see Tuovinen et al. (2019) and Jenkins and Richards (2019).

45 For 'It doesn't matter what you call it' see Hillman (2018). Goward has formulated a definition of lichens that takes account of these recent findings: 'The enduring physical byproduct of lichenisation defined as a process whereby a nonlinear system comprising an unspecified number of fungal, algal and bacterial taxa give rise to a thallus [the shared body of the lichen] viewed as an emergent property of its constituent parts' (Goward 2009c).

46 For lichens as microbial reservoirs see Grube et al. (2015), Aschenbrenner et al. (2016), and Cernava et al. (2019).

47 For queer theory for lichens see Griffiths (2015).

48 See Gilbert et al. (2012) for a more detailed breakdown of how microbes confuse different definitions of biological individuality. For more on microbes and immunity see McFall-Ngai (2007) and Lee and Mazmanian (2010). Some propose alternative definitions of biological individuals based on the 'common fate' of the living system. For instance, Frédéric Bouchard proposes that 'A biological individual is a functionally integrated entity whose integration is linked to the common fate of the system when faced with selective pressures from the environment' (Bouchard 2018).

49 Gordon et al. (2013); Bordenstein and Theis (2015).

50 For infections caused by gut bacteria see Van Tyne et al. (2019).

51 Gilbert et al. (2012).

Chapter 4: Mycelial Minds

Epigraph: Sabina, M., from recording by Gordon Wasson. Quoted in Schultes et al. (2001), p. 156.

1 For brief summary of clinical studies into psychedelics see Winkelman (2017); for extended discussion see Pollan (2018).

2 Hughes et al. (2016).

3 For timing and height of ants' death grips see Hughes et al. (2011) and Hughes (2013); for orientation see Chung et al. (2017). There are many different species of *Ophiocordyceps* fungi, and many different species of carpenter ant, but each ant is host to only one species of fungus, and each species of fungus can only control one species of ant (de Bekker et al., 2014). Different fungus–ant pairings are particular about their choice of death site. Some fungi cause their insect avatars to bite onto twigs, some onto bark, and some onto leaves (Andersen et al., 2009, Chung et al., 2017).

4 For fungal proportion of ant biomass see Mangold et al. (2019); for visualisation of fungal network within ant bodies see Fredericksen et al. (2017).

5 For hypothesis that fungal manipulation takes place by chemical means see Fredericksen et al. (2017); for chemicals produced by *Ophiocordyceps* see de Bekker et al. (2014); for discussion of *Ophiocordyceps* and ergot alkaloids see Mangold et al (2019).

6 For fossilised leaf scars see Hughes et al. (2011).

7 For Mckenna quote see Letcher (2006), p. 258.

8 Schultes et al. (2001), p. 9. For wide-ranging if sometimes uncritical discussions of intoxication in the animal world see Siegel (2005) and Samorini (2002).

9 For discussion of *Amanita muscaria* see Letcher (2006), ch. 7–9. Some hypothesise that the accusers in the Salem witch trials were afflicted by convulsive ergotism (Caporael, 1976, and Matossian, 1982), although their arguments have been robustly countered by Spanos and Gottleib (1976). Known in the Middle Ages and Renaissance as Saint Anthony's fire, ergot-induced visions and psycho-spiritual anguish are thought to have inspired contemporary visions of hell. For Heironymus Bosch see Dixon (1984). Livestock, too, are vulnerable to ergot poisoning. 'Sleepy grass', 'drunk grass' and the 'ryegrass staggers' are all named after their effects on cattle, horses and sheep (Clay, 1988). Ergot fungi also have powerful medicinal effects, and have been used for hundreds of years by midwives to stop post-partem bleeding. Henry Wellcome, the entrepreneur whose endowment founded the Wellcome Trust, researched reports into the medicinal effects of ergot, the grain fungus. He recorded that ergot was regarded by midwives in sixteenth-century Scotland, Germany and France 'to be of remarkable and certain efficacy' in inducing uterine contractions and controlling bleeding after childbirth. It was from these herb-wives or midwives that – male – physicians learned of the therapeutic properties of ergot, which form the basis of the drug ergometrine, still used today to treat heavy bleeding following childbirth (Dugan, 2011, pp. 20–1). It was for their reputation as obstetric drugs that Albert Hofmann began investigating them at Sandoz Laboratories in the 1930s, a research programme that led to the synthesis of LSD in 1938. For a discussion of ergot alkaloids, their history and uses, see Wasson et al. (2009), ch. 2: 'A Challenging Question and My Answer'.

10 For discussion of the history of psilocybin mushroom use in Mexico see Letcher (2006), ch. 5, Schultes (1940) and Schultes et al. (2001), 'Little Flowers of the Gods'; for Sahagún quote see Schultes (1940).

11 Letcher (2006), p. 76.

12 For McKenna and Tassili painting and quote see McKenna (1992), ch. 6; for discussion of McKenna and the Tassili painting see Metzner (2005), pp. 42–3; for a more critical discussion see Letcher (2006), pp. 37–8.

13 A paper published in 2019 analysed the residues inside a pouch made from a fox snout found in a ritual bundle excavated in Bolivia, dating from over 1,000 years ago. The researchers found traces of multiple psychoactive compounds – including cocaine (from coca), DMT, harmine and bufotenine. The analysis provided tentative evidence of psilocin – a psychoactive breakdown product of psilocybin – which, if true, would suggest that psilocybin mushrooms had been present in the ritual bundle (Miller et al., 2019). The Eleusinian Mysteries – a celebration of Demeter, the goddess of grain and harvest, and her daughter Persephone – were one of the major religious festivals in ancient Greece. As part of the celebrations, initiates drank a cup of a liquid known as *kykeon*. Following the drink, initiates experienced ghostly apparitions, and awe-inspiring ecstatic and visionary states. Many described being permanently changed by their experience (Wasson et al.,

2009, ch. 3). Although the identity of *kykeon* remained a carefully guarded secret, it is very likely to have been a mind-altering brew – a notorious scandal erupted when it was found that Athenian aristocrats had been drinking *kykeon* at home with their guests at dinner parties (Wasson et al., 1986, p. 155). There were no registration lists for the rites of Eleusis, and so there is some uncertainty about exactly who attended. However, most Athenian citizens were initiates, and many notable figures are thought to have attended, including Euripides, Sophocles, Pindar and Aeschylus. Plato wrote about the experience of mystery initiations in some detail in his *Symposium* and in *Phaedrus*, using language that refers clearly to the rites at Eleusis (Burkett, 1987, pp. 91–3). Aristotle did not refer explicitly to the mysteries at Eleusis, but did refer to mystery initiations – a reference that is probably compatible with the Eleusinian Mysteries, given the pre-eminence of the Eleusinian rites by the mid fourth century BCE. Hofmann, along with Gordon Wasson and Carl Ruck, hypothesised that *kykeon* was made from ergot fungi growing on grain, somehow purified to avoid the dreadful symptoms associated with its accidental consumption (Wasson et al., 2009). Terence McKenna (1992, ch. 8) speculated that the priests at Eleusis distributed psilocybin mushrooms. Others have suggested a preparation made from opium poppies. There are other examples of possible use of mushrooms in ancient religious contexts. In central Asia a religious cult sprang up around the use of a mind-altering preparation called 'soma'. Soma induced ecstatic states, and devotional hymns to soma are recorded in the Rig Veda, an ancient text dating from around 1500 BCE. Like *kykeon*, the identity of the drink remains unknown. Some – most notably Gordon Wasson – have argued that it was the red and white spotted mushroom, *Amanita muscaria* (for a discussion see Letcher, 2008, ch. 8). Terence McKenna – true to form – suggested that psilocybin mushrooms are more likely candidates. Others have suggested cannabis. There is no unequivocal evidence either way.

14 For reference to the fictional monsters see Yong (2017). In 2018, researchers at the University of the Ryukyus in Japan discovered that several species of cicada had domesticated *Ophiocordyceps* fungi that lived within their body (Matsuura et al., 2018). Like many insects that live mostly on sap, cicadas depend on symbiotic bacteria to produce several essential nutrients and vitamins, without whom they can't survive. But in a number of Japanese species of cicada, the bacteria have been replaced by a species of *Ophiocordyceps*. It is the last thing one would expect. *Ophiocordyceps* are brutally effective killers that have honed their abilities over tens of millions of years. Yet somehow, over the course of their long history together, *Ophiocordyceps* have become essential life partners of the cicadas. What's more, it has happened at least three times in three separate lineages of cicada. Domesticated *Ophiocordyceps* are a reminder that the distinction between 'beneficial' and 'parasitic' microbes is not always clear cut.

15 For immunosuppressant drugs see *State of the World's Fungi* (2018), 'Useful Fungi'; for nostrum for eternal youth see Adachi and Chiba (2007).

16 Coyle et al. (2018); for 'whackiest discovery' see //twitter.com/mbeisen/status/1019655132940627969 [accessed 29 October 2019].

17 For description of the behaviour of infected flies see Hughes et al. (2016) and Cooley et al. (2018); for 'flying saltshakers of death' see Yong (2018).

18 For Kasson study see Boyce et al. (2019) and discussion in Yong (2018). It isn't the first report that insect-manipulating fungi might control their hosts using chemicals that can also alter human minds; cousins of *Ophiocordyceps* fungi

are eaten alongside psilocybin mushrooms in some indigenous ceremonies in Mexico (Guzmán et al., 1998).

19 Cathinone has been reported to increase aggression in ants, and it might be responsible for the hyperactive behaviours observed in infected cicadas (Boyce et al., 2019).

20 See Ovid (1958), p. 186; for Amazonian shamanism see Viveiros de Castro (2004); for Yukaghir people see Willerslev (2007).

21 For 'fungus in ant's clothing' see Hughes et al. (2016). Neuromicrobiology is a relatively new field, and understanding of gut microbes' influence on animal behaviours, cognition and psychological states remains patchy (Hooks et al., 2018). Nonetheless, some patterns are starting to emerge. Mice, for example, require a healthy gut microflora to develop a functional nervous system in the first place (Bruce-Keller et al., 2018). If one knocks out the microbiome of adolescent mice before they have had the chance to develop a functional nervous system, they develop cognitive defects. These include memory problems, and difficulty identifying objects (de la Fuente-Nunez et al., 2017). The most dramatic demonstrations come from studies that swap the microbiota between different mouse lines. When 'timid' mouse strains are given faecal transplants from 'normal' strains, they lose their caution. Likewise, if 'normal' strains are inoculated with the microbes of the 'timid' strains, they acquire 'exaggerated caution and hesitancy' (Bruce-Keller et al., 2018). Differences in gut microbiota in mice affect the ability of mice to forget the experience of pain (Pennisi, 2019a, Chu et al., 2019). Many gut microbes produce chemicals that influence the activity of the nervous system, including neurotransmitters and short chain fatty acids (SCFAs). More than 90 per cent of the serotonin in our bodies – the neurotransmitter that when abundant makes us feel happy, and when depleted makes us feel depressed – is produced in our gut, and gut microbes play a major role in regulating its production (Yano et al., 2015). Two studies have investigated the effect of transplanting the faecal microbiota of depressed human patients into germ-free mice and rats. The animals developed symptoms of depression including anxiety and a loss of interest in pleasurable behaviours. These studies suggest that not only can imbalances in the gut microbiota result in depression, but that the same imbalances may be responsible for depressed behaviour in mice and in humans (Zheng et al., 2016, Kelly et al., 2016). Further studies on humans have shown that certain probiotic treatments can reduce symptoms of depression, anxiety and the occurrence of negative thoughts (Mohajeri et al., 2018, Valles-Colomer et al., 2019). However, a multi-billion-dollar probiotics industry hovers around the field of neuromicrobiology, and a number of researchers have pointed out the tendency to overhype findings. Gut communities are complex, and manipulating them is a challenge. There are so many variables involved that few studies are able to identify causal links between the action of specific microbes and specific behaviours (Hooks et al., 2018).

22 For full exposition of the 'extended phenotype' see Dawkins (1982); for 'tightly limited speculation' see Dawkins (2004); for discussion of fungal manipulation of insect behaviour in terms of extended phenotypes see Andersen et al. (2009), Hughes (2013 and 2014), Cooley et al. (2018).

23 For a discussion of the 'first wave' of psychedelic research in the 1950s and 1960s see Dyke (2008) and Pollan (2018), ch. 3.

24 For Johns Hopkins study see Griffiths et al. (2016); for NYU study see Ross et al. (2016); for interview with Griffiths see *Fantastic Fungi: The Magic Beneath Us*, directed by Louis Schwartzberg; for general discussion including the record 'treatment effect' sizes see Pollan (2018), ch. 1.

25 For study on psilocybin-occasioned mystical experience see Griffiths et al. (2008); for role of awe in psychedelic-assisted psychotherapy see Hendricks (2018).

26 For role of psilocybin in treating tobacco addiction see Johnson et al. (2014, 2015); for psilocybin-induced 'openness' and life satisfaction see MacLean et al. (2011); for general discussion of the role of psychedelics in treating addiction see Pollan (2018), ch. 6, pt 2; for sense of connection with the natural world see Lyons and Carhart-Harris (2018) and Studerus et al. (2011). There is a long tradition of Native American communities using the psychedelic cactus peyote as a treatment for alcoholism. Between the 1950s and 1970s, a number of studies investigated the possibility that psilocybin and LSD could be used to treat drug addiction. Several reported positive effects. In 2012, a meta-analysis pooled the data from the most rigorously controlled trials. It reported that a single dose of LSD had a beneficial effect on alcohol misuse that lasted up to six months (Krebs and Johansen 2012). In an online survey designed to investigate the 'natural ecology' of the phenomenon, Matthew Johnson and his colleagues (2017) analysed accounts from more than 300 people who reported that they had reduced their tobacco intake or stopped entirely following an experience with psilocybin or LSD.

27 For 'started out stone cold' see Pollan (2018), ch. 4; for non-material reality as basis for religious belief see Pollan (2018), ch. 2. Even sitters who guide and observe the sessions at Johns Hopkins have reported unexpected changes in their world views. One guide who had sat through dozens of psilocybin sessions described their experience:

> I started out on the atheist side, but I began seeing things every day in my work that were at odds with this belief. My world became more and more mysterious as I sat with people on psilocybin. (Pollan, 2018, ch. 1).

28 For influence of psychedelics on the growth and architecture of neurones see Ly et al. (2018).

29 For psilocybin and the DMN see Carhart-Harris et al. (2012) and Petri et al. (2014); for effects of LSD on brain connectivity see Carhart-Harris et al. (2016b).

30 For Hoffer quote see Pollan (2018), ch. 3.

31 For Johnson quote see Pollan (2018), ch. 6; for role of psilocybin in treating 'rigid pessimism' of depression see Carhart-Harris et al. (2012).

32 For discussion of ego dissolution and 'merging' see Pollan (2018), prologue, ch. 5.

33 For 'cool night of the mind' and 'baroque' see McKenna and McKenna (1976), pp. 8–9.

34 For Whitehead quote see Russell (1956), p. 39; for 'tightly limited' speculation see Dawkins (2004).

35 It isn't straightforward to estimate when exactly the first mushrooms became 'magic'. The simplest approach is to assume that the ability to make psilocybin originated in the most recent common ancestor of all the fungi that make psilocybin. However, this doesn't work because:

1) psilocybin has been horizontally transferred between fungal lineages (Reynolds et al., 2018), and
2) psilocybin biosynthesis has evolved more than once (Awan et al., 2018).

Jason Slot, a researcher at Ohio State University, made the estimate of 75 million years based on a hypothesis that the genes needed to make psilocybin first clustered in an ancestor of the genera *Gymnopilus* and *Psilocybe*. Slot suspects this to be the case because the other occurrences of the psilocybin gene cluster have been shown to have arisen through horizontal gene transfer.

36 For horizontal gene transfer of psilocybin gene cluster see Reynolds et al. (2018); for multiple origins of psilocybin biosynthesis see Awan et al. (2018).
37 Some relationships between insects and fungi involve more ambiguous manipulation, like 'cuckoo fungi', which capitalise on the social behaviour of termites by producing small balls that look like termite eggs, and produce a pheromone found in real termite eggs. Termites carry the fake eggs into their nest, where they tend to them. When they fail to germinate, the fungal 'eggs' are thrown into waste piles. Surrounded with a nutrient-rich compost, the cuckoo fungi sprout and are able to live free from competition with other fungi (Matsuura et al., 2009).
38 For leafcutter ants foraging for psilocybin mushrooms see Masiulionis et al. (2013); for gnats and other insects that eat psilocybin mushrooms, and the 'lure' hypothesis of psilocybin, see Awan et al. (2018). Pure crystalline psilocybin is expensive, and heavy regulation makes research hard. There is some evidence that psilocybin impedes the behaviour of insects and other invertebrates. In a well-known series of experiments in the 1960s, researchers gave a range of drugs to spiders to study the webs that they spun. High doses of psilocybin prevented web-building altogether. Spiders on lower doses spun looser webs, behaving 'as if they were heavier'. By contrast, LSD caused the spiders to produce 'unusually regular' webs (Witt, 1971). More recently, studies have found that fruit flies given metitepine, a chemical which blocks the serotonin receptors that psilocybin stimulates, lost their appetites. This has led some to suggest that psilocybin may serve to *increase* the appetite of flies – possibly serving to disperse fungal spores (Awan et al., 2018). Michael Beug, a biochemist and mycologist at Evergreen State College, is among the researchers that argue against the psilocybin-as-deterrent hypothesis. Mushrooms are a fruit. Just as an apple tree makes its fruit conspicuous to facilitate the dispersal of its seeds, so fungi produce mushrooms to facilitate the dispersal of their spores. Psilocybin, as Beug points out, is found in high concentrations in mushrooms of psilocybin-producing species, but negligible quantities in the mycelium of most psilocybin-producing species (though not all: *Psilocybe caerulescens* and *Psilocybe hoogshagenii/semperviva* are reported to contain significant concentrations of psilocybin in their mycelium). Yet it is the mycelium, not the mushrooms, that is in most need of defence. Why would psilocybin mushrooms go to the trouble to defend their fruit while leaving their mycelium unprotected (Pollan, 2018, ch. 2)?
39 Other mammals, too, are known to eat species of psilocybin mushroom without ill effect. Michael Beug, the biochemist and mycologist in charge of the poisoning reports filed with the North American Mycological Association, has received many such accounts. 'With horses or cows, it may or may not be accidental,' Beug told me. In some cases, however, animals do seem to seek them out. 'Some dogs will see their owners picking psilocybin mushrooms and take an interest – and

then will eat the mushrooms again and again with effects that appear familiar to the human observer.' Only once has he dealt with reports of a cat, 'who ate mushrooms repeatedly, and appeared to become quite "be-mushroomed"'.

40 Schultes (1940).

41 For a discussion of Wasson's article in *Life* magazine and its reach see Pollan (2018), ch. 2, and Davis (1996), ch. 4.

42 For 'trailing our mother' see McKenna (2012). Possibly 'the' first account of a trip in a widely read organ was written by the journalist Sidney Katz, who published an article in the popular Canadian magazine *Maclean's* entitled 'My Twelve Hours as a Madman'. For a discussion see Pollan (2018), ch. 3.

43 For a discussion of Leary's 'visionary voyage' and the Harvard Psilocybin Project see Letcher (2006), pp. 198–201 and Pollan (2018), ch. 3. For Leary quote see Leary (2005).

44 Letcher (2006), pp. 201, 254–5; Pollan (2018), ch. 3.

45 For discussion of the growing interest in magic mushrooms see Letcher (2006), ch. 13, 'Underground, Overground'; for discussion of the development of cultivation techniques see Letcher (2006), ch. 15, 'Muck and Brass'; for grower's guide see McKenna and McKenna (1976).

46 For discussion of *The Mushroom Cultivator,* and the Dutch and English magic mushroom scenes, see Letcher (2006), ch. 15, 'Muck and Brass'.

47 In Central American pastures mushrooms grow readily, and there is nothing to suggest that people actively cultivated them.

48 For psilocybin-containing lichen see Schmull et al. (2014); for global distribution of psilocybin mushrooms see Stamets (1996, 2005); for 'occur in abundance' see Allen and Arthur (2005); for an account of the discovery of psilocybin mushrooms around the world see Letcher (2006), pp. 221–5; for 'parks, housing developments ...' see Stamets (2005).

49 Schultes et al. (2001), p. 23.

50 See James (2002), p. 300.

Chapter 5: Before Roots

Epigraph: Kathleen Brennan & Tom Waits, 'Green Grass', on *Real Gone* (2004).

1 For evolution of land plants see Lutzoni et al. (2018), Delwiche and Cooper (2015), and Pirozynski and Malloch (1975); for biomass of plants see Bar-On et al. (2018).

2 For early biocrusts see Beerling (2019), p. 15 and Wellman and Strother (2015); for Ordovician life see: web.archive.org/web/20071221094614/ http://www.palaeos. com/Paleozoic/Ordovician/Ordovician.htm#Life [accessed 29 October 2019].

3 For incentives of life on land for the ancestors of plants see Beerling (2019), p. 155. Perhaps unsurprisingly, there has not always been consensus on this topic. The idea was first proposed by Kris Pirozynski and David Malloch in 1975 in their paper 'The origin of land plants: a matter of mycotropism'. In it they made the claim that 'land plants never had any independence [from fungi], for if they had, they could never have colonised the land'. It was a radical idea at the time because it posited that symbiosis had been a major force in one of the most significant evolutionary developments in the history of life. Lynn Margulis ran with the idea and described symbiosis as 'the moon that pulled the tide of life from its oceanic depths to dry land and up into the air' (Beerling 2019, pp. 126–7). For

discussion of fungi and their role in the evolution of land plants see Lutzoni et al. (2018), Hoysted et al. (2018), Selosse et al. (2015), Strullu-Derrien et al. (2018).

4 For proportion of plant species that form mycorrhizal associations see Brundrett and Tedersoo (2018). The 7 per cent of land plant species that don't form mycorrhizal associations have evolved alternative strategies, like parasitism, or carnivory. This figure may be even less than 7 per cent: recent studies have found that plants that are traditionally thought of as 'non-mycorrhizal' – those in the cabbage family, for example – form relationships with non-mycorrhizal fungi that provide benefit to the plant as mycorrhizal associations do (van der Heijden et al., 2017, Cosme et al., 2018, Hiruma et al., 2018).

5 For fungi in seaweeds – 'mycophycobiosis' – see Selosse and Tacon (1998); for 'soft green balls' see Hom and Murray (2014).

6 A group of living plants called liverworts are thought to be the earliest-diverging lineage of land plants, and may stretch back more than 400 million years. Liverworts in the genera *Treubia* and *Haplomitrium* may provide us with the best glimpse at early plant life (Beerling 2019, p. 25). There are a number of lines of evidence besides fossils. The genetic apparatus responsible for the chemical signals used by plants to communicate with mycorrhizal fungi is identical in all living plant groups, implying that it was present in the common ancestor of all plants (Wang et al., 2010, Bonfante and Selosse, 2010, Delaux et al., 2015). The surviving ancestors of the earliest land plants – the liverworts – form relationships with the most ancient lineages of mycorrhizal fungi (Pressel et al., 2010). Furthermore, the most recent estimates of the timings suggest that fungi made the transition to land earlier than the ancestors of modern land plants, indicating that it would have been nearly impossible for early plants not to have encountered fungi (Lutzoni et al., 2018).

7 For evolution of roots see Brundrett et al. (2002) and Brundrett and Tedersoo (2018).

8 For evolution of thinner, more opportunistic roots see Ma et al. (2018). The diameter of fine roots varies, but is typically between 100 and 500 micrometres. Among one of the most ancient lineages of mycorrhizal fungi – the arbuscular mycorrhizal fungi – transport hyphae are around 20–30 micrometres in diameter, and their fine absorptive hyphae are as thin as 2–7 micrometres (Leake et al., 2004).

9 For a third to a half of soil biomass see Johnson et al. (2013); for estimates of lengths of mycorrhizal fungi in the top 10 centimetres of soil see Leake and Read (2017). These estimates are based on the lengths of mycorrhizal mycelium found in different ecosystems, and take into account mycorrhizal type and land use type. Data comes from Leake et al. (2004).

10 For Frank's work on mycorrhizal fungi see Frank (2005); for discussion of Frank's work see Trappe (2005).

11 One of Frank's most vocal critics was the botanist and later dean of Harvard Law School, Roscoe Pound, who denounced his propositions as 'decidedly fishy'. Pound took the side of more 'sober' authors who maintained that mycorrhizal fungi were 'probably injurious by taking nourishment properly belonging to the tree'. 'In all cases', Pound thundered, symbiosis 'results advantageously to one of the parties, and we can never be sure that the other would not have been nearly as well off, if left to itself' (Sapp, 2004).

12 For description of Frank's experiments see Beerling (2019), p. 129.

13 Tolkien (2014): for 'For you, little gardener …' see Vol. II, ch. 8: 'Farewell to Lórien'; for 'Sam Gamgee planted …' see Vol. III, ch. 9: 'The Grey Havens'.

14 For rapid evolution in the Devonian see Beerling (2019), pp. 152, 155; for drop in carbon dioxide see Johnson et al. (2013) and Mills et al. (2017). There are alternative hypotheses regarding the triggers for the drop in atmospheric carbon dioxide. For example, carbon dioxide and other greenhouse gasses are emitted by volcanism and other tectonic activity. If levels of volcanic carbon dioxide emissions fell, then levels of atmospheric carbon dioxide would also fall, potentially triggering a period of global cooling (McKenzie et al., 2016).

15 For mycorrhizal assistance to the plant boom in the Devonian see Beerling (2019), p. 162; for discussion of weathering in light of mycorrhizal activity see Taylor et al. (2009).

16 Mills used the COPSE model (Carbon, Oxygen, Phosphorus, Sulphur and Evolution), which examines the cycling of all these elements over long periods of evolutionary time in relation to a 'simplified representation of the land biota, atmosphere, oceans and sediments' (Mills et al., 2017).

17 Mills et al. (2017); for Katie Field's experiments on mycorrhizal responses to ancient climates see Field et al. (2012).

18 For a general discussion of mycorrhizal evolution see Brundrett and Tedersoo (2018). The group of fungi that helped plants onto land, and which thrive in grasslands and tropical forests – arbuscular mycorrhizal fungi – are thought to have evolved only once. Arbuscular mycorrhizal fungi are the ones that grow into feathery lobes within plant cells. The type that dominates in temperate forests – ectomycorrhizal fungi – has arisen on more than sixty separate occasions (Hibbett et al., 2000). These fungi – which include truffles – weave themselves into mycelial sleeves around plant root tips, as Frank observed in the late nineteenth century. Orchids have their own type of mycorrhizal relationship, with its own evolutionary history. So do plants in the blueberry family, or *Ericaceae* (Martin et al., 2017). Katie Field and her colleagues are studying a completely different group of mycorrhizal fungus that was only discovered in the late 2000s, known as the *Mucoromycotina*. It occurs across the plant kingdom and is thought to be as old as the earliest land plants, but had gone entirely unnoticed despite decades of study. There may well be more hiding in plain sight (van der Heijden et al., 2017, Cosme et al., 2018, Hiruma et al., 2018, Selosse et al., 2018).

19 For strawberry experiments see Orrell (2018); for a further study on the influence of mycorrhizal fungi on plant–pollinator interactions see Davis et al. (2019).

20 For basil see Copetta et al. (2006); for tomatoes see Copetta et al. (2011), Rouphael et al. (2015); for mint see Gupta et al. (2002); for lettuce see Baslam et al. (2011); for artichokes see Ceccarelli et al. (2010); for St John's wort and echinacea see Rouphael et al. (2015); for bread see Torri et al. (2013).

21 Rayner (1945).

22 For the social function of intellect see Humphrey (1976).

23 For 'reciprocal rewards' see Kiers et al. (2011). Kiers and her colleagues were able to be so precise because she used an artificial system. The plants weren't normal plants but root 'organ cultures' – disembodied roots that grow without shoots or leaves. Nonetheless, the ability of plants and fungi to preferentially transfer nutrients or carbon to more favourable partners has been demonstrated with whole plants growing in soil (Bever et al., 2009, Fellbaum et al., 2014, Zheng et al., 2015). Exactly how plants and fungi are able to regulate these fluxes is not well understood, but it appears to be a general feature of the relationship (Werner and Kiers, 2015).

24 Not all plant and fungal species are able to control their exchange to the same degree. Some species of plant inherit an ability to preferentially supply carbon to favourable fungal partners. Some species just don't have this talent (Grman, 2012). Some plants depend more on their fungal partners than others. Some species, like those that produce dust seeds, won't germinate without a fungus present; many plants will. Some plants don't give anything back to the fungus when they're young but start to reward the fungus when they get older, a life-style that Katie Field calls the 'take now, pay later' approach (Field et al., 2015).

25 For study on resource inequality see Whiteside et al. (2019).

26 Kiers and her colleagues measured the speed of transport through the network, observing maximum speeds of more than fifty micrometres per second – roughly a hundred times faster than passive diffusion – as well as regular changes, or oscillations, in the direction of flow through the network (Whiteside et al., 2019).

27 For the role of context in mycorrhizal associations see Hoeksema et al. (2010) and Alzarhani et al. (2019); for impact of phosphorus on plant 'pickiness' see Ji and Bever (2016). Even within plant and fungal species, there is a large amount of variation between the behaviour of individual plants and fungi (Mateus et al., 2019).

28 For estimate of number of trees on Earth see Crowther et al. (2015).

29 For a discussion of knowledge gaps in mycorrhizal research see Lekberg and Helgason (2018).

30 For discussion of plant and fungal exchange and how it is controlled see Wipf et al. (2019). In one study, a single fungus connected to two different species of plant at the same time – flax and sorghum – supplied more nutrients to flax, even though sorghum supplied the fungus with more carbon. Based on a cost–benefit analysis, one would expect the fungus to supply more nutrients to sorghum (Walder et al., 2012, see also Hortal et al., 2017). Some species of plant are even more extreme, and don't provide any carbon to their mycorrhizal partners at all. In these instances, exchange between partners appears not to be based on recipro-cal rewards exchanged tit for tat. Of course, there may be many other benefits and costs that aren't being taken into account, but it's hard to measure so many variables at once. For this reason, most studies focus on a small number of easily manipulated parameters, like carbon and phosphorus. This provides fine detail but makes it difficult to extend the findings to complex real-world scenarios (Walder and van der Heijden, 2015, van der Heijden and Walder, 2016).

31 For influence of mycorrhizal fungi on forest dynamics on a continental scale see Phillips et al. (2013), Bennett et al. (2017), Averill et al. (2018), Zhu et al. (2018), Steidinger et al. (2019), Chen et al. (2019); for migration of trees following the retreat of the Laurentide Ice Sheet see Pither et al. (2018).

32 For study at University of British Columbia see Pither et al. (2018) and com-mentary by Zobel (2018); for a study on mycorrhiza-mediated encroachment of plants onto heathlands see Collier and Bidartondo (2009); for co-migration of plants and their mycorrhizal partners see Peay (2016).

33 Rodriguez et al. (2009).

34 Osborne et al. (2018), with commentary by Geml and Wagner (2018).

35 For involution see Hustak and Myers (2012).

36 For discussion of the role of plant–fungal relations in adaptation to climate change see Pickles et al. (2012), Giauque and Hawkes (2013), Kivlin et al. (2013), Mohan et al. (2014), Fernandez et al. (2017), Terrer et al. (2016); for 'alarm-ing deterioration' see Sapsford et al. (2017) and van der Linde et al. (2018). Mycorrhizal relationships can pattern the above-ground world in a number

of ways, for example through their influence on soil nutrient cycles. One can think of soil nutrient cycles as chemical weather systems. The chemical 'climate' set up by different types of fungi helps to determine what kind of plants grow where. The influence of different plants, in turn, feeds back on the behaviour of mycorrhizal fungi. Arbuscular mycorrhizal (AM) fungi – the ancient lineage that grows inside plant cells – steer chemical weather systems in a completely different direction than ectomycorrhizal (EM) fungi – the type that has evolved multiple times, and grows around plant roots in a mycelial sleeve. Unlike AM fungi, EM fungi have descended from free-living decomposer fungi. As a result, they are better at decomposing organic matter than arbuscular mycorrhizal fungi. On an ecosystem scale, this makes a big difference. EM fungi thrive in colder climates where decomposition is slower. AM thrive in warmer, wetter climates where decomposition is faster. EM tend to compete with free-living decomposers and reduce the rate at which carbon cycles. AM tend to promote the activity of free-living decomposers and increase the rate at which carbon cycles. EM cause more carbon to become immobilised in the upper soil layers. AM cause more carbon to trickle down into lower soil layers and become immobilised there (Phillips et al., 2013, Craig et al., 2018, Zhu et al., 2018, Steidinger et al., 2019). Mycorrhizal relationships can also influence the ways in which plants interact with one another. In some situations, mycorrhizal fungi increase the diversity of plant life by easing competitive interactions between plants, allowing less dominant plant species to establish themselves (van der Heijden et al., 2008, Bennett and Cahill, 2016, Bachelot et al., 2017, Chen et al., 2019). In others, they reduce diversity by allowing plants to exclude competitors. In some cases, plant feedbacks with mycorrhizal communities span generations, sometimes known as 'legacy effects' (Mueller et al., 2019). A study into the effects of the deadly pine beetle on the west coast of North America found that the survival of young pine seedlings varied depending on where their mycorrhizal communities came from. If grown with mycorrhizal fungi taken from areas where adult pines had been killed by pine beetles, seedlings had higher rates of mortality. Mycorrhizal communities allowed the effects of pine beetles to cascade through generations of trees (Karst et al., 2015).

37 For 'marriage …' see Howard (1945), ch. 2; for 'living fungous threads' see Howard (1945), ch. 1; for 'can mankind regulate … ?' see Howard (1940), ch. 1.

38 For doubling of crop production see Tilman et al. (2002); for agricultural emissions and plateau of crop yields see Foley et al. (2005) and Godfray et al. (2010); for dysfunction of use of phosphorus fertiliser see Elser and Bennett (2011); for loss of crops see King et al. (2017); for thirty football fields see Arsenault (2014); for projections of global food demand see Tilman et al. (2011).

39 For a study of traditional agricultural practices in China see King (1911); for Howard's concern about the 'life of the soil' see Howard (1940); for damage to soil microbial communities by agriculture see Wagg et al. (2014), de Vries et al. (2013), Toju et al. (2018).

40 For Agroscope study see Banerjee et al. (2019); for impact of ploughing on mycorrhizal communities see Helgason et al. (1998); for comparison of organic and inorganic practices on mycorrhizal communities see Verbruggen et al. (2010), Manoharan et al (2017) and Rillig et al. (2019).

41 For 'ecosystem engineers' see Banerjee et al. (2018); for role of mycorrhizal fungi in soil stability see Leifheit et al. (2014), Mardhiah et al. (2016), Delavaux et al. (2017), Lehmann et al. (2017), Powell and Rillig (2018), Chen et al. (2018);

for impact of mycorrhizal fungi on soil water absorption see Martínez-García et al. (2017); for carbon stored in soil see Swift (2001) and Scharlemann et al. (2014); for analysis of soil carbon bound up in fungi see Clemmensen et al. (2013), Lehmann et al. (2017); for estimates of number of organisms in the soil see Berendsen et al. (2012); and for estimate of number of people that have ever lived see www.prb.org/howmanypeoplehaveeverlivedonearth/ [accessed 29 October 2019].

42 For impact of mycorrhizal fungi on plant resistance to stress see Zabinski and Bunn (2014); Delavaux et al. (2017), Brito et al. (2018), Rillig et al. (2018), Chialva et al. (2018). Other studies have found that by inoculating crops with the endophytic fungi that live in plants' shoots, they can dramatically increase crops' tolerance to drought and heat stress (Redman and Rodriguez 2017).

43 For unpredictable outcomes of mycorrhizal associations on crop yields see Ryan and Graham (2018) (but see Rillig et al., 2019 and Zhang et al., 2019); for Katie Field's studies on crop responses to mycorrhizal fungi see Thirkell et al. (2017); for variability of mycorrhizal response between crop varieties see Thirkell et al. (2019).

44 For discussion of effectiveness of commercial mycorrhizal products see Hart et al. (2018) and Kaminsky et al. (2018). There is a growing number of products which use fungal endophytes of plants to protect crops. In 2019 the US Environmental Protection Agency gave approval to a fungal pesticide designed to be delivered to plants by bees (Fritts 2019).

45 For Kiers' approach see Kiers and Denison (2014).

46 For 'complete scientific explanation' see Howard (1940), ch. 11.

47 Bateson (1987), ch. 4.94; Merleau-Ponty (2002), pt 1, ch. 3, 'The Spatiality of One's Own Body and Motility'.

Chapter 6: Wood Wide Webs

Epigraph: von Humboldt (1845), vol. 1, p. 33. Translation: Anna Westermeier. The sentence containing the phrase 'net-like, entangled fabric' (*Eine allgemeine Verkettung, nicht in einfacher linearer Richtung, sondern in netzartig verschlugenem Gewebe, … stellt sich allmählich dem forschenden Natursinn dar*) does not occur in the published English translation of 1849.

1 The Russian botanist was F. Kamienski, who published his speculation about *Monotropa* in 1882 (Trappe, 2015); for study with radioactive glucose see Björkman (1960).

2 For discussion of Humboldt's 'net-like, entangled fabric' see Wulf (2015), ch. 18, 'Humboldt's Cosmos'.

3 For Read's study with radioactive carbon dioxide see Francis and Read (1984). In 1988, E.I. Newman, the author of a classic review on the subject of shared mycorrhizal networks, commented that 'if this phenomenon is widespread, it could have profound implications for the functioning of ecosystems.' Newman identified five routes by which shared mycorrhizal networks might make an impact:

1) Seedlings may quickly become linked into a large hyphal network and begin to benefit from it at an early stage.

2) One plant may receive organic materials [such as energy-rich carbon compounds] from another via hyphal links, perhaps sufficient to increase the receiver's growth and chance of survival.

3) The balance of competition between plants may be altered if they are obtaining mineral nutrients from a common mycelial network, rather than separately taking them up from the soil.

4) Mineral nutrients may pass from one plant to another, thus perhaps reducing competitive dominance.

5) Nutrients released from dying roots may pass directly via hyphal links to living roots, without ever entering the soil solution.

4   Simard et al. (1997). Simard grew seedlings of three species of tree in a forest in the Canadian province of British Columbia. Two of the species – paper birch and Douglas fir – form relationships with the same type of mycorrhizal fungus. The third species – Western red cedar – forms relationships with a quite unrelated type of mycorrhizal fungus. This meant that she could be fairly sure that the birch and fir shared a network, while the cedar just shared root space with no direct fungal connections (although this approach does not show beyond doubt that the plants remain unconnected – a point for which her study was later criticised). In an important twist on Read's previous studies, Simard exposed pairs of tree seedlings to carbon dioxide labelled with two different radioactive isotopes of carbon. With only a single isotope, it's impossible to follow the *bidirectional* movement of carbon between plants. One might well find that a receiver plant has taken up labelled carbon from a donor plant. But the donor plant might have taken up just as much carbon from the receiver and one would have no way of knowing. Simard's approach allowed her to calculate the net movement between plants.

5   Read (1997).

6   For root grafts see Bader and Leuzinger (2019); for 'we should place ...' see Read (1997). Root grafts have received comparatively little attention in the last few decades, yet they account for a number of interesting phenomena, such as 'living stumps', which continue to survive long after they have been cut. Root grafts can occur between roots of a single individual, individuals of the same species, and even individuals of different species.

7   Barabási (2001).

8   For study of the World Wide Web see Barabási and Albert (1999); for general discussion of developments in network science in the mid 1990s see Barabási (2014); for 'more in common ...' see Barabási (2001); for 'cosmic web' and network structure of the universe see accessible summary by Ferreira (2019), also Gott (2016), ch. 9, Govoni et al. (2019), and Umehata et al. (2019) with commentary from Hamden (2019).

9   For summary of the studies that have found biologically meaningful transfer of resources between plants see Simard et al. (2015). For '280 kilograms' see Klein et al. (2016) and commentary by van der Heijden (2016). The study by Klein et al. (2016) was unusual in measuring the transfer of carbon between mature trees in a forest. The trees were of similar age, meaning there were no obvious source–sink gradients between them.

10 For studies that report little or variable benefit see van der Heijden et al. (2009), and Booth (2004). On the whole, experiments that have found clear benefits to plants have looked at species that form relationships with a group known

as ectomycorrhizal fungi. Studies that have found more ambiguous effects have examined one of the oldest groups, the arbuscular mycorrhizal fungi.

11 For discussion of the variety of opinions within the research community and the differences in interpretation of the evidence see Hoeksema (2015). Part of the problem is that experimenting on shared mycorrhizal networks is complicated in controlled lab conditions, let alone in wild soils. To start with, it's very difficult to show that two plants are connected by the very same fungus. Living systems are leaky. There are countless ways in which a radioactive label applied to one plant could end up in another. What's more, any experiment on networks must compare networked with non-networked plants. The problem is that networks are the default mode. Some researchers sever the fungal ties between plants by moving the position of fine mesh barriers between them. Others dig trenches to separate plants, but it's hard to know whether these interventions are causing collateral damage.

12 For multiple origins of mycoheterotrophy see Merckx (2013). Charles Darwin was a great orchid enthusiast, and spent time puzzling over how orchids could survive with such small seeds. In 1863, in a letter to Joseph Hooker, the director of Kew Gardens, Darwin wrote that although he had 'not a fact to go on', he had a 'firm conviction' that germinating orchid seeds 'are parasites in early youth on cryptogams [or fungi]'. It was not until three decades later that fungi were shown to be crucial for the germination of orchid seeds (Beerling, 2019, p. 141).

13 For the 'snow plant' see Muir (1912), ch. 8; for the 'thousand invisible cords' see Wulf (2015), ch. 23. This was a recurring theme for Muir, who also wrote of 'innumerable unbreakable cords', besides his more well-known line: 'When we try to pick out anything by itself, we find it hitched to everything else in the universe.'

14 For a discussion of *Allotropa* and matsutake see Tsing (2015), 'Interlude. Dancing'.

15 Source–sink dynamics regulate plant photosynthesis. When the products of photosynthesis accumulate, the rate of photosynthesis is reduced. Mycorrhizal fungal networks increase the rate of plant photosynthesis by acting as a carbon sink, thereby preventing the build-up of the products of photosynthesis which would normally slow the process (Gavito et al., 2019).

16 For Simard shading fir seedlings see Simard et al. (1997); for dying plants see Eason et al. (1991).

17 For switching of direction of carbon flow see Simard et al. (2015).

18 For discussion of the evolutionary puzzle see Wilkinson (1998) and Gorzelak et al. (2015).

19 For sharing of surplus resources as a 'public good' see Walder and van der Heijden (2015). Another possibility is that the receiver plants harbour a multitude of different fungal species. Plant A might benefit from plant B's community of fungi when conditions change. Diverse fungal communities offer insurance against environmental uncertainty (Moeller and Neubert, 2016).

20 For kin selection mediated by shared mycorrhizal connections see Gorzelak et al. (2015), Pickles et al. (2017), Simard (2018). A number of species of fern have employed a form of kin selection, or parental 'care', using shared mycorrhizal networks, and have probably done so for millions of years (Beerling 2019, pp. 138–40). These fern species (in the genera *Lycopodium, Huperzia, Psilotum, Botrychium* and *Ophioglossum*) have two phases of their life cycle. Spores germinate into a structure called a 'gametophyte'. Gametophytes are small underground structures that don't photosynthesise. They are where fertilisation takes place. Once a gametophyte has been fertilised, it develops into the above-ground 'adult' phase called a 'sporophyte'. The sporophyte is where photosynthesis takes place.

Gametophytes are only able to survive underground because they are supplied with carbon via mycorrhizal networks, shared with the adult sporophytes. It is a case of 'take now, pay later'.

21 For bidirectional transport see Lindahl et al. (2001) and Schmieder et al. (2019).

22 For studies showing benefits of plant participation in shared mycorrhizal networks see Booth (2004), McGuire (2007), Bingham and Simard (2011), and Simard et al. (2015).

23 For study showing no benefit of participation in shared mycorrhizal networks see Booth (2004); for amplification of competition by shared mycorrhizal networks see Weremijewicz et al. (2016) and Jakobsen and Hammer (2015).

24 For 'fungal fast lane' and fungal transport of poisons see Barto et al. (2011, 2012), and Achatz and Rillig (2014).

25 For hormones see Pozo et al. (2015); for nuclear transport through mycorrhizal fungal networks see Giovannetti et al. (2004, 2006); for transport of RNA between a parasitic plant and its host see Kim et al. (2014); for RNA-mediated interaction between plants and fungal pathogens see Cai et al. (2018).

26 For bacterial use of fungal networks see Otto et al. (2017), Berthold et al. (2016), Zhang et al. (2018); for influence of 'endohyphal' bacteria on fungal metabolism see Vannini et al. (2016), Bonfante and Desirò (2017), Deveau et al. (2018); for bacterial farming in thick-footed morel see Pion et al. (2013) and Lohberger et al. (2019).

27 Babikova et al. (2013).

28 Babikova et al. (2013).

29 For plant–plant information transfer between tomato plants see Song and Zeng (2010); for stress-signalling between Douglas fir and pine seedlings see Song et al. (2015a); for transfer between Douglas fir and pine seedlings see Song et al. (2015b).

30 For electrical signalling in plants see Mousavi et al. (2013), Toyota et al. (2018), and commentary by Muday and Brown-Harding (2018); for plant electrical response to herbivory see Salvador-Recatalà et al. (2014). Many questions remain about the chemical conversations that take place between plant roots and fungi that allow them to form their relationships in the first place. David Read once tried to grow the mycohetertrophic snow plant – John Muir's 'glowing pillar of fire' – and made some progress before hitting 'a brick wall'. 'It was fascinating,' Read recalled:

> the fungus grew towards the seed and showed huge excitement and interest – it fluffed up and 'said hi'. There's clearly signalling going on. The sadness is that we never had plants big enough to let it go further. These signalling questions are questions that the next generation of researchers will have to work on.

31 David Read is of a similar opinion. As he explained to me: 'Somebody from a radio programme wanted to interview me a couple of weeks ago about plants talking to each other and all that kind of nonsense.'

32 Beiler et al. (2009, 2015). Other studies have looked at the architecture of shared mycorrhizal networks based on which species interact, but these have not been explicit about the spatial arrangement of trees within an ecosystem. These include Southworth et al. (2005), Toju et al. (2014, 2016), Toju and Sato (2018).

33 If one drew lines between the trees in Beiler's forest plot randomly, each tree would end up with a similar number of links between them. Trees with an exceptionally high or exceptionally low number of links would be rare. One could calculate an average number of links per tree, and most trees would fall somewhere around

this number. In network language this characteristic node would represent the network's 'scale'. In reality we see something different. In Beiler's plots, Barabási's map of the Web, or a network of aeroplane routes, a few highly connected hubs account for the vast majority of connections in the network. The nodes in this type of network differ from one another so greatly that there is no such thing as a characteristic node. The networks have no 'scale', and are described as 'scale-free'. Barabási's discovery of scale-free networks in the late 1990s helped to provide a framework to model the behaviour of complex systems. For difference between well-connected and poorly connected hubs see Barabási (2014), ch. 'The Sixth Link: The 80/20 Rule'; for vulnerability of scale-free networks see Albert et al. (2000) and Barabási (2001); for a discussion of scale-free networks in the natural world see Bascompte (2009).

34 For discussion of different types of shared mycorrhizal networks and their contrasting architectures see Simard et al. (2012); for discussion of fusion between different arbuscular mycorrhizal networks see Giovannetti et al. (2015). Just because two trees are linked, it doesn't mean that they are linked in the same way. Some types of alder tree, for instance, associate with a very low number of fungal species, which in turn tend not to associate with plants other than alder. This means that alders have an isolationist tendency, and form closed, inward-facing networks with one another. In terms of the overall architecture of a patch of forest, an alder grove would be a 'module' – well connected inside, but only sparsely *inter*-linked (Kennedy et al. 2015). We are used to this idea. Plot a network of your acquaintances on a piece of paper. Then consider that each link is a relationship. How many of your relationships are equivalent? What do you forfeit when you count your relationship with your sister, your third cousin, your friend from work and your landlord as equivalent links in your social network? The network scientists Nicholas Christakis and James Fowler describe how influential a given link in a social network is in terms of its contagion. You may have a social link between your sister and your landlord, but the amount of influence, the 'contagion', each of these links carries will differ. Christakis and Fowler have a theory known as 'three degrees of influence' to describe how social influence drops off after three degrees of separation (Christakis and Fowler, 2009, ch. 1).

35 Prigogine and Stengers (1984), ch. 1.

36 For ecosystems as complex adaptive systems see Levin (2005); for the dynamic non-linear behaviour of ecosystems see Hastings et al. (2018).

37 For Simard's parallels between shared mycorrhizal networks and neural networks see Simard (2018). Researchers in other fields share this view. Manicka and Levin (2019) argue that tools so far only used to study brain function should be transferred to other biological arenas to overcome the problem of 'thematic silos' that segregate fields of biological enquiry. In neuroscience, a 'connectome' is a map of neural connections within a brain. Would it be possible to plot the mycorrhizal connectome of an ecosystem? 'If I had unlimited funding,' Beiler told me, 'I would sample the hell out of a forest. Then you could get a very precise view of the network – who exactly is associating with whom and *where* – and also a broad view of the system as a whole.' For an example of a study in neuroscience that takes an analogous approach see Markram et al. (2015).

38 Simard (2018).

39 'Many fungi interact with roots in a loose way,' Marc-André Selosse explained to me. 'Take truffles, for example. Of course, you can find truffle mycelium growing on the roots of its official "host" trees. But you can also find it in the roots of surrounding plants which aren't its normal hosts and don't usually form

mycorrhizal associations at all. These casual relationships aren't strictly mycor-rhizal, but they nonetheless exist.' For more on non-mycorrhizal fungi that link different plants see Toju and Sato (2018).

Chapter 7: *Radical Mycology*

Epigraph: Le Guin (2017).

1   Many of these early plants – classed as lycophytes and pteridophytes – produced comparatively little 'true' wood, and are thought to have been made up mostly of a bark-like material known as 'periderm' (Nelsen et al., 2016).

2   For three trillion trees see Crowther et al. (2015). The current best estimates of global biomass distributions put plants at around 80 per cent of the total biomass on earth. Around 70 per cent of this plant fraction is estimated to be 'woody' stem and trunk, making wood around 60 per cent of global biomass (Bar-On et al., 2018).

3   For composition of wood and relative abundances of lignin and cellulose see Moore (2013a), ch. 1.

4   For an introduction to wood decomposition and enzymatic combustion see Moore et al. (2011), ch. 10.7, and Watkinson et al. (2015), ch. 5.

5   For 85 gigatonnes see Hawksworth (2009); for 2018 global carbon budget see Quéré et al. (2018). The other major group of decomposing fungi are 'brown rot' fungi, so called because they cause wood to turn a brown colour. Brown rot fungi largely digest the cellulose component of wood. But they are able, too, to use radical chemistry to accelerate the breakdown of lignin. Their approach is slightly different to white rotters. Rather than use free radicals to break apart lignin molecules, they produce radicals that react with lignin and make it vulner-able to bacterial decay (Tornberg and Olsson, 2002).

6   How so much wood could go un-rotted for such a long time has been the subject of considerable discussion. In a paper published in *Science* in 2012, a team headed by David Hibbett argued that the evolution of lignin peroxidases in the white rot fungi approximately coincided with the 'sharp decrease' in carbon burial at the end of the Carboniferous period, suggesting that the Carboniferous deposits may have arisen because fungi hadn't yet evolved the ability to degrade lignin (Floudas et al., 2012, with commentary by Hittinger, 2012). This finding supported the hypothesis first proposed by Jennifer Robinson (1990). In 2016, Matthew Nelsen et al. published a paper refuting this hypothesis, on several grounds:

> 1) Many of the plants that formed the Carboniferous deposits responsible for large amounts of carbon burial were not major lignin producers.
> 2) Lignin-degrading fungi and bacteria may have been present before the Carboniferous period.
> 3) Significant coal seams have formed after the point at which white rot fungi are estimated to have evolved lignin-degrading enzymes.
> 4) If there had been no degradation of lignin before the Carboniferous period, all the carbon dioxide in the atmosphere would have been removed in less than a million years (Nelsen et al., 2016, with commentary by Montañez 2016).

The case isn't clear cut. The relative rates of decomposition versus carbon burial are difficult to measure, and it is hard to imagine that the ability of white rot fungi to degrade lignin and other tough components of wood, such as crystalline cellulose, would have no impact on the global levels of carbon burial (Hibbett et al., 2016).

7 For fungal degradation of coal see Singh (2006), pp. 14–15; the 'kerosene fungus' is a yeast, *Candida keroseneae* (Buddie et al., 2011).

8 Hawksworth (2009). See also Rambold et al. (2013), who argue that 'mycology should be recognised as a field in biology at eye level with other major disciplines'.

9 For mycology in ancient China see Yun-Chang (1985); for the state of mycology in modern China and global production of mushrooms see *State of the World's Fungi* (2018); for deaths by mushroom poisoning see Marley (2010).

10 *State of the World's Fungi* (2018); Hawksworth (2009).

11 For a discussion of the recent history of citizen science and the 'zooniverse' – a digital platform that allows people to participate in research projects across a wide number of fields – see Lintott (2019), reviewed by West (2019); for a classic discussion of 'lay experts' with regard to the AIDS crisis see Epstein (1995); for a discussion of modern crowdsourced participation in science see Kelty (2010); for citizen science in ecology see Silvertown (2009); for a discussion of the history of experimental 'thrifty' science as conducted at home see Werrett (2019). The work of Charles Darwin is a notable example. For most of his life, he conducted almost all of his work at home. He bred orchids on the windowsills, apples in the orchard, racing pigeons, and earthworms on the terrace. Much of the evidence Darwin mobilised in support of his theory of evolution came from networks of amateur animal and plant breeders, and he maintained a large volume of correspondence with well-organised networks of hobbyist collectors and backyard enthusiasts. (Boulter, 2010). Today, digital platforms open new possibilities. In late 2018, a low-frequency seismic hum travelled around the world, evading mainstream earthquake detection systems. Its trajectory and identity were pieced together in an impromptu collaboration between academic and 'citizen' seismologists interacting on Twitter (Sample 2018).

12 For a history of the DIY mycology see Steinhardt (2018).

13 McCoy (2016), p. xx.

14 For figures on agricultural waste see Moore et al. (2011), ch. 11.6; for diapers in Mexico City see Espinosa-Valdemar et al. (2011) – when the plastic was left on, the mass loss was still an impressive 70 per cent. For agricultural waste in India see Prasad (2018).

15 For fungal proliferation at the Cretaceous–Tertiary extinction see Vajda and McLoughlin (2004); for matsutake after Hiroshima see Tsing (2015), 'Prologue'. Tsing writes in her notes that the source of this story is difficult to pin down.

16 For a video of *Pleurotus* on Cigarette butts see: https://web.archive.org/web/20200429100059/https://www.youtube.com/watch?v=fCAX9P50SNU

17 For discussion of non-specific fungal enzymes and the potential for breaking down toxins see Harms et al. (2011).

18 In 2015, Stamets was given an award by the Mycological Society of America. In the official announcement, he was described as a 'highly original, self-trained member of the mycological community who has had a huge and sustained impact on the field of Mycology' (fungi.com/blogs/articles/paul-receives-the-gordon-and-tina-wasson-award [accessed 29 October 2019]). In a 2018 interview with Tim Ferris, Stamets explained that he had been given the award for 'bringing more students into mycology than anyone in history'

(tim.blog/2018/10/15/the-tim-ferriss-show-transcripts-paul-stamets/ [accessed 29 October 2019]).

19 For DMMP see Stamets (2011), 'Part II: Mycorestoration'. Note that *Psilocybe azurescens* is not mentioned here – Stamets told me about this in person.

20 For summary of fungal ability to break down toxins see Harms et al. (2011); for broader discussion of mycoremediation see McCoy (2016), ch. 10.

21 For mycelial highways see Harms et al. (2011); for mycofiltration of *E. coli* see Taylor et al. (2015); for Finnish company reclaiming gold with mycelium see: https://web.archive.org/web/20200429095819/https://phys.org/news/2014-04-filter-recover-gold-mobile-scrap.html. A number of studies reported mushrooms enriched in the radioactive heavy metal caesium following the nuclear fallout at Chernobyl (Oolbekkink and Kuyper, 1989, Kammerer et al., 1994, Nikolova et al., 1997).

22 For discussion of additional fungal needs see Harms et al. (2011); for challenges see McCoy (2016), ch. 10.

23 For CoRenewal see corenewal.org [accessed 29 October 2019]; for fungal clean-up after California fires see: newfoodeconomy.org/mycoremediation-radical-mycology-mushroom-natural-disaster-pollution-clean-up/ [accessed 29 October 2019]; for *Pleurotus* booms in the Danish harbour see: www.sailing.org/news/87633.php#.XCkcIc9KiOE [accessed 29 October 2019].

24 For polyurethane-digesting fungus see Khan et al. (2017); for another example of a plastic-digesting fungus see Brunner et al. (2018). The mycologist Tradd Cotter at the organisation Mushroom Mountain runs a crowdsourced initiative to collect strains of fungi from unusual places: newfoodeconomy.org/mycoremediation-radical-mycology-mushroom-natural-disaster-pollution-clean-up/ [accessed 29 October 2019].

25 For Mary Hunt see Bennett and Chung (2001). The 'crowd' need not always be 'non-scientists'. In 2017, a study published by the Earth Microbiome Project in *Nature* attracted attention for its unusual methodology. Researchers put out a call to scientists around the world for well-preserved environmental samples for inclusion in the survey of global microbial diversity (Raes, 2017).

26 Every year Charles Darwin competed with his cousin, a vicar, as to who could grow the largest pears by crossing the latest varieties. It was a contest that became a source of much family entertainment. See Boulter (2010), p. 31.

27 For Wu San Kwung see McCoy (2016), p. 71; for 'Paris' mushrooms see Monaco (2017); for a general history of cultivation in Europe see Ainsworth (1976), ch. 4. There is a modern twist in the story of underground mushroom-growing in Paris. Car ownership in Paris is falling, and several underground car parks have been converted into successful edible mushroom farms; see www.bbc.co.uk/news/av/business-49928362/turning-paris-s-underground-car-parks-into-mushrooms-farms [accessed 29 October 2019].

28 The preparation of mushrooms is certainly not limited to humans. Several species of North American squirrel are known to dry mushrooms and cache them for later (O'Regan et al., 2016).

29 For age of *Macrotermes* mounds see Erens et al. (2015); for complexity of *Macrotermes* societies see Aanen et al. (2002).

30 For discussion of *Macrotermes* digestion and prolific metabolisms see Aanen et al. (2002), Poulsen et al. (2014) and Yong (2014).

31 For termites eating 'private property' see Margonelli (2018), ch. 1; for termites eating banknotes see www.bbc.co.uk/news/world-south-asia-13194864 [accessed 29 October 2019]; for discussion of Stamets' insect-killing fungal products see Stamets (2011), ch. 8: 'Mycopesticides'. A study published in *Science* in 2019

reported that a genetically modified strain of *Metarhizium* eliminated nearly all of the mosquitoes in an experimental 'near-natural environment' in Burkina Faso. The authors propose the use of the modified strain of *Metarhizium* to fight the spread of malaria (Lovett et al., 2019).

32 For 'waking up' the soil see Fairhead and Scoones (2005); for benefits of termite earths see Fairhead (2016); for destruction of the French garrison see Fairhead and Leach (2003).

33 For spiritual hierarchies see Fairhead (2016). In parts of Guinea, people plaster the walls of houses with earth harvested from the inside of *Macrotermes* mounds (Fairhead 2016).

34 For discussion of materials made from fungi see Haneef et al. (2017), Jones et al. (2019); for portabello batteries see Campbell et al. (2015); for fungal skin substitutes see Suarato et al. (2018).

35 For termite-resistant mycomaterials see: phys.org/news/2018–06-scientists-material-fungus-rice-glass.html [accessed 29 October 2019]. Mycelial building materials have been used in a number of high-profile exhibits, including the 2014 PS1 gallery pavilion at the Museum of Modern Art in New York, and the Shell Mycelium Installation in Kochi in India.

36 For NASA growing structures in space see: www.nasa.gov/directorates/spacetech/niac/2018_Phase_I_Phase_II/Myco-architecture_off_planet/[accessed 29 October 2019]; for 'self-healing' concrete using fungi see Luo et al. (2018).

37 To make the wood-mycelium composite, sawdust and corn are mixed into a damp slurry. The mixture is inoculated with fungal mycelium and loaded into plastic moulds. The mycelium 'runs' through the substrate, forming a cast made of an interlocked mass of mycelium and partially digested wood. It is a different story for the leather and soft foam. Rather than pack the inoculated substrate into moulds, it is spread on flat sheets. By controlling the growth conditions, the mycelium is persuaded to grow upwards into the air. In less than a week, the spongy layer can be harvested. When compressed and tanned, it produces a material that feels remarkably like leather. If dried as it is, it forms a foam.

38 Eben Bayer's longer-term goal is to understand the biophysics of how mycelium creates physical structures. 'I think about fungi as nanotech assemblers that put molecules in place,' he explained. 'We're trying to understand how the 3D orientation of the microfibres influences the properties of the materials; their strength, durability and flexibility.' Bayer's vision is to develop genetically programmable fungi. With this level of control, he explained, 'we'll be able to dial in a different material. You could even have it excrete a plasticising compound like glycerin. Then you'd have something that's naturally more flexible and water-resistant. There's so much you could do.' 'Could' is the operative word. Fungal genetics are byzantine and poorly understood. To insert a gene and have the fungus express it is one thing. To insert a gene and have the fungus express it in a stable and predictable way is another. To programme fungal behaviour by issuing a stream of genetic commands is yet another.

39 There is no precedent for building with fungi, so a lot of the research has to be done from scratch. This is a bigger focus for Bayer than straightforward production. Over the last ten years, they have invested $30 million in research. To work with mycelium in these ways requires new methods, new ways to persuade the fungus to grow, to behave differently.

40 For FUNGAR see: info.uwe.ac.uk/news/uwenews/news.aspx?id=3970 [accessed 29 October 2019] and www.theregister.co.uk/2019/09/17/like_computers_love_fungus/ [accessed 29 October 2019].

41 Stamets et al. (2018).

42 For importance of pollinators and pollinator decline see Klein et al. (2007) and Potts et al. (2010); for problems caused by varroa mites see Stamets et al. (2018).

43 For review of fungal antiviral compounds see Linnakoski et al. (2018); for discussion of Project BioShield see Stamets (2011), ch. 4. Stamets told me that the fungi found to have the strongest antiviral activity were agarikon (*Lacrifomes officionalis*), chaga (*Inonotus obliquus*), reishi (*Ganoderma* spp.), birch polypore (*Fomitopsis betulina*) and turkey tail (*Trametes versicolor*). The most richly documented histories of fungal cures come from China, where medicinal mushrooms have occupied a central place in the pharmacopoeia for at least 2,000 years. The classic herbal dating from around 200 CE, the *Shennong Ben Cao*, thought to be a compilation of far older orally transmitted traditions, includes several fungi still in medicinal use today, including reishi (*Ganoderma lucidum*) and the umbrella polypore (*Polyporus umbellatus*). Reishi was one of the most venerated, and can be found depicted in countless paintings, carvings and embroidery (Powell 2014).

44 Stamets et al. (2018).

Chapter 8: Making Sense of Fungi

Epigraph: Haraway (2016), ch. 4.

1   For yeasts in the human microbiome see Huffnagle and Noverr (2013).

2   For sequencing of yeast genome see Goffeau et al. (1996); for Nobel Prizes on yeast see *State of the World's Fungi* (2018), ch. 'Useful Fungi'.

3   For discussion of evidence for early brewing practices see Money (2018), ch. 2.

4   Lévi-Strauss (1973), p. 473.

5   For domestication of yeast see Money (2018), ch. 1 and Legras et al. (2007); for bread-before-beer see Wadley and Hayden (2015) and Dunn (2012). The development of agriculture affected a number of human relationships with fungi. Many fungal pathogens of plants are considered to have evolved in parallel with domesticated crops. As is the case today, domestication and cultivation provide fungal pathogens of plants with new opportunities (Dugan, 2008, p. 56).

6   I was inspired by the excellent book *Sacred Herbal and Healing Beers* (Buhner, 1998).

7   For Sumerians and Egyptian *Book of the Dead* see Katz (2003), ch. 2; for Ch'orti' see Aasved (1988), p. 757; for Dionysus see Kerényi (1976) and Paglia (2001), ch. 3.

8   For discussion of yeast in biotechnology see Money (2018), ch. 5; for Sc.2.0 see syntheticyeast.org/sc2-0/introduction/[accessed 29 October 2019].

9   For rhapsodic verse see Yun-Chang (1985); Yamaguchi Sodo quoted in Tsing (2015), 'Prologue'; Albertus Magnus quoted in Letcher (2006), p. 50; John Gerard quoted in Letcher (2006), p. 49.

10 Wasson and Wasson (1957), vol. II, ch. 18. The Wassons divided up much of the world into their categories. The United States (Wasson was American) was mycophobic, along with Anglo-Saxons and Scandinavians. Russia (Valentina was Russian) was mycophilic, along with Slavs and Catalans. 'The Greeks', the Wassons observed disdainfully, 'have always been mycophobes … From beginning to end in the writings of the ancient Greeks we find not one enthusiastic word for mushrooms.' Of course, things are rarely so straightforward. The Wassons confected a binary system and were the first to dissolve its hard edges. They

observed that Finns were 'by tradition mycophobes', but in areas where the Russians used to go on vacation had learned to 'know and love many species'. Where exactly the reformed Finns fell between the two poles of their system the Wassons neglected to say.

11 For reclassification of fungi and bacteria see Sapp (2009), p. 47.

12 For a discussion of the history of fungal taxonomy see Ainsworth (1976), ch. 10.

13 For Theophrastus see Ainsworth (1976), p. 35; for association of fungi with lightning strikes and general discussion of European understanding of fungi see Ainsworth (1976), ch. 2; for 'The order of Fungi' and a good general history of fungal taxonomy see Ramsbottom (1953), ch. 3.

14 Money (2013).

15 Raverat (1952), p. 136.

16 One of the first recorded taxonomic attempts to order the fungi was made in 1601, and divided mushroom species into categories of 'edible' and 'poisonous': that is, the potential relationship they would have with a human body (Ainsworth, 1976, p. 183). These judgements are seldom meaningful. Brewer's yeast can be used to make bread and alcohol, yet can cause a life-threatening infection if it gets into your blood.

17 The word 'mutualism' was explicitly political for the first decades of its life, describing a school of early anarchist thought. The concept of the 'organism', too, was understood in explicitly political terms by German biologists of the late nineteenth century. Rudolf Virchow understood the organism to be made up of a community of co-operating cells, each working for the good of the whole, just as a population of interdependent co-operating citizens underpinned the operation of a healthy nation state (Ball, 2019, ch. 1).

18 For 'close to the margins' see Sapp (2004). The relationship between Darwin's theory of evolution by natural selection, Thomas Malthus' analysis of food supply and human populations and Adam Smith's theory of the market has received considerable scholarly attention. See for example Young (1985).

19 Sapp (1994), ch. 2.

20 Sapp (2004).

21 For Needham see Haraway (2004), p. 106; Lewontin (2000), p. 3.

22 Toby Kiers, professor at Vrije Universiteit Amsterdam, is one of the leading proponents of applying 'biological market frameworks' to plant and fungal interactions. Biological markets are not themselves a new idea – they have been used to think about animal behaviour for decades. But Kiers and her colleagues are the first to apply them to organisms that don't have brains (see for example: Werner et al., 2014, Wyatt et al., 2014, Kiers et al., 2016, Noë and Kiers, 2018). For Kiers, economic metaphors underpin economic models, which are helpful investigative tools. 'It's not about trying to make analogies to human markets,' she told me. Instead, 'it allows us to make more testable predictions'. Rather than sweep the dizzyingly variable world of plant and fungal exchange into vague notions of 'complexity' or 'context-dependency', economic models make it possible to break down dense webs of interactions and test basic hypotheses. Kiers became interested in biological markets after she found that plants and mycorrhizal fungi use 'reciprocal rewards' to regulate their exchange of carbon and phosphorus. Plants that receive more phosphorus from a fungus provide it with more carbon; fungi that receive more carbon provide the plant with more phosphorus (Kiers et al., 2011). In Kiers' view, market models provide a way to understand how these 'strategic trading behaviours' might have evolved, and how they might change in different conditions. 'So far it's been a very useful tool, even in the way that it

allows us to set up different experiments,' she explained. 'We might say, "Theory suggests that as we increase the number of partners the trade strategy is going to change in a certain way depending on those resources." That allows us to set up an experiment: let's try changing the number of partners and see if this strategy actually changes. It's a sounding board rather than a strict protocol.' In this case, the market frameworks are a tool, a set of stories based on human interactions that help to formulate questions about the world, to generate new perspectives. It is not to say, as Kropotkin did, that humans should base their behaviour on the behaviour of non-human organisms. Nor is it to say that plants and fungi are actually capitalist individuals making rational decisions. Of course, even if they were it is unlikely that their behaviour would fit perfectly within a given human economic model. As any economist will admit, human markets don't behave like 'ideal' markets in practice. The messy complexity of human economic life spills out of the models built to house it. And in fact, fungal lives don't fit neatly into biological market theory either. For a start, biological markets depend – like the human capitalist markets from which they derive – on being able to identify individual 'traders' that act in their own interest. The truth of the matter is that it is not clear what counts as an individual 'trader' (Noë and Kiers, 2018). The mycelium of a 'single' mycorrhizal fungus might fuse with another and end up with several different types of nucleus – several different genomes – travelling around its network. What counts as an individual? An individual nucleus? A single interconnected network? One tuft of a network? Kiers is straightforward about these challenges. 'If biological market theory is not a useful way to study interactions between plants and fungi, then we'll stop using it.' Market frameworks are tools whose utility is not known in advance. Nonetheless, biological markets are a problem for some researchers in the field. As Kiers remarked, 'this debate can get emotional with no particular reason for it to get emotional.' Perhaps it is the fact that biological market frameworks touch a socio-political nerve. Human economic systems are many and diverse. Yet the body of theory known as biological market frameworks bears a striking resemblance to free-market capitalism. Would it help to compare the value of economic models drawn from different cultural systems? There are many ways to attribute value. There might be other currencies that haven't been taken into account.

23 The Internet and World Wide Web are more of a self-organising system than many human technologies (in Barabási's words, the World Wide Web appears to have 'more in common with a cell or an ecological system than with a Swiss watch'). Nonetheless, these networks are built from machines and protocols which are not self-organising, and which would cease to function without constant human attention.

24 Jan Sapp told me a story that illustrates how easy it is for biologists' metaphors to become a flash point. He noticed that many portrayed larger, more complex organisms, like animals and plants, as more 'successful' than the bacteria or fungi that they partnered with. Sapp gave this argument short shrift. 'By what definition of success? The last time I looked, the world was primarily microbial. This planet belongs to microbes. Microbes were at the beginning and they will be at the end, long after complex "higher" animals are gone. They created the atmosphere and life as we know it, they make up most of our bodies.' Sapp explained how he had observed the evolutionary biologist John Maynard Smith playing down microbes by changing a metaphor. If a microbe was gaining from a relationship, Maynard Smith called it a 'microbial parasite', and the large organism the 'host'. However, if the large organism was manipulating the microbe, Maynard Smith

didn't call the big organism the parasite. He changed metaphors, and called the big organism the master, and the microbe the slave. Sapp's concern lay in the fact that the microbe was either a parasite or a slave, but for Maynard Smith it could never be understood as a dominant partner manipulating the 'host'. The microbe could never be the one in control.

25 For 'puhpowee' see Kimmerer (2013), ch: 'Learning the Grammar of Animacy' and 'Allegiance to Gratitude'. The Dutch primatologist Frans de Waal, frustrated at people using the charge of 'anthropomorphism' to defend human exceptionalism, complains of 'anthropodenial': 'the *a priori* rejection of shared characteristics between humans and animals when in fact they may exist' (de Waal 1999).

26 Hustak and Myers (2012).

27 Ingold (2003) asks how human thought would look different if fungi, not animals, had been taken as the 'paradigmatic instance of a life form'. He explores the implications of adopting a 'fungal model' of life, arguing that humans are no less embedded in networks: it is just that our 'pathways of relationship' are more difficult to see than those of fungi.

28 For 'Sharing resources …' see Waller et al. (2018).

29 Deleuze and Guattari (2005), p. 11.

30 Carrigan et al. (2015). Alcohol dehydrogenase is different from acetaldehyde dehydrogenase, another enzyme responsible for alcohol metabolism that varies in between human populations and can cause people to have trouble metabolising alcohol.

31 For the 'drunken monkey' hypothesis see Dudley (2014). Fungal infestations have been shown to boost fruit aroma and removal by animals and birds (Peris et al., 2017).

32 Wiens et al. (2008), Money (2018), ch. 2.

33 For consequences of biofuel production in the United States see Money (2018), ch. 5; for land use change and biofuels see Wright and Wimberly (2013); for subsidies and carbon release see Lu et al. (2018).

34 Stukeley (1752).

Epilogue: This Compost

Epigraph: Saint Francis of Assisi (1181/2–1216). In Ladinsky (2002).

Bibliography

Aanen, D. K., Eggleton, P., Rouland-Lefevre, C., Guldberg-Froslev, T., Rosendahl, S., Boomsma, J. J., 'The evolution of fungus-growing termites and their mutualistic fungal symbionts', *Proceedings of the National Academy of Sciences*, 99 (2002), pp. 14887–92.

Aasved, M. J., *Alcohol, drinking and intoxication in preindustrial societies: Theoretical, nutritional, and religious considerations*, PhD thesis, University of California at Santa Barbara (1988).

Abadeh, A., Lew. R. R., 'Mass flow and velocity profiles in *Neurospora* hyphae: partial plug flow dominates intra-hyphal transport', *Microbiology*, 159 (2013), pp. 2386–94.

Achatz, M., Rillig, M. C., 'Arbuscular mycorrhizal fungal hyphae enhance transport of the allelochemical juglone in the field', *Soil Biology and Biochemistry*, 78 (2014), pp. 76–82.

Adachi, K., Chiba, K., 'FTY720 story. Its discovery and the following accelerated development of sphingosine 1-phosphate receptor agonists as immunomodulators based on reverse pharmacology', *Perspectives in Medicinal Chemistry*, 1 (2007), pp. 11–23.

Adamatzky, A., *Advances in Physarum Machines* (Springer International Publishing, 2016).

Adamatzky, A., 'Towards fungal computer', *Journal of the Royal Society Interface Focus*, 8 (2018a), 20180029.

Adamatzky, A., 'On spiking behaviour of oyster fungi *Pleurotus djamor*', *Scientific Reports*, 8 (2018b), 7873.

Adamatzky, A., 'A brief history of liquid computers', *Philosophical Transactions of the Royal Society B*, 374 (2019), 20180372.

Ahmadjian, V., Heikkilä, H., 'The culture and synthesis of *Endocarpon pusillum* and *Staurothele clopima*', *Lichenologist*, 4 (1970), pp. 259–67.

Ahmadjian, V., 'Lichens are more important than you think,' *BioScience*, 45 (1995), p. 123.

Ainsworth, G. C., *Introduction to the History of Mycology* (Cambridge, Cambridge University Press, 1976).

Albert, R., Jeong, H., Barabási, A-L., 'Error and attack tolerance of complex networks', *Nature*, 406 (2000), pp. 378–82.

Alberti, S., 'Don't go with the cytoplasmic flow,' *Developmental Cell*, 34 (2015), pp. 381–2.

Alim, K., Andrew, N., Pringle, A., Brenner, M. P., 'Mechanism of signal propagation in *Physarum polycephalum*', *Proceedings of the National Academy of Sciences*, 114 (2017), pp. 5136–41.

Alim, K., 'Fluid flows shaping organism morphology', *Philosophical Transactions of the Royal Society B*, 373 (2018), 20170112.

Allaway, W., Ashford, A., 'Motile tubular vacuoles in extramatrical mycelium and sheath hyphae of ectomycorrhizal systems', *Protoplasma*, 215 (2001), pp. 218–25.

Allen, J., Arthur, J., 'Ethnomycology and Distribution of Psilocybin Mushrooms', in ed. R. Metzner, *Sacred Mushroom of Visions: Teonanacatl*, Rochester, VT: Park Street Press (2005), pp. 49–68.

Alpert, C., 'Unraveling the Mysteries of the Canadian Whiskey Fungus', *Wired* (2011), www.wired.com/2011/05/ff-angelsshare/ [accessed 29 October 2019].

Alpi, A., Amrhein, N., Bertl, A., Blatt, M. R., Blumwald, E., Cervone, F., Dainty, J., Michelis, M., Epstein, E., Galston, A. W. et al., 'Plant neurobiology: no brain, no gain?' *Trends in Plant Science*, 12 (2007), pp. 135–6.

Aly, A., Debbab, A., Proksch, P., 'Fungal endophytes: unique plant inhabitants with great promises', *Applied Microbiology and Biotechnology*, 90 (2011), pp. 1829–45.

Alzarhani, K. A., Clark, D. R., Underwood, G. J., Ford, H., Cotton, A. T., Dumbrell, A. J., 'Are drivers of root-associated fungal community structure context specific?' *ISME Journal*, 13 (2019), pp. 1330–44.

Andersen, S. B., Gerritsma, S., Yusah, K. M., Mayntz, D., Hywel Jones, N. L., Billen, J., Boomsma, J. J., Hughes, D. P., 'The life of a dead ant: the expression of an adaptive extended phenotype', *American Naturalist*, 174 (2009), pp. 424–33.

Anderson, J. B., Bruhn, J. N., Kasimer, D., Wang, H., Rodrigue, N., Smith, M. L., 'Clonal evolution and genome stability in a 2,500-year-old fungal individual', *Proceedings of the Royal Society B*, 285 (2018), 20182233.

Araldi-Brondolo, S. J., Spraker, J., Shaffer, J. P., Woytenko, E. H., Baltrus, D. A., Gallery, R. E., Arnold, E. A., 'Bacterial endosymbionts: master modulators of fungal phenotypes', *Microbiology Spectrum*, 5 (2017), FUNK-0056-2016.

Arnaud-Haond, S., Duarte, C. M., Diaz-Almela, E., Marbà, N., Sintes, T., Serrão, E. A., 'Implications of extreme life span in clonal organisms: millenary clones in meadows of the threatened seagrass *Posidonia oceanica*', *PLOS ONE*, 7 (2012), e30454.

Arnold, E. A., Mejía, L., Kyllo, D., Rojas, E. I., Maynard, Z., Robbins, N., Herre, E., 'Fungal endophytes limit pathogen damage in a tropical tree', *Proceedings of the National Academy of Sciences*, 100 (2003), pp. 15649–54.

Arnold, E. A., Miadlikowska, J., Higgins, L. K., Sarvate, S. D., Gugger, P., Way, A., Hofstetter, V., Kauff, F., Lutzoni, F., 'A phylogenetic estimation of trophic transition networks for ascomycetous fungi: are lichens cradles of symbiotrophic fungal diversification?', *Systematic Biology*, 58 (2009), pp. 283–97.

Arsenault, C., 'Only 60 Years of Farming Left if Soil Degradation Continues', *Scientific American* (2014), www.scientificamerican.com/article/only-60-years-of-farming-left-if-soil-degradation-continues/ [accessed 29 October 2019].

Aschenbrenner, I. A., Cernava, T., Berg, G., Grube, M., 'Understanding microbial multi-species symbioses', *Frontiers in Microbiology*, 7 (2016), 180.

Asenova, E., Lin, H.-Y., Fu, E., Nicolau, D. V., 'Optimal fungal space searching algorithms', *IEEE Transactions on NanoBioscience*, 15 (2016), pp. 613–18.

Ashford, A. E., Allaway, W. G., 'The role of the motile tubular vacuole system in mycorrhizal fungi', *Plant and Soil*, 244 (2002), pp. 177–87.

Averill, C., Dietze, M. C., Bhatnagar, J. M., 'Continental-scale nitrogen pollution is shifting forest mycorrhizal associations and soil carbon stocks', *Global Change Biology*, 24 (2018), pp. 4544–53.

Awan, A. R., Winter, J. M., Turner, D., Shaw, W. M., Suz, L. M., Bradshaw, A. J., Ellis, T., Dentinger, B., 'Convergent evolution of psilocybin biosynthesis by psychedelic mushrooms', *bioRxiv* (2018), 374199.

Babikova, Z., Gilbert, L., Bruce, T. J., Birkett, M., Caulfield, J. C., Woodcock, C., Pickett, J. A., Johnson, D., 'Underground signals carried through common mycelial networks warn neighbouring plants of aphid attack', *Ecology Letters*, 16 (2013), 835–43.

Bachelot, B., Uriarte, M., McGuire, K. L., Thompson, J., Zimmerman, J., 'Arbuscular mycorrhizal fungal diversity and natural enemies promote coexistence of tropical tree species', *Ecology*, 98 (2017), pp. 712–20.

Bader, M.K.-F., Leuzinger, S., 'Hydraulic coupling of a leafless kauri tree remnant to conspecific hosts', *iScience* 19 (2019), pp. 1238–43.

Bahn, Y.-S., Xue, C., Idnurm, A., Rutherford, J. C., Heitman, J., Cardenas, M. E., 'Sensing the environment: lessons from fungi', *Nature Reviews Microbiology*, 5 (2007), pp. 57–69.

Bain, N., Bartolo, D., 'Dynamic response and hydrodynamics of polarized crowds', *Science*, 363 (2019) pp. 46–9.

Ball, P., *How to Grow a Human* (London, William Collins, 2019).

Banerjee, S., Schlaeppi, K., van der Heijden, M. G., 'Keystone taxa as drivers of microbiome structure and functioning', *Nature Reviews Microbiology*, 16 (2018), pp. 567–76.

Banerjee, S., Walder, F., Büchi, L., Meyer, M., Held, A. Y., Gattinger, A., Keller, T., Charles, R., van der Heijden, M. G., 'Agricultural intensification reduces microbial network complexity and the abundance of keystone taxa in roots', *ISME Journal*, 13 (2019), pp. 1722–36.

Bar-On, Y. M., Phillips, R., Milo, R., 'The biomass distribution on Earth', *Proceedings of the National Academy of Sciences*, 115 (2018), pp. 6506–11.

Barabási, A.-L., Albert, R., 'Emergence of scaling in random networks', *Science*, 286 (1999), pp. 509–12.

Barabási, A. L., 'The physics of the Web', *Physics World*, 14 (2001), pp. 33–8, physicsworld.com/a/the-physics-of-the-web/ [accessed 29 October 2019].

Barabási, A.-L., *Linked: How Everything is Connected to Everything Else and What It Means for Business, Science, and Everyday Life* (New York, Basic Books, 2014).

Barbey, A. K., 'Network neuroscience theory of human intelligence', *Trends in Cognitive Sciences*, 22 (2018), pp. 8–20.

Barto, K. E., Hilker, M., Müller, F., Mohney, B. K., Weidenhamer, J. D., Rillig, M. C., 'The fungal fast lane: common mycorrhizal networks extend bioactive zones of allelochemicals in soils', *PLOS ONE*, 6 (2011), e27195.

Barto, K. E., Weidenhamer, J. D., Cipollini, D., Rillig, M. C., 'Fungal superhighways: do common mycorrhizal networks enhance below ground communication?' *Trends in Plant Science*, 17 (2012), pp. 633–7.

Bascompte, J., 'Mutualistic networks', *Frontiers in Ecology and the Environment*, 7 (2009), pp. 429–36.

Baslam, M., Garmendia, I., Goicoechea, N., 'Arbuscular mycorrhizal fungi (AMF) improved growth and nutritional quality of greenhouse-grown lettuce', *Journal of Agricultural and Food Chemistry*, 59 (2011), pp. 5504–15.

Bass, D., Howe, A., Brown, N., Barton, H., Demidova, M., Michelle, H., Li, L., Sanders, H., Watkinson, S. C., Willcock, S. et al., 'Yeast forms dominate fungal diversity in the deep oceans', *Proceedings of the Royal Society B*, 274 (2007), pp. 3069–77.

Bassett, D. S., Sporns, O., 'Network neuroscience', *Nature Neuroscience*, 20 (2017), pp. 353–64.

Bassett, E., Keith, M. S., Armelagos, G., Martin, D., Villanueva, A., 'Tetracycline-labeled human bone from ancient Sudanese Nubia (AD 350)', *Science*, 209 (1980), pp. 1532–4.

Bateson, B., *William Bateson, Naturalist* (Cambridge, Cambridge University Press, 1928).

Bateson, G., *Steps to an Ecology of Mind* (Northvale, NJ, Jason Aronson Inc., 1987).

Bebber, D. P., Hynes, J., Darrah, P. R., Boddy, L., Fricker, M. D., 'Biological solutions to transport network design', *Proceedings of the Royal Society B*, 274 (2007), pp. 2307–15.

Beck, A., Divakar, P., Zhang, N., Molina, M., Struwe, L., 'Evidence of ancient horizontal gene transfer between fungi and the terrestrial alga *Trebouxia*', *Organisms Diversity & Evolution*, 15 (2015), pp. 235–48.

Beerling, D., *Making Eden* (Oxford, Oxford University Press, 2019).

Beiler, K. J., Durall, D. M., Simard, S. W., Maxwell, S. A., Kretzer, A. M., 'Architecture of the wood-wide web: *Rhizopogon* spp. genets link multiple Douglas-fir cohorts', *New Phytologist*, 185 (2009), pp. 543–53.

Beiler, K. J., Simard, S. W., Durall, D. M., 'Topology of tree-mycorrhizal fungus interaction networks in xeric and mesic Douglas-fir forests', *Journal of Ecology*, 103 (2015), pp. 616–28.

Bengtson, S., Rasmussen, B., Ivarsson, M., Muhling, J., Broman, C., Marone, F., Stampanoni, M., Bekker, A., 'Fungus-like mycelial fossils in 2.4-billion-year-old vesicular basalt', *Nature Ecology & Evolution*, 1 (2017), 0141.

Bennett, J. A., Cahill, J. F., 'Fungal effects on plant–plant interactions contribute to grassland plant abundances: evidence from the field', *Journal of Ecology*, 104 (2016), pp. 755–64.

Bennett, J. A., Maherali, H., Reinhart, K. O., Lekberg, Y., Hart, M. M., Klironomos, J., 'Plant – soil feedbacks and mycorrhizal type influence temperate forest population dynamics', *Science*, 355 (2017), pp. 181–4.

Bennett, J. W., Chung, K. T., 'Alexander Fleming and the discovery of penicillin', *Advances in Applied Microbiology*, 49 (2001), pp. 163–84.

Berendsen, R. L., Pieterse, C. M. J., Bakker, P. A., 'The rhizosphere microbiome and plant health', *Trends in Plant Science*, 17 (2012), pp. 478–86.

Bergson, H., *Creative Evolution* (New York, Henry Holt and Company, 1911).

Berthold, T., Centler, F., Hübschmann, T., Remer, R., Thullner, M., Harms, H., Wick, L. Y., 'Mycelia as a focal point for horizontal gene transfer among soil bacteria', *Scientific Reports*, 6 (2016), 36390.

Bever, J. D., Richardson, S. C., Lawrence, B. M., Holmes, J., Watson, M., 'Preferential allocation to beneficial symbiont with spatial structure maintains mycorrhizal mutualism', *Ecology Letters*, 12 (2009), pp. 13–21.

Bingham, M. A., Simard, S. W., 'Mycorrhizal networks affect ectomycorrhizal fungal community similarity between conspecific trees and seedlings', *Mycorrhiza*, 22 (2011), pp. 317–26.

Björkman, E., '*Monotropa hypopitys* L. – an epiparasite on tree roots', *Physiologia Plantarum*, 13 (1960), pp. 308–27.

Boddy, L., Hynes, J., Bebber, D. P., Fricker, M. D., 'Saprotrophic cord systems: dispersal mechanisms in space and time', *Mycoscience*, 50 (2009), pp. 9–19.

Bonfante, P., 'The future has roots in the past: the ideas and scientists that shaped mycorrhizal research', *New Phytologist*, 220 (2018), pp. 982–95.

Bonfante, P., Desirò, A., 'Who lives in a fungus? The diversity, origins and functions of fungal endobacteria living in Mucoromycota', *ISME Journal*, 11 (2017), pp. 1727–35.

Bonfante, P., Selosse, M.-A., 'A glimpse into the past of land plants and of their mycorrhizal affairs: from fossils to evo-devo', *New Phytologist*, 186 (2010), pp. 267–70.

Bonifaci, V., Mehlhorn, K., Varma, G., '*Physarum* can compute shortest paths', *Journal of Theoretical Biology*, 309 (2012), pp. 121–33.

Booth, M. G., 'Mycorrhizal networks mediate overstorey–understorey competition in a temperate forest', *Ecology Letters*, 7 (2004), pp. 538–46.

Bordenstein, S. R., Theis, K. R., 'Host biology in light of the microbiome: ten principles of holobionts and hologenomes', *PLOS Biology*, 13 (2015), e1002226.

Bouchard, F., 'Symbiosis, Transient Biological Individuality, and Evolutionary Process', in eds. J. Dupré and J. Nicholson, *Everything Flows: Towards a Processual Philosophy of Biology* (Oxford, Oxford University Press, 2018), pp. 186–98.

Boulter, M., *Darwin's Garden: Down House and the Origin of Species* (London, Counterpoint, 2010).

Boyce, G. R., Gluck-Thaler, E., Slot, J. C., Stajich, J. E., Davis, W. J., James, T. Y., Cooley, J. R., Panaccione, D. G., Eilenberg, J., Licht, H. H. et al., 'Psychoactive plant- and mushroom-associated alkaloids from two behaviour-modifying cicada pathogens', *Fungal Ecology*, 41 (2019), pp. 147–64.

Brand, A., Gow, N. A., 'Mechanisms of hypha orientation of fungi', *Current Opinion in Microbiology*, 12 (2009), pp. 350–7.

Brandt, A., de Vera, J. P., Onofri, S., Ott, S., 'Viability of the lichen *Xanthoria elegans* and its symbionts after 18 months of space exposure and simulated Mars conditions on the ISS', *International Journal of Astrobiology*, 14 (2014), pp. 411–25.

Brandt, A., Meeßen, J., Jänicke, R. U., Raguse, M., Ott, S., 'Simulated space radiation: impact of four different types of high-dose ionizing radiation on the lichen *Xanthoria elegans*', *Astrobiology*, 17 (2017), pp. 136–44.

Bringhurst, R., *Everywhere Being Is Dancing* (Berkeley, CA, Counterpoint, 2009).

Brito, I., Goss, M. J., Alho, L., Brígido, C., van Tuinen, D., Félix, M. R., Carvalho, M., 'Agronomic management of AMF functional diversity to overcome biotic and abiotic stresses – the role of plant sequence and intact extraradical mycelium', *Fungal Ecology*, 40 (2018), pp. 72–81.

Bruce-Keller, A. J., Salbaum, M. J., Berthoud, H.-R., 'Harnessing gut microbes for mental health: getting from here to there', *Biological Psychiatry*, 83 (2018), pp. 214–23.

Bruggeman, F. J., van Heeswijk, W. C., Boogerd, F. C., Westerhoff, H. V., 'Macromolecular intelligence in micro-organisms', *Biological Chemistry*, 381 (2000), pp. 965–72.

Brundrett, M. C., 'Co-evolution of roots and mycorrhizas of land plants', *New Phytologist*, 154 (2002), pp. 275–304.

Brundrett, M. C., Tedersoo, L., 'Evolutionary history of mycorrhizal symbioses and global host plant diversity', *New Phytologist*, 220 (2018), pp. 1108–15.

Brunet, T., Arendt, D., 'From damage response to action potentials: early evolution of neural and contractile modules in stem eukaryotes', *Philosophical Transactions of the Royal Society B*, 371 (2015), 20150043.

Brunner, I., Fischer, M., Rüthi, J., Stierli, B., Frey, B., 'Ability of fungi isolated from plastic debris floating in the shoreline of a lake to degrade plastics', *PLOS ONE*, 13 (2018), e0202047.

Bublitz, D. C., Chadwick, G. L., Magyar, J. S., Sandoz, K. M., Brooks, D. M., Mesnage, S., Ladinsky, M. S., Garber, A. I., Bjorkman, P. J., Orphan, V. J. et al., 'Peptidoglycan production by an insect-bacterial mosaic', *Cell*, 179 (2019), pp. 1–10.

Buddie, A. G., Bridge, P. D., Kelley, J., Ryan, M. J., '*Candida keroseneae* sp. nov., a novel contaminant of aviation kerosene', *Letters in Applied Microbiology*, 52 (2011), pp. 70–5.

Büdel, B., Vivas, M., Lange, O. L., 'Lichen species dominance and the resulting photosynthetic behavior of Sonoran Desert soil crust types (Baja California, Mexico)', *Ecological Processes*, 2 (2013), p. 6.

Buhner, S. H., *Sacred Herbal and Healing Beers* (Boulder, CO, Siris Books, 1998).

Buller, A. H. R., *Researches on Fungi*, vol. 4 (London, Longmans, Green, and Co., 1931).

Büntgen, U., Egli, S., Schneider, L., von Arx, G., Rigling, A., Camarero, J. J., Sangüesa-Barreda, G., Fischer, C. R., Oliach, D., Bonet, J. A. et al., 'Long-term irrigation effects on Spanish holm oak growth and its black truffle symbiont', *Agriculture, Ecosystems & Environment*, 202 (2015), pp. 148–59.

Burford, E. P., Kierans, M., Gadd, G. M., 'Geomycology: fungi in mineral substrata', *Mycologist*, 17 (2003), pp. 98–107.

Burkett, W., *Ancient Mystery Cults* (Cambridge, MA, Harvard University Press, 1987).

Burr, C., *The Emperor of Scent* (New York, Random House, 2012).

Bushdid, C., Magnasco, M., Vosshall, L., Keller, A., 'Humans can discriminate more than 1 trillion olfactory stimuli', *Science*, 343 (2014), pp. 1370–2.

Cai, Q., Qiao, L., Wang, M., He, B., Lin, F.-M., Palmquist, J., Huang, S.-D., Jin, H., 'Plants send small RNAs in extracellular vesicles to fungal pathogen to silence virulence genes', *Science*, 360 (2018), pp. 1126–9.

Calvo Garzón, P., Keijzer, F., 'Plants: adaptive behavior, root-brains, and minimal cognition', *Adaptive Behavior*, 19 (2011), pp. 155–71.

Campbell, B., Ionescu, R., Favors, Z., Ozkan, C. S., Ozkan, M., 'Bio-derived, binderless, hierarchically porous carbon anodes for Li-ion batteries', *Scientific Reports*, 5 (2015), 14575.

Caporael, L., 'Ergotism: the Satan loosed in Salem?', *Science*, 192 (1976), pp. 21–6.

Carhart-Harris, R. L., Bolstridge, M., Rucker, J., Day, C. M., Erritzoe, D., Kaelen, M., Bloomfield, M., Rickard, J. A., Forbes, B., Feilding, A. et al., 'Psilocybin with psychological support for treatment-resistant depression: an open-label feasibility study', *Lancet Psychiatry*, 3 (2016a), 619–27.

Carhart-Harris, R. L., Erritzoe, D., Williams, T., Stone, J., Reed, L. J., Colasanti, A., Tyacke, R. J., Leech, R., Malizia, A. L., Murphy, K. et al., 'Neural correlates of the psychedelic state as determined by fMRI studies with psilocybin', *Proceedings of the National Academy of Sciences*, 109 (2012), pp. 2138–43.

Carhart-Harris, R. L., Muthukumaraswamy, S., Roseman, L., Kaelen, M., Droog, W., Murphy, K., Tagliazucchi, E., Schenberg, E. E., Nest, T., Orban, C. et al., 'Neural correlates of the LSD experience revealed by multimodal neuroimaging', *Proceedings of the National Academy of Sciences*, 113 (2016b), pp. 4853–8.

Carrigan, M. A., Uryasev, O., Frye, C. B., Eckman, B. L., Myers, C. R., Hurley, T. D., Benner, S. A., 'Hominids adapted to metabolize ethanol long before human-directed fermentation', *Proceedings of the National Academy of Sciences*, 112 (2015), pp. 458–63.

Casadevall, A., 'Fungi and the rise of mammals', *Pathogens*, 8 (2012), e1002808.

Casadevall, A., Cordero, R. J., Bryan, R., Nosanchuk, J., Dadachova, E., 'Melanin, radiation, and energy transduction in fungi', *Microbiology Spectrum*, 5 (2017), FUNK-0037-2016.

Casadevall, A., Kontoyiannis, D. P., Robert, V., 'On the emergence of *Candida auris*: climate change, azoles, swamps and birds', *mBio*, 10 (2019), e01397-19.

Ceccarelli, N., Curadi, M., Martelloni, L., Sbrana, C., Picciarelli, P., Giovannetti, M., 'Mycorrhizal colonization impacts on phenolic content and antioxidant properties of artichoke leaves and flower heads two years after field transplant', *Plant and Soil*, 335 (2010), pp. 311-23.

Cepelewicz, J., 'Bacterial Complexity Revises Ideas about "Which Came First?"', *Quanta* (2019), www.quantamagazine.org/bacterial-organelles-revise-ideas-about-which-came-first-20190612/ [accessed 29 October 2019].

Cerdá-Olmedo, E., '*Phycomyces* and the biology of light and color', *FEMS Microbiology Reviews*, 25 (2001), pp. 503-12.

Cernava, T., Aschenbrenner, I., Soh, J., Sensen, C. W., Grube, M., Berg, G., 'Plasticity of a holobiont: desiccation induces fasting-like metabolism within the lichen microbiota', *SME Journal*, 13 (2019), pp. 547-56.

Chen, J., Blume, H., Beyer, L., 'Weathering of rocks induced by lichen colonization – a review', *Catena*, 39 (2000), pp. 121-46.

Chen, L., Swenson, N. G., Ji, N., Mi, X., Ren, H., Guo, L., Ma, K., 'Differential soil fungus accumulation and density dependence of trees in a subtropical forest', *Science*, 366 (2019), pp. 124-8.

Chen, M., Arato, M., Borghi, L., Nouri, E., Reinhardt, D., 'Beneficial services of arbuscular mycorrhizal fungi – from ecology to application', *Frontiers in Plant Science*, 9 (2018), 1270.

Chialva, M., di Fossalunga, A., Daghino, S., Ghignone, S., Bagnaresi, P., Chiapello, M., Novero, M., Spadaro, D., Perotto, S., Bonfante, P., 'Native soils with their microbiotas elicit a state of alert in tomato plants', *New Phytologist*, 220 (2018), pp. 1296-1308.

Chrisafis, A., 'French truffle farmer shoots man he feared was trying to steal "black diamonds"', *Guardian* (2010), www.theguardian.com/world/2010/dec/22/french-truffle-farmer-shoots-trespasser [accessed 29 October 2019].

Christakis, N. A., Fowler, J. H., *Connected: The Surprising Power of Our Social Networks and How They Shape Our Lives* (London, HarperPress, 2009).

Chu, C., Murdock, M. H., Jing, D., Won, T. H., Chung, H., Kressel, A. M., Tsaava, T., Addorisio, M. E., Putzel, G. G., Zhou, L. et al., 'The microbiota regulate neuronal function and fear extinction learning', *Nature*, 574 (2019), pp. 543-8.

Chung, T.-Y., Sun, P.-F., Kuo, J.-I., Lee, Y.-I., Lin, C.-C., Chou, J.-Y., 'Zombie ant heads are oriented relative to solar cues', *Fungal Ecology*, 25 (2017), pp. 22-8.

Cixous, H. *The Book of Promethea* (Lincoln, University of Nebraska Press, 1991).

Claus, R., Hoppen, H., Karg, H., 'The secret of truffles: A steroidal pheromone?' *Experientia*, 37 (1981), pp. 1178-9.

Clay, K. 'Fungal endophytes of grasses: a defensive mutualism between plants and fungi', *Ecology*, 69 (1988), pp. 10-16.

Clemmensen, K., Bahr, A., Ovaskainen, O., Dahlberg, A., Ekblad, A., Wallander, H., Stenlid, J., Finlay, R., Wardle, D., Lindahl, B., 'Roots and associated fungi drive long-term carbon sequestration in boreal forest', *Science*, 339 (2013), pp. 1615-18.

Cockell, C. S., 'The interplanetary exchange of photosynthesis', *Origins of Life and Evolution of Biospheres*, 38 (2008), pp. 87-104.

Cohen, R., Jan, Y., Matricon, J., Dellbrück, M., 'Avoidance response, house response, and wind responses of the sporangiophore of *Phycomyces*', *Journal of General Physiology*, 66 (1975), pp. 67-95.

Collier, F. A., Bidartondo, M. I., 'Waiting for fungi: the ectomycorrhizal invasion of lowland heathlands', *Journal of Ecology*, 97 (2009), pp. 950–63.

Collinge, A., Trinci, A., 'Hyphal tips of wild-type and spreading colonial mutants of *Neurospora crassa*', *Archive of Microbiology*, 99 (1974), pp. 353–68.

Cooke, M., *Fungi: Their Nature and Uses* (New York, D. Appleton and Company, 1875).

Cooley, J. R., Marshall, D. C., Hill, K. B. R., 'A specialized fungal parasite (*Massospora cicadina*) hijacks the sexual signals of periodical cicadas (Hemiptera: Cicadidae: Magicicada), *Scientific Reports*, 8 (2018), 1432.

Copetta, A., Bardi, L., Bertolone, E., Berta, G., 'Fruit production and quality of tomato plants (*Solanum lycopersicum* L.) are affected by green compost and arbuscular mycorrhizal fungi', *Plant Biosystems*, 145 (2011), pp. 106–15.

Copetta, A., Lingua, G., Berta, G., 'Effects of three AM fungi on growth, distribution of glandular hairs, and essential oil production in *Ocimum basilicum* L. var. Genovese', *Mycorrhiza*, 16 (2006), pp. 485–94.

Corbin, A., *The Foul and the Fragrant: Odor and the French Social Imagination* (Leamington Spa, Berg, 1986).

Cordero, R. J., 'Melanin for space travel radioprotection', *Environmental Microbiology*, 19 (2017), pp. 2529–32.

Corrales, A., Mangan, S. A., Turner, B. L., Dalling, J. W., 'An ectomycorrhizal nitrogen economy facilitates monodominance in a neotropical forest', *Ecology Letters*, 19 (2016), pp. 383–92.

Corrochano, L. M., Galland, P., 'Photomorphogenesis and Gravitropism in Fungi', in ed. J. Wendland, *Growth, Differentiation, and Sexuality* (Springer International Publishing, 2016), pp. 235–66.

Cosme, M., Fernández, I., van der Heijden, M. G., Pieterse, C., 'Non-mycorrhizal plants: the exceptions that prove the rule', *Trends in Plant Science*, 23 (2018), pp. 577–87.

Costello, E. K., Lauber, C. L., Hamady, M., Fierer, N., Gordon, J. I., Knight, R., 'Bacterial community variation in human body habitats across space and time', *Science*, 326 (2009), pp. 1694–7.

Cottin, H., Kotler, J., Billi, D., Cockell, C., Demets, R., Ehrenfreund, P., Elsaesser, A., d'Hendecourt, L., van Loon, J. J., Martins, Z. et al., 'Space as a tool for astrobiology: review and recommendations for experimentations in earth orbit and beyond', *Space Science Reviews*, 209 (2017) pp. 83–181.

Coyle, M. C., Elya, C. N., Bronski, M. J., Eisen, M. B., 'Entomophthovirus: An insect-derived iflavirus that infects a behavior manipulating fungal pathogen of dipterans', *bioRxiv* (2018), 371526.

Craig, M. E., Turner, B. L., Liang, C., Clay, K., Johnson, D. J., Phillips, R. P., 'Tree mycorrhizal type predicts within-site variability in the storage and distribution of soil organic matter', *Global Change Biology*, 24 (2018), pp. 3317–30.

Crowther, T., Glick, H., Covey, K., Bettigole, C., Maynard, D., Thomas, S., Smith, J., Hintler, G., Duguid, M., Amatulli, G. et al., 'Mapping tree density at a global scale', *Nature*, 525 (2015), pp. 201–68.

Currie, C. R., Poulsen, M., Mendenhall, J., Boomsma, J. J., Billen, J., 'Coevolved crypts and exocrine glands support mutualistic bacteria in fungus-growing ants', *Science*, 311 (2006), pp. 81–3.

Currie, C. R., Scott, J. A., Summerbell, R. C., Malloch, D., 'Fungus-growing ants use antibiotic-producing bacteria to control garden parasites', *Nature*, 398 (1999), pp. 701–4.

Dadachova, E., Casadevall, A., 'Ionizing radiation: how fungi cope, adapt and exploit with the help of melanin', *Current Opinion in Microbiology*, 11 (2008), pp. 525–31.

Dance, A., 'Inner workings: the mysterious parentage of the coveted black truffle', *Proceedings of the National Academy of Sciences*, 115 (2018), pp. 10188–90.

Darwin, C., Darwin, F., *The Power of Movement in Plants* (London, John Murray, 1880).

Davis, J., Aguirre, L., Barber, N., Stevenson, P., Adler, L., 'From plant fungi to bee parasites: mycorrhizae and soil nutrients shape floral chemistry and bee pathogens', *Ecology*, 100 (2019), e02801.

Davis, W., *One River: Explorations and Discoveries in the Amazon Rainforest* (New York, Simon and Schuster, 1996).

Dawkins, R., *The Extended Phenotype* (Oxford, Oxford University Press, 1982).

Dawkins, R., 'Extended phenotype – but not too extended. A reply to Laland, Turner and Jablonka', *Biology and Philosophy*, 19 (2004), pp. 377–96.

de Bekker, C., Quevillon, L. E., Smith, P. B., Fleming, K. R., Ghosh, D., Patterson A. D., Hughes, D. P., 'Species-specific ant brain manipulation by a specialized fungal parasite', *BMC Evolutionary* Biology, 14 (2014), p. 166.

de Gonzalo, G., Colpa, D. I., Habib, M., Fraaije, M. W., 'Bacterial enzymes involved in lignin degradation', *Journal of Biotechnology*, 236 (2016), pp. 110–19.

de Jong, E., Field, J. A., Spinnler, H. E., Wijnberg, J. B., de Bont, J. A., 'Significant biogenesis of chlorinated aromatics by fungi in natural environments', *Applied and Environmental Microbiology*, 60 (1994) pp. 264–70.

de la Fuente-Nunez, C., Meneguetti, B., Franco, O., Lu, T. K., 'Neuromicrobiology: how microbes influence the brain', *ACS Chemical Neuroscience*, 9 (2017), pp. 141–50.

de la Torre, R., Miller, A., Cubero, B., Martín-Cerezo, L. M., Raguse, M., Meeßen, J., 'The effect of high-dose ionizing radiation on the astrobiological model lichen *Circinaria gyrosa*', *Astrobiology*, 17 (2017), pp. 145–53.

de la Torre Noetzel, R., Miller, A. Z., de la Rosa, J. M., Pacelli, C., Onofri, S., Sancho, L., Cubero, B., Lorek, A., Wolter, D., de Vera, J. P., 'Cellular responses of the lichen *Circinaria gyrosa* in Mars-like conditions', *Frontiers in Microbiology*, 9 (2018), 308.

de los Ríos, A., Sancho, L., Grube, M., Wierzchos, J., Ascaso, C., 'Endolithic growth of two *Lecidea* lichens in granite from continental Antarctica detected by molecular and microscopy techniques', *New Phytologist*, 165 (2005), pp. 181–90.

de Vera, J. P., Alawi, M., Backhaus, T., Baqué, M., Billi, D., Böttger, U., Berger, T., Bohmeier, M., Cockell, C., Demets, R. et al., 'Limits of life and the habitability of Mars: the ESA space experiment BIOMEX on the ISS', *Astrobiology*, 19 (2019), pp. 145–57.

de Vries, F. T., Thébault, E., Liiri, M., Birkhofer, K., Tsiafouli, M. A., Bjørnlund, L., Jørgensen, H., Brady, M., Christensen, S., de Ruiter, P. C. et al., 'Soil food web properties explain ecosystem services across European land use systems', *Proceedings of the National Academy of Sciences*, 110 (2013), pp. 14296–301.

de Waal, F. B. M., 'Anthropomorphism and anthropodenial', *Philosophical Topics*, 27 (1999), pp. 255–80.

Delaux, P.-M., Radhakrishnan, G. V., Jayaraman, D., Cheema, J., Malbreil, M., Volkening, J. D., Sekimoto, H., Nishiyama, T., Melkonian, M., Pokorny, L. et al., 'Algal ancestor of land plants was preadapted for symbiosis', *Proceedings of the National Academy of Sciences*, 112 (2015), pp. 13390–5.

Delavaux, C. S., Smith-Ramesh, L., Kuebbing, S. E., 'Beyond nutrients: a meta-analysis of the diverse effects of arbuscular mycorrhizal fungi on plants and soils', *Ecology*, 98 (2017), pp. 2111–19.

Deleuze, G., Guattari, F., *A Thousand Plateaus: Capitalism and Schizophrenia* (Minneapolis, University of Minnesota Press, 2005).

Delwiche, C., Cooper, E., 'The evolutionary origin of a terrestrial flora', *Current Biology*, 25 (2015), pp. R899–R910.

Deng, Y., Qu, Z., Naqvi, N. I., 'Twilight, a novel circadian-regulated gene, integrates phototropism with nutrient and redox homeostasis during fungal development', *PLOS Pathogens*, 11 (2015), e1004972.

Deveau, A., Bonito, G., Uehling, J., Paoletti, M., Becker, M., Bindschedler, S., Hacquard, S., Hervé, V., Labbé, J., Lastovetsky, O. et al., 'Bacterial–fungal interactions: ecology, mechanisms and challenges', *FEMS Microbiology Reviews*, 42 (2018), pp. 335–52.

di Fossalunga, A., Lipuma, J., Venice, F., Dupont, L., Bonfante, P., 'The endobacterium of an arbuscular mycorrhizal fungus modulates the expression of its toxin–antitoxin systems during the life cycle of its host', *ISME Journal*, 11 (2017), pp. 2394–8.

Diamant, L., *Chaining the Hudson: The Fight for the River in the American Revolution* (New York, Fordham University Press, 2004).

Ditengou, F. A., Müller, A., Rosenkranz, M., Felten, J., Lasok, H., van Doorn, M., Legué, V., Palme, K., Schnitzler, J.-P., Polle, A., 'Volatile signalling by sesquiterpenes from ectomycorrhizal fungi reprogrammes root architecture', *Nature Communications*, 6 (2015), 6279.

Dixon, L.S., 'Bosch's *St Anthony Triptych* – An apothecary's apotheosis', *Art Journal*, 44 (1984), pp. 119–31.

Donoghue, P. C., Antcliffe, J. B., 'Early life: origins of multicellularity', *Nature*, 466 (2010), p. 41.

Doolittle, F. W., Booth, A., 'It's the song, not the singer: an exploration of holobiosis and evolutionary theory', *Biology & Philosophy*, 32 (2017), pp. 5–24.

Dressaire, E., Yamada, L., Song, B., Roper, M., 'Mushrooms use convectively created airflows to disperse their spores', *Proceedings of the National Academy of Sciences*, 113 (2016), pp. 2833–8.

Dudley, R., *The Drunken Monkey: Why We Drink and Abuse Alcohol* (Berkeley, University of California Press, 2014).

Dugan, F. M., *Fungi in the Ancient World* (St Paul, MN, American Phytopathological Society, 2008).

Dugan, F. M., *Conspectus of World Ethnomycology* (St Paul, MN, American Phytopathological Society, 2011).

Dunn, R., 'A Sip for the Ancestors', *Scientific American* (2012), blogs.scientificamerican.com/guest-blog/a-sip-for-the-ancestors-the-true-story-of-civilizations-stumbling-debt-to-beer-and-fungus/ [accessed 29 October 2019].

Dupré, J., Nicholson, D. J., 'A manifesto for a processual biology', in eds. J. Dupré and D. J. Nicholson, *Everything Flows: Towards a Processual Philosophy of Biology* (Oxford, Oxford University Press, 2018), pp. 3–48.

Dyke, E., *Psychedelic Psychiatry: LSD from Clinic to Campus* (Baltimore, MD, Johns Hopkins University Press, 2008).

Eason, W., Newman, E., Chuba, P., 'Specificity of interplant cycling of phosphorus: the role of mycorrhizas', *Plant and Soil*, 137 (1991), pp. 267–74.

Elser, J. and Bennett, E., 'A broken biogeochemical cycle', *Nature* (2011), www.nature.com/articles/478029a [accessed 29 October 2019].

Eltz, T., Zimmermann, Y., Haftmann, J., Twele, R., Francke, W., Quezada-Euan, J. J. G., Lunau, K., 'Enfleurage, lipid recycling and the origin of perfume collection in orchid bees', *Proceedings of the Royal Society B*, 274 (2007), pp. 2843–8.

Eme, L., Spang, A., Lombard, J., Stairs, C. W., Ettema, T. J. G., 'Archaea and the origin of eukaryotes', *Nature Reviews Microbiology*, 15 (2017), pp. 711–23.

Engelthaler, D. M., Casadevall, A., 'On the Emergence of *Cryptococcus gattii* in the Pacific Northwest: ballast tanks, tsunamis, and black swans', *mBio*, 10 (2019), e02193–19.

Ensminger, P. A., *Life under the Sun* (New Haven, CT, Yale Scholarship Online, 2001).

Epstein, S., 'The construction of lay expertise: AIDS activism and the forging of credibility in the reform of clinical trials', *Science, Technology, Human Values*, 20 (1995), pp. 408–37.

Erens, H., Boudin, M., Mees, F., Mujinya, B., Baert, G., Strydonck, M., Boeckx, P., Ranst, E., 'The age of large termite mounds – radiocarbon dating of *Macrotermes falciger* mounds of the Miombo woodland of Katanga, DR Congo', *Palaeogeography, Palaeoclimatology, Palaeoecology*, 435 (2015), pp. 265–71.

Espinosa-Valdemar, R., Turpin-Marion, S., Delfín-Alcalá, I., Vázquez-Morillas, A., 'Disposable diapers biodegradation by the fungus *Pleurotus ostreatus*', *Waste Management*, 31 (2011), pp. 1683–8.

Fairhead, J., Leach, M., 'Termites, Society and Ecology: Perspectives from West Africa', in eds. E. Motte-Florac and J. Thomas, *Insects in Oral Literature and Traditions* (Leuven, Belgium, Peeters, 2003).

Fairhead, J., Scoones, I., 'Local knowledge and the social shaping of soil investments: critical perspectives on the assessment of soil degradation in Africa', *Land Use Policy*, 22 (2005), pp. 33–41.

Fairhead, J. R., 'Termites, mud daubers and their earths: a multispecies approach to fertility and power in West Africa', *Conservation and Society*, 14 (2016), pp. 359–67.

Farahany, N. A., Greely, H. T., Hyman, S., Koch, C., Grady, C., Paşca, S. P., Sestan, N., Arlotta, P., Bernat, J. L., Ting, J. et al., 'The ethics of experimenting with human brain tissue', *Nature*, 556 (2018), pp. 429–32.

Ferreira, B., 'There's growing evidence that the universe is connected by giant structures', *Vice* (2019), www.vice.com/en_us/article/zmj7pw/theres-growing-evidence-that-the-universe-is-connected-by-giant-structures [accessed 16 November 2019].

Fellbaum, C. R., Mensah, J. A., Cloos, A. J., Strahan, G. E., Pfeffer, P. E., Kiers, T. E., Bücking, H., 'Fungal nutrient allocation in common mycorrhizal networks is regulated by the carbon source strength of individual host plants', *New Phytologist*, 203 (2014) pp. 646–56.

Ferguson, B. A., Dreisbach, T., Parks, C., Filip, G., Schmitt, C., 'Coarse-scale population structure of pathogenic *Armillaria* species in a mixed-conifer forest in the Blue Mountains of northeast Oregon', *Canadian Journal of Forest Research*, 33 (2003), pp. 612–23.

Fernandez, C. W., Nguyen, N. H., Stefanski, A., Han, Y., Hobbie, S. E., Montgomery, R. A., Reich, P. B., Kennedy, P. G., 'Ectomycorrhizal fungal response to warming is linked to poor host performance at the boreal-temperate ecotone', *Global Change Biology*, 23 (2017), pp. 1598–609.

Field, K. J., Cameron, D. D., Leake, J. R., Tille, S., Bidartondo, M. I., Beerling, D. J., 'Contrasting arbuscular mycorrhizal responses of vascular and non-vascular plants to a simulated Palaeozoic CO_2 decline', *Nature Communications*, 3 (2012), 835.

Field, K. J., Leake, J. R., Tille, S., Allinson, K. E., Rimington, W. R., Bidartondo, M. I., Beerling, D. J., Cameron, D. D., 'From mycoheterotrophy to mutualism: mycorrhizal specificity and functioning in *Ophioglossum vulgatum* sporophytes', *New Phytologist*, 205 (2015), pp. 1492–1502.

Fisher, M. C., Hawkins, N. J., Sanglard, D., Gurr, S. J., 'Worldwide emergence of resistance to antifungal drugs challenges human health and food security', *Science*, 360 (2018), pp. 739–42.

Fisher, M. C., Henk, D. A., Briggs, C. J., Brownstein, J. S., Madoff, L. C., McCraw, S. L., Gurr, S. J., 'Emerging fungal threats to animal, plant and ecosystem health', *Nature*, 484 (2012), pp. 186–94.

Floudas, D., Binder, M., Riley, R., Barry, K., Blanchette, R. A., Henrissat, B., Martínez, A. T., Otillar, R., Spatafora, J. W., Yadav, J. S. et al., 'The Paleozoic origin of enzymatic lignin decomposition reconstructed from 31 fungal genomes', *Science*, 336 (2012), pp. 1715–19.

Foley, J. A., DeFries, R., Asner, G. P., Barford, C., Bonan, G., Carpenter, S. R., Chapin, S. F., Coe, M. T., Daily, G. C., Gibbs, H. K. et al., 'Global consequences of land use', *Science*, 309 (2005), pp. 570–4.

Francis, R., Read, D. J., 'Direct transfer of carbon between plants connected by vesicular–arbuscular mycorrhizal mycelium', *Nature*, 307 (1984), pp. 53–6.

Frank, A. B., 'On the nutritional dependence of certain trees on root symbiosis with belowground fungi (an English translation of A. B. Frank's classic paper of 1885)', *Mycorrhiza*, 15 (2005), pp. 267–75.

Fredericksen, M. A., Zhang, Y., Hazen, M. L., Loreto, R. G., Mangold, C. A., Chen, D. Z., Hughes, D. P., 'Three-dimensional visualization and a deep-learning model reveal complex fungal parasite networks in behaviorally manipulated ants', *Proceedings of the National Academy of Sciences*, 114 (2017), pp. 12590–5.

Fricker, M. D., Heaton, L. L., Jones, N. S., Boddy, L., 'The mycelium as a network', *Microbiology Spectrum*, 5 (2017), FUNK-0033-2017.

Fricker, M. D., Boddy, L., Bebber, D. P., 'Network Organisation of Mycelial Fungi', in eds. R. J. Howard and N. A. R. Gow, *Biology of the Fungal Cell* (Springer International Publishing, 2007a), pp. 309–30.

Fricker, M. D., Lee, J., Bebber, D., Tlalka, M., Hynes, J., Darrah, P., Watkinson, S., Boddy, L., 'Imaging complex nutrient dynamics in mycelial networks', *Journal of Microscopy*, 231 (2008), pp. 317–31.

Fricker, M. D., Lee, J., Tlalka, M., Bebber, D., Tagaki, S., Watkinson, S. C., Darrah, P. R., 'Fourier-based spatial mapping of oscillatory phenomena in fungi', *Fungal Genetics and Biology*, 44 (2007b), pp. 1077–84.

Fries, N., '*Untersuchungen über Sporenkeimung und Mycelentwicklung bodenbewohneneder Hymenomyceten*', *Symbolae Botanicae Upsaliensis*, 6 (1943), pp. 633–64.

Fritts, R., 'A new pesticide is all the buzz', *Ars Technica* (2019), arstechnica.com/science/2019/10/now-available-in-the-us-a-pesticide-delivered-by-bees/ [accessed 29 October 2019].

Fröhlich-Nowoisky, J., Pickersgill, D. A., Després, V. R., Pöschl, U., 'High diversity of fungi in air particulate matter', *Proceedings of the National Academy of Sciences*, 106 (2009), pp. 12814–9.

Fukusawa Y., Savoury M., Boddy L., 'Ecological memory and relocation decisions in fungal mycelial networks: responses to quantity and location of new resources', *ISME Journal* (2019) 10.1038/s41396-018-0189-7.

Galland, P., 'The sporangiophore of *Phycomyces blakesleeanus*: a tool to investigate fungal gravireception and graviresponses', *Plant Biology*, 16 (2014), pp. 58–68.

Gavito, M. E., Jakobsen, I., Mikkelsen, T. N., Mora, F., 'Direct evidence for modulation of photosynthesis by an arbuscular mycorrhiza-induced carbon sink strength', *New Phytologist*, 223 (2019), pp. 896–907.

Geml, J., Wagner, M. R., 'Out of sight, but no longer out of mind – towards an increased recognition of the role of soil microbes in plant speciation', *New Phytologist*, 217 (2018), pp. 965–7.

Giauque, H., Hawkes, C. V., 'Climate affects symbiotic fungal endophyte diversity and performance', *American Journal of Botany*, 100 (2013), pp. 1435–44.

Gilbert, C. D., Sigman, M., 'Brain states: top-down influences in sensory processing', *Neuron*, 54 (2007), pp. 677–96.

Gilbert, J. A., Lynch, S. V., 'Community ecology as a framework for human microbiome research', *Nature Medicine*, 25 (2019), pp. 884–9.

Gilbert, S. F., Sapp, J., Tauber, A. I., 'A symbiotic view of life: we have never been individuals', *Quarterly Review of Biology*, 87 (2012), pp. 325–41.

Giovannetti, M., Avio, L., Fortuna, P., Pellegrino, E., Sbrana, C., Strani, P., 'At the root of the Wood Wide Web', *Plant Signaling & Behavior*, 1 (2006), pp. 1–5.

Giovannetti, M., Avio, L., Sbrana, C., 'Functional Significance of Anastomosis in Arbuscular Mycorrhizal Networks', in ed. T. Horton, *Mycorrhizal Networks* (Springer International Publishing, 2015), pp. 41–67.

Giovannetti, M., Sbrana, C., Avio, L., Strani, P., 'Patterns of below-ground plant interconnections established by means of arbuscular mycorrhizal networks', *New Phytologist*, 164 (2004), pp. 175–81.

Gluck-Thaler, E., Slot, J. C., 'Dimensions of horizontal gene transfer in eukaryotic microbial pathogens', *PLOS Pathogens*, 11 (2015), e1005156.

Godfray, C. H., Beddington, J. R., Crute, I. R., Haddad, L., Lawrence, D., Muir, J. F., Pretty, J., Robinson, S., Thomas, S. M., Toulmin, C., 'Food security: the challenge of feeding 9 billion people', *Science*, 327 (2010), pp. 812–18.

Godfrey-Smith, P., *Other Minds: The Octopus and the Evolution of Intelligent Life* (London, William Collins, 2017).

Goffeau, A., Barrell, B., Bussey, H., Davis, R., Dujon, B., Feldmann, H., Galibert, F., Hoheisel, J., Jacq, C., Johnston, M. et al., 'Life with 6000 genes', *Science*, 274 (1996), pp. 546–67.

Gogarten, P. J., Townsend, J. P., 'Horizontal gene transfer, genome innovation and evolution', *Nature Reviews Microbiology*, 3 (2005), pp. 679–87.

Gond, S. K., Kharwar, R. N., White, J. F., 'Will fungi be the new source of the blockbuster drug Taxol?' *Fungal Biology Reviews*, 28 (2014), pp. 77–84.

Gontier, N., 'Reticulate Evolution Everywhere', in ed. N. Gontier, *Reticulate Evolution* (Springer International Publishing, 2015a).

Gontier, N., 'Historical and Epistemological Perspectives on What Horizontal Gene Transfer Mechanisms Contribute to Our Understanding of Evolution', in ed. N. Gontier, *Reticulate Evolution* (Springer International Publishing, 2015b).

Gordon, J., Knowlton, N., Relman, D. A., Rohwer, F., Youle, M., 'Superorganisms and holobionts', *Microbe* 8 (2013), pp. 152–3.

Goryachev, A. B., Lichius, A., Wright, G. D., Read, N. D., 'Excitable behavior can explain the "ping-pong" mode of communication between cells using the same chemoattractant', *BioEssays*, 34 (2012), pp. 259–66.

Gorzelak, M. A., Asay, A. K., Pickles, B. J., Simard, S. W., 'Inter-plant communication through mycorrhizal networks mediates complex adaptive behaviour in plant communities', *AoB PLANTS*, 7 (2015), plv050.

Gott, J. R., *The Cosmic Web: Mysterious Architecture of the Universe* (Princeton, NJ, Princeton University Press, 2016).

Govoni, F., Orrù, E., Bonafede, A., Iacobelli, M., Paladino, R., Vazza, F., Murgia, M., Vacca, V., Giovannini, G., Feretti, L., et al., 'A radio ridge connecting two galaxy clusters in a filament of the cosmic web', *Science*, 364 (2019), pp. 981–4.

Gow, N. A. R., Morris, B. M., 'The electric fungus', *Botanical Journal of Scotland*, 47 (2009), pp. 263–77.

Goward, T. 'Here for a long time, not a good time.', *Nature Canada*, 24: 9 (1995), www.waysofenlichenment.net/public/pdfs/Goward_1995_Here_for_a_good_time_not_a_long_time.pdf [accessed 29 October 2019].

Goward, T., 'Twelve readings on the lichen Thallus VII – species', *Evansia*, 26 (2009a), pp. 153–62, www.waysofenlichenment.net/ways/readings/essay7 [accessed 29 October 2019].

Goward, T., 'Twelve readings on the lichen Thallus IV – re-emergence', *Evansia* 26 (2009b), pp. 1–6, www.waysofenlichenment.net/ways/readings/essay4 [accessed 29 October 2019].

Goward, T., 'Twelve readings on the lichen Thallus V – conversational', *Evansia*, 26 (2009c), pp. 31–7. www.waysofenlichenment.net/ways/readings/essay5 [accessed 29 October 2019].

Goward, T., 'Twelve readings on the lichen Thallus VIII – theoretical', *Evansia* 27 (2010), pp. 2–10, www.waysofenlichenment.net/ways/readings/essay8 [accessed 29 October 2019].

Gregory, P. H., 'Fairy rings; free and tethered', *Bulletin of the British Mycological Society*, 16 (1982), pp. 161–3.

Griffiths, D., 'Queer theory for lichens', *UnderCurrents*, 19 (2015), pp. 36–45.

Griffiths, R., Johnson, M., Carducci, M., Umbricht, A., Richards, W., Richards, B., Cosimano, M., Klinedinst, M., 'Psilocybin produces substantial and sustained decreases in depression and anxiety in patients with life-threatening cancer: a randomized double-blind trial', *Journal of Psychopharmacology*, 30 (2016), pp. 1181–97.

Griffiths, R., Richards, W., Johnson, M., McCann, U., Jess, R., 'Mystical-type experiences occasioned by psilocybin mediate the attribution of personal meaning and spiritual significance 14 months later', *Journal of Psychopharmacology*, 22 (2008), pp. 621–32.

Grman, E., 'Plant species differ in their ability to reduce allocation to non-beneficial arbuscular mycorrhizal fungi', *Ecology*, 93 (2012), pp. 711–18.

Grube, M., Cernava, T., Soh, J., Fuchs, S., Aschenbrenner, I., Lassek, C., Wegner, U., Becher, D., Riedel, K., Sensen, C. W. et al., 'Exploring functional contexts of symbiotic sustain within lichen-associated bacteria by comparative omics', *ISME Journal*, 9 (2015), pp. 412–24.

Gupta, M., Prasad, A., Ram, M., Kumar, S., 'Effect of the vesicular–arbuscular mycorrhizal (VAM) fungus *Glomus fasciculatum* on the essential oil yield related characters and nutrient acquisition in the crops of different cultivars of menthol mint (*Mentha arvensis*) under field conditions', *Bioresource Technology*, 81 (2002), pp. 77–9.

Guzmán, G., Allen, J. W., Gartz, J., 'A worldwide geographical distribution of the neurotropic fungi, an analysis and discussion', *Annali del Museo Civico di Rovereto: Sezione Archeologia, Storia, Scienze Naturali*, 14 (1998), pp. 189–280, www.museocivico.rovereto.tn.it/UploadDocs/104_art09-Guzman%20&%20C.pdf [accessed 29 October 2019].

Hague, T., Florini, M., Andrews, P., 'Preliminary *in vitro* functional evidence for reflex responses to noxious stimuli in the arms of *Octopus vulgaris*', *Journal of Experimental Marine Biology and Ecology*, 447 (2013), pp. 100–5.

Hall, I. R., Brown, G. T., Zambonelli, A., *Taming the Truffle* (Portland, OR, Timber Press, 2007).

Hamden, E., 'Observing the cosmic web', *Science*, 366 (2019), pp. 31–2.

Haneef, M., Ceseracciu, L., Canale, C., Bayer, I. S., Heredia-Guerrero, J. A., Athanassiou, A., 'Advanced materials from fungal mycelium: fabrication and tuning of physical properties', *Scientific Reports*, 7 (2017), 41292.

Hanson, K. L., Nicolau, D. V., Filipponi, L., Wang, L., Lee, A. P., Nicolau, D. V., 'Fungi use efficient algorithms for the exploration of microfluidic networks', *Small*, 2 (2006), pp. 1212–20.

Haraway, D. J., *Crystals, Fabrics, and Fields* (Berkeley, CA, North Atlantic Books, 2004).

Haraway, D. J., *Staying with the Trouble: Making Kin in the Chthulucene* (Durham, NC, Duke University Press, 2016).

Harms, H., Schlosser, D., Wick, L. Y., 'Untapped potential: exploiting fungi in bio-remediation of hazardous chemicals', *Nature Reviews Microbiology*, 9 (2011), pp. 177–92.

Harold, F. M., Kropf, D. L., Caldwell, J. H., 'Why do fungi drive electric currents through themselves?' *Experimental Mycology*, 9 (1985), pp. 183–6.

Hart, M. M., Antunes, P. M., Chaudhary, V., Abbott, L. K., 'Fungal inoculants in the field: is the reward greater than the risk?' *Functional Ecology*, 32 (2018), 126–35.

Hastings, A., Abbott, K. C., Cuddington, K., Franci, T., Gellner, G., Lai, Y.-C., Morozov, A., Petrovskii, S., Scranton, K., Zeeman, M., 'Transient phenomena in ecology', *Science*, 361 (2018), eaat6412.

Hawksworth, D., 'The magnitude of fungal diversity: the 1.5 million species estimate revisited', *Mycological Research*, 12 (2001), pp. 1422–32.

Hawksworth, D., 'Mycology: A Neglected Megascience', in eds. M. Rai and P. D. Bridge, *Applied Mycology* (Oxford, CABI, 2009), pp. 1–16.

Hawksworth, D. L., Lücking, R., 'Fungal diversity revisited: 2.2 to 3.8 million species', *Microbiology Spectrum*, 5 (2017), FUNK-00522016.

Heads, S. W., Miller, A. N., Crane, L. J., Thomas, J. M., Ruffatto, D. M., Methven, A. S., Raudabaugh, D. B., Wang, Y., 'The oldest fossil mushroom', *PLOS ONE*, 12 (2017), e0178327.

Hedger, J., 'Fungi in the tropical forest canopy', *Mycologist*, 4 (1990), pp. 200–2.

Held, M., Edwards, C., Nicolau, D., 'Fungal intelligence; or on the behaviour of micro-organisms in confined micro-environments', *Journal of Physics: Conference Series*, 178 (2009), 012005.

Held, M., Edwards, C., Nicolau, D. V., 'Probing the growth dynamics of *Neurospora crassa* with microfluidic structures', *Fungal Biology*, 115 (2011), pp. 493–505.

Held, M., Kašpar, O., Edwards, C., Nicolau, D. V., 'Intracellular mechanisms of fungal space searching in microenvironments', *Proceedings of the National Academy of Sciences*, 116 (2019), pp. 13543–52.

Held, M., Lee, A. P., Edwards, C., Nicolau, D. V., 'Microfluidics structures for prob-ing the dynamic behaviour of filamentous fungi', *Microelectronic Engineering*, 87 (2010), pp. 786–9.

Helgason, T., Daniell, T., Husband, R., Fitter, A., Young, J., 'Ploughing up the wood-wide web?' *Nature*, 394 (1998), pp. 431.

Hendricks, P. S., 'Awe: a putative mechanism underlying the effects of classic psy-chedelic-assisted psychotherapy', *International Review of Psychiatry*, 30 (2018), pp. 1–12.

Hibbett, D., Blanchette, R., Kenrick, P., Mills, B., 'Climate, decay and the death of the coal forests', *Current Biology*, 26 (2016), pp. R563–7.

Hibbett, D., Gilbert, L., Donoghue, M., 'Evolutionary instability of ectomycorrhizal symbioses in basidiomycetes', *Nature*, 407 (2000), pp. 506–8.

Hickey, P. C., Dou, H., Foshe, S., Roper, M., 'Anti-jamming in a fungal transport network' (2016), arXiv:1601.06097v1 [physics.bio-ph].

Hickey, P. C., Jacobson, D., Read, N. D., Glass, L. N., 'Live-cell imaging of vegetative hyphal fusion in *Neurospora crassa*', *Fungal Genetics and Biology*, 37 (2002), 109–19.

Hillman, B., *Extra Hidden Life, among the Days* (Middletown, CT, Wesleyan University Press, 2018).

Hiruma, K., Kobae, Y., Toju, H., 'Beneficial associations between Brassicaceae plants and fungal endophytes under nutrient-limiting conditions: evolutionary origins and host–symbiont molecular mechanisms', *Current Opinion in Plant Biology*, 44 (2018), pp. 145–54.

Hittinger, C., 'Endless rots most beautiful', *Science*, 336 (2012), pp. 1649–50.

Hoch, H. C., Staples, R. C., Whitehead, B., Comeau, J., Wolf, E. D., 'Signaling for growth orientation and cell differentiation by surface topography in *Uromyces*', *Science*, 235 (1987), pp. 1659–62.

Hoeksema, J., 'Experimentally Testing Effects of Mycorrhizal Networks on Plant – Plant Interactions and Distinguishing among Mechanisms', in ed. T. Horton, *Mycorrhizal Networks* (Springer International Publishing, 2015), pp. 255–77.

Hoeksema, J. D., Chaudhary, V. B., Gehring, C. A., Johnson, N. C., Karst, J., Koide, R. T., Pringle, A., Zabinski, C., Bever, J. D., Moore, J. C. et al., 'A meta-analysis of context-dependency in plant response to inoculation with mycorrhizal fungi', *Ecology Letters*, 13 (2010), pp. 394–407.

Hom, E. F., Murray, A. W., 'Niche engineering demonstrates a latent capacity for fungal–algal mutualism', *Science*, 345 (2014), pp. 94–8.

Honegger, R., 'Simon Schwendener (1829–1919) and the dual hypothesis of lichens, *Bryologist*, 103 (2000), pp. 307–13.

Honegger, R., Edwards, D., Axe, L., 'The earliest records of internally stratified cyanobacterial and algal lichens from the Lower Devonian of the Welsh Borderland', *New Phytologist*, 197 (2012), pp. 264–75.

Honegger, R., Edwards, D., Axe, L., Strullu-Derrien, C., 'Fertile *Prototaxites taiti*: a basal ascomycete with inoperculate, polysporous asci lacking croziers', *Philosophical Transactions of the Royal Society B*, 373 (2018), 20170146.

Hooks, K. B., Konsman, J., O'Malley, M. A., 'Microbiota-gut-brain research: a critical analysis', *Behavioral and Brain Sciences*, 42 (2018), e60.

Horie, M., Honda, T., Suzuki, Y., Kobayashi, Y., Daito, T., Oshida, T., Ikuta, K., Jern, P., Gojobori, T., Coffin, J. M. et al., 'Endogenous non-retroviral RNA virus elements in mammalian genomes', *Nature*, 463 (2010), pp. 84–7.

Hortal, S., Plett, K., Plett, J., Cresswell, T., Johansen, M., Pendall, E., Anderson, I., 'Role of plant–fungal nutrient trading and host control in determining the competitive success of ectomycorrhizal fungi', *ISME Journal*, 11 (2017), pp. 2666–76.

Howard, A., *An Agricultural Testament* (Oxford, Oxford University Press, 1940), www.journeytoforever.org/farm_library/howardAT/ATtoc.html#contents [accessed 29 October 2019].

Howard A., *Farming and Gardening for Health and Disease* (London, Faber and Faber, 1945), journeytoforever.org/farm_library/howardSH/SHtoc.html [accessed 29 October 2019].

Howard, R., Ferrari, M., Roach, D., Money, N., 'Penetration of hard substrates by a fungus employing enormous turgor pressures', *Proceedings of the National Academy of Sciences*, 88 (1991), pp. 11281–4.

Hoysted, G. A., Kowal, J., Jacob, A., Rimington, W. R., Duckett, J. G., Pressel, S., Orchard, S., Ryan, M. H., Field, K. J., Bidartondo, M. I., 'A mycorrhizal revolution', *Current Opinion in Plant Biology*, 44 (2018), pp. 1–6.

Hsueh, Y.-P., Mahanti, P., Schroeder, F. C., Sternberg, P. W., 'Nematode-trapping fungi eavesdrop on nematode pheromones', *Current Biology*, 23 (2013), pp. 83–6.

Huffnagle, G. B., Noverr, M. C., 'The emerging world of the fungal microbiome', *Trends in Microbiology*, 21 (2013), pp. 334–41.

Hughes, D. P., 'On the origins of parasite-extended phenotypes', *Integrative and Comparative Biology*, 54 (2014), pp. 210–7.

Hughes, D. P., Araújo, J., Loreto, R., Quevillon, L., de Bekker, C., Evans, H., 'From so simple a beginning: the evolution of behavioural manipulation by fungi', *Advances in Genetics*, 94 (2016), pp. 437–69.

Hughes, D. P., 'Pathways to understanding the extended phenotype of parasites in their hosts', *Journal of Experimental Biology*, 216 (2013), pp. 142–7.

Hughes, D. P., Wappler, T., Labandeira, C. C., 'Ancient death-grip leaf scars reveal ant–fungal parasitism', *Biology Letters*, 7 (2011), pp. 67–70.

Humphrey, N., 'The Social Function of Intellect', in eds. P. Bateson and R. A. Hindle, *Growing Points in Ethology* (Cambridge, Cambridge University Press, 1976), pp. 303–17.

Hustak, C., Myers, N., 'Involutionary momentum: affective ecologies and the sciences of plant/insect encounters', *Differences*, 23 (2012), pp. 74–118.

Hyde, K., Jones, E., Leano, E., Pointing, S., Poonyth, A., Vrijmoed, L., 'Role of fungi in marine ecosystems', *Biodiversity and Conservation*, 7 (1998), pp. 1147–61.

Ingold, T., 'Two Reflections on Ecological Knowledge', in eds. G. Sanga and G. Ortall, *Nature Knowledge: Ethnoscience, Cognition, and Utility* (Oxford, Berghahn Books, 2003), pp. 301–11.

Islam, F., Ohga, S., 'The response of fruit body formation on *Tricholoma matsutake in situ* condition by applying electric pulse stimulator', *ISRN Agronomy* (2012), pp. 1–6.

Jackson, S., Heath, I., 'UV microirradiations elicit Ca2+-dependent apex-directed cytoplasmic contractions in hyphae', *Protoplasma*, 70 (1992), pp. 46–52.

Jacobs, L. F., Arter, J., Cook, A., Sulloway, F. J., 'Olfactory orientation and navigation in humans', *PLOS ONE*, 10 (2015), e0129387.

Jacobs, R., *The Truffle Underground* (New York, Clarkson Potter, 2019).

Jakobsen, I., Hammer, E., 'Nutrient Dynamics in Arbuscular Mycorrhizal Networks', in ed. T. Horton, *Mycorrhizal Networks* (Springer International Publishing, 2015), pp. 91–131.

James, W., *The Varieties of Religious Experience: A Study in Human Nature (Centenary Edition)* (London, Routledge, 2002).

Jedd, G., Pieuchot, L., 'Multiple modes for gatekeeping at fungal cell-to-cell channels', *Molecular Microbiology*, 86 (2012), pp. 1291–4.

Jenkins, B., Richards, T. A., 'Symbiosis: wolf lichens harbour a choir of fungi', *Current Biology*, 29 (2019), R88–90.

Ji, B., Bever, J. D., 'Plant preferential allocation and fungal reward decline with soil phosphorus: implications for mycorrhizal mutualism', *Ecosphere*, 7 (2016), e01256.

Johnson, D., Gamow, R., 'The avoidance response in *Phycomyces*', *Journal of General Physiology*, 57 (1971), pp. 41–9.

Johnson, M. W., Garcia-Romeu, A., Cosimano, M. P., Griffiths, R. R., 'Pilot study of the 5-HT 2AR agonist psilocybin in the treatment of tobacco addiction', *Journal of Psychopharmacology*, 28 (2014), pp. 983–92.

Johnson, M. W., Garcia-Romeu, A., Griffiths, R. R., 'Long-term follow-up of psilocybin-facilitated smoking cessation', *American Journal of Drug and Alcohol Abuse*, 43 (2015), pp. 55–60.

Johnson, M. W., Garcia-Romeu, A., Johnson, P. S., Griffiths, R. R., 'An online survey of tobacco smoking cessation associated with naturalistic psychedelic use', *Journal of Psychopharmacology*, 31 (2017), pp. 841–50.

Johnson, N. C., Angelard, C., Sanders, I. R., Kiers, T. E., 'Predicting community and ecosystem outcomes of mycorrhizal responses to global change', *Ecology Letters*, 16 (2013), pp. 140–53.

Jolivet, E., L'Haridon, S., Corre, E., Forterre, P., Prieur, D., '*Thermococcus gammatolerans* sp. nov., a hyperthermophilic archaeon from a deep-sea hydrothermal vent that resists ionizing radiation', *International Journal of Systematic and Evolutionary Microbiology*, 53 (2003), pp. 847–51.

Jones, M. P., Lawrie, A. C., Huynh, T. T., Morrison, P. D., Mautner, A., Bismarck, A., John, S., 'Agricultural by-product suitability for the production of chitinous composites and nanofibers', *Process Biochemistry*, 80 (2019), pp, 95–102.

Jönsson, K. I., Rabbow, E., Schill, R. O., Harms-Ringdahl, M., Rettberg, P., 'Tardigrades survive exposure to space in low Earth orbit', *Current Biology*, 18 (2008), R729–31.

Jönsson, K. I., Wojcik, A., 'Tolerance to X-rays and heavy ions (Fe, He) in the tardigrade *Richtersius coronifer* and the bdelloid rotifer *Mniobia russeola*', *Astrobiology*, 17 (2017), pp. 163–7.

Kaminsky, L. M., Trexler, R. V., Malik, R. J., Hockett, K. L., Bell, T. H., 'The inherent conflicts in developing soil microbial inoculants', *Trends in Biotechnology*, 37 (2018), pp. 140–51.

Kammerer, L., Hiersche, L., Wirth, E., 'Uptake of radiocaesium by different species of mushrooms', *Journal of Environmental Radioactivity*, 23 (1994), pp. 135–50.

Karst, J., Erbilgin, N., Pec, G. J., Cigan, P. W., Najar, A., Simard, S. W., Cahill, J. F., 'Ectomycorrhizal fungi mediate indirect effects of a bark beetle outbreak on secondary chemistry and establishment of pine seedlings', *New Phytologist*, 208 (2015), pp. 904–14.

Katz, S. E., *Wild Fermentation* (White River Junction, VT, Chelsea Green Publishing Company, 2003).

Kavaler, L., *Mushrooms, Moulds and Miracles: The Strange Realm of Fungi* (London, Harrap, 1967).

Keijzer, F. A., 'Evolutionary convergence and biologically embodied cognition', *Journal of the Royal Society Interface Focus*, 7 (2017), 20160123.

Keller, E.F., *A Feeling for the Organism* (New York, Times Books, 1984).

Kelly, J. R., Borre, Y., O'Brien, C., Patterson, E., El Aidy, S., Deane, J., Kennedy, P. J., Beers, S., Scott, K., Moloney, G. et al., 'Transferring the blues: depression-associated gut microbiota induces neurobehavioural changes in the rat', *Journal of Psychiatric Research*, 82 (2016), pp. 109–18.

Kelty, C., 'Outlaws, hackers, Victorian amateurs: diagnosing public participation in the life sciences today', *Journal of Science Communication*, 9 (2010).

Kendi, I. X., *Stamped from the Beginning* (New York, Nation Books, 2017).

Kennedy, P. G., Walker, J. K. M., Bogar, L. M., 'Interspecific Mycorrhizal Networks and Non-networking Hosts: Exploring the Ecology of the Host Genus *Alnus*', in ed. T. Horton, *Mycorrhizal Networks* (Springer International Publishing, 2015) pp. 227–54.

Kerényi, C., *Dionysus: Archetypal Image of Indestructible Life* (Princeton, NJ, Princeton University Press, 1976).

Kern, V. D., 'Gravitropism of basidiomycetous fungi – on Earth and in microgravity', *Advances in Space Research*, 24 (1999), pp. 697–706.

Khan, S., Nadir, S., Shah, Z., Shah, A., Karunarathna, S. C., Xu, J., Khan, A., Munir, S., Hasan, F., 'Biodegradation of polyester polyurethane by *Aspergillus tubingensis*', *Environmental Pollution*, 225 (2017), pp. 469–80.

Kiers, E. T., Denison, R. F., 'Inclusive fitness in agriculture', *Philosophical Transactions of the Royal Society B*, 369 (2014), 20130367.

Kiers, T. E., Duhamel, M., Beesetty, Y., Mensah, J. A., Franken, O., Verbruggen, E., Fellbaum, C., Fellbaum, C. R., Kowalchuk, G. A. et al., 'Reciprocal rewards stabilize co-operation in the mycorrhizal symbiosis', *Science*, 333 (2011), pp. 880–2.

Kiers, T. E., West, S. A., Wyatt, G. A., Gardner, A., Bücking, H., Werner, G. D., 'Misconceptions on the application of biological market theory to the mycorrhizal symbiosis', *Nature Plants*, 2 (2016), 16063.

Kim, G., LeBlanc, M. L., Wafula, E. K., dePamphilis, C. W., Westwood, J. H., 'Genomic-scale exchange of mRNA between a parasitic plant and its hosts', *Science*, 345 (2014), pp. 808–11.

Kimmerer, R. W., *Braiding Sweetgrass* (Minneapolis, MN, Milkweed Editions, 2013).

King, A., 'Technology: the future of agriculture', *Nature*, 544 (2017), pp. S21–3.

King, F. H., *Farmers of Forty Centuries* (Emmaus, PA, Organic Gardening Press, 1911), soilandhealth.org/wp-content/uploads/01aglibrary/010122king/ffc.html [accessed 29 October 2019].

Kivlin, S. N., Emery, S. M., Rudgers, J. A., 'Fungal symbionts alter plant responses to global change', *American Journal of Botany*, 100 (2013), pp. 1445–57.

Klein, A.-M., Vaissière, B. E., Cane, J. H., Steffan-Dewenter, I., Cunningham, S. A., Kremen, C., Tscharntke, T., 'Importance of pollinators in changing landscapes for world crops', *Proceedings of the Royal Society B*, 274 (2007), pp. 303–13.

Klein, T., Siegwolf, R. T., Körner, C., 'Below-ground carbon trade among tall trees in a temperate forest', *Science*, 352 (2016), pp. 342–4.

Kozo-Polyanksy, B. M., *Symbiogenesis: A New Principle of Evolution* (Cambridge, MA, Harvard University Press, 2010).

Krebs, T. S., Johansen, P.-Ø., 'Lysergic acid diethylamide (LSD) for alcoholism: meta-analysis of randomised controlled trials', *Journal of Psychopharmacology*, 26 (2012), pp. 994–1002.

Kroken, S., '"Miss Potter's First Love" – a rejoinder', *Inoculum*, 58 (2007), p. 14.

Kusari, S., Singh, S., Jayabaskaran, C., 'Biotechnological potential of plant-associated endophytic fungi: hope versus hype', *Trends in Biotechnology*, 32 (2014), pp. 297–303.

Ladinsky, D., *Love Poems from God* (New York, Penguin, 2002).

Ladinsky, D., *A Year with Hafiz: Daily Contemplations* (New York, Penguin, 2010).

Lai, J., Koh, C., Tjota, M., Pieuchot, L., Raman, V., Chandrababu, K., Yang, D., Wong, L., Jedd, G., 'Intrinsically disordered proteins aggregate at fungal cell-to-cell channels and regulate intercellular connectivity', *Proceedings of the National Academy of Sciences*, 109 (2012), pp. 15781–6.

Lalley, J., Viles, H., 'Terricolous lichens in the northern Namib Desert of Namibia: distribution and community composition', *Lichenologist*, 37 (2005), pp. 77–91.

Lanfranco, L., Fiorilli, V., Gutjahr, C., 'Partner communication and role of nutrients in the arbuscular mycorrhizal symbiosis', *New Phytologist*, 220 (2018), pp. 1031–46.

Latty, T., Beekman, M., 'Irrational decision-making in an amoeboid organism: transitivity and context-dependent preferences', *Proceedings of the Royal Society B*, 278 (2011), pp. 307–12.

Le Guin, U., 'Deep in Admiration', in eds. A. Tsing, H. Swanson, E. Gan and N. Bubandt, *Arts of Living on a Damaged Planet: Ghosts of the Anthropocene* (Minneapolis, University of Minnesota Press, 2017), pp. M15–21.

Leake, J., Johnson, D., Donnelly, D., Muckle, G., Boddy, L., Read, D., 'Networks of power and influence: the role of mycorrhizal mycelium in controlling plant communities and agroecosystem functioning', *Canadian Journal of Botany*, 82 (2004), pp. 1016–45.

Leake, J., Read, D., 'Mycorrhizal Symbioses and Pedogenesis Throughout Earth's History', in eds. N. Johnson, C. Gehring and J. Jansa, *Mycorrhizal Mediation of Soil: Fertility, Structure, and Carbon Storage* (Oxford, Elsevier, 2017), pp. 9–33.

Leary, T., 'The Initation of the "High Priest", in ed. R. Metzner, *Sacred Mushrooms of Visions: Teonanacatl* (Rochester, VT, Park Street Press, 2005), pp. 160–78.

Lederberg. J., 'Cell genetics and hereditary symbiosis', *Physiological Reviews*, 32 (1952), pp. 403–30.

Lederberg, J., Cowie, D., 'Moondust; the study of this covering layer by space vehicles may offer clues to the biochemical origin of life', *Science*, 127 (1958), pp. 1473–5.

Ledford, H., 'Billion-year-old fossils set back evolution of earliest fungi', *Nature* (2019), www.nature.com/articles/d41586-019-01629-1 [accessed 29 October 2019].

Lee, N. N., Friz, J., Fries, M. D., Gil, J. F., Beck, A., Pellinen-Wannberg, A., Schmitz, B., Steele, A., Hofmann, B. A., 'The Extreme Biology of Meteorites: Their Role in Understanding the Origin and Distribution of Life on Earth and in the Universe', in eds. H. Stan-Lotter and S. Fendrihan, *Adaptation of Microbial Life to Environmental Extremes* (Springer International Publishing, 2017), pp. 283–325.

Lee, Y., Mazmanian, S. K., 'Has the microbiota played a critical role in the evolution of the adaptive immune system?' *Science*, 330 (2010), pp. 1768–73.

Legras, J., Merdinoglu, D., Couet, J., Karst, F., 'Bread, beer and wine: *Saccharomyces cerevisiae* diversity reflects human history', *Molecular Ecology*, 16 (2007), pp. 2091–2102.

Lehmann, A., Leifheit, E. F., Rillig, M. C., 'Mycorrhizas and Soil Aggregation', in eds. N. Johnson, C. Gehring and J. Jansa, *Mycorrhizal Mediation of Soil: Fertility, Structure, and Carbon Storage* (Oxford, Elsevier, 2017), pp. 241–62.

Leifheit, E. F., Veresoglou, S. D., Lehmann, A., Morris, K. E., Rillig, M. C., 'Multiple factors influence the role of arbuscular mycorrhizal fungi in soil aggregation – a meta-analysis', *Plant and Soil*, 374 (2014), pp. 523–37.

Lekberg, Y., Helgason, T., '*In situ* mycorrhizal function – knowledge gaps and future directions', *New Phytologist*, 220 (2018), pp. 957–62.

Leonhardt, Y., Kakoschke, S., Wagener, J., Ebel, F., 'Lah is a transmembrane protein and requires Spa10 for stable positioning of Woronin bodies at the septal pore of *Aspergillus fumigatus*', *Scientific Reports*, 7 (2017), 44179.

Letcher, A., *Shroom: A Cultural History of the Magic Mushroom* (London, Faber and Faber, 2006).

Lévi-Strauss, C., *From Honey to Ashes: Introduction to a Science of Mythology*, 2 (New York, Harper and Row, 1973).

Levin, M., 'The wisdom of the body: future techniques and approaches to morphogenetic fields in regenerative medicine, developmental biology and cancer', *Regenerative Medicine*, 6 (2011), pp. 667–73.

Levin, M., 'Morphogenetic fields in embryogenesis, regeneration, and cancer: non-local control of complex patterning', *Biosystems*, 109 (2012), pp. 243–61.

Levin, S. A., 'Self-organization and the emergence of complexity in ecological systems', *BioScience*, 55 (2005), pp. 1075–9.

Lewontin, R., *The Triple Helix: Gene, Organism, and Environment* (Cambridge, MA, Harvard University Press, 2000).

Lewontin, R., *It Ain't Necessarily So: The Dream of the Human Genome and Other Illusions* (New York, New York Review of Books, 2001).

Li, N., Alfiky, A., Vaughan, M. M., Kang, S., 'Stop and smell the fungi: fungal volatile metabolites are overlooked signals involved in fungal interaction with plants', *Fungal Biology Reviews*, 30 (2016), pp. 134–44.

Li, Q., Yan, L., Ye, L., Zhou, J., Zhang, B., Peng, W., Zhang, X., Li, X., 'Chinese black truffle (*Tuber indicum*) alters the ectomycorrhizosphere and endoectomycosphere microbiome and metabolic profiles of the host tree *Quercus aliena*', *Frontiers in Microbiology*, 9 (2018), 2202.

Lindahl. B., Finlay, R., Olsson, S., 'Simultaneous, bidirectional translocation of 32P and 33P between wood blocks connected by mycelial cords of *Hypholoma fasciculare*', *New Phytologist*, 150 (2001), pp. 189–94.

Linnakoski, R., Reshamwala, D., Veteli, P., Cortina-Escribano, M., Vanhanen, H., Marjomäki, V., 'Antiviral agents from fungi: diversity, mechanisms and potential applications', *Frontiers in Microbiology*, 9 (2018), 2325.

Lintott, C., *The Crowd and the Cosmos: Adventures in the Zooniverse* (Oxford, Oxford University Press, 2019).

Lipnicki, L. I., 'The role of symbiosis in the transition of some eukaryotes from aquatic to terrestrial environments', *Symbiosis*, 65 (2015), pp. 39–53.

Liu, J., Martinez-Corral, R., Prindle, A., Lee, D.-Y. D., Larkin, J., Gabalda-Sagarra, M., Garcia-Ojalvo, J., Süel, G. M., 'Coupling between distant biofilms and emergence of nutrient time-sharing', *Science*, 356 (2017), pp. 638–42.

Lohberger, A., Spangenberg, J. E., Ventura, Y., Bindschedler, S., Verrecchia, E. P., Bshary, R., Junier, P., 'Effect of organic carbon and nitrogen on the interactions of *Morchella* spp. and bacteria dispersing on their mycelium', *Frontiers in Microbiology*, 10 (2019), 124.

Löpez-Franco, R., Bracker, C. E., 'Diversity and dynamics of the Spitzenkörper in growing hyphal tips of higher fungi', *Protoplasma*, 195 (1996), pp. 90–111.

Loron, C. C., François, C., Rainbird, R. H., Turner, E. C., Borensztajn, S., Javaux, E. J., 'Early fungi from the Proterozoic era in Arctic Canada', *Nature*, 570 (2019), pp. 232–5.

Lovett, B., Bilgo, E., Millogo, S., Ouattarra, A., Sare, I., Gnambani, E., Dabire, R. K., Diabate, A., Leger, R.J., 'Transgenic *Metarhizium* rapidly kills mosquitoes in a malaria-endemic region of Burkina Faso', *Science*, 364 (2019), pp. 894–7.

Lu, C., Yu, Z., Tian, H., Hennessy, D. A., Feng, H., Al-Kaisi, M., Zhou, Y., Sauer, T., Arritt, R., 'Increasing carbon footprint of grain crop production in the US Western Corn Belt', *Environmental Research Letters*, 13 (2018), 124007.

Luo, J., Chen, X., Crump, J., Zhou, H., Davies, D. G., Zhou, G., Zhang, N., Jin, C., 'Interactions of fungi with concrete: significant importance for bio-based self-healing concrete', *Construction and Building Materials*, 164 (2018), pp. 275–85.

Lutzoni, F., Nowak, M. D., Alfaro, M. E., Reeb, V., Miadlikowska, J., Krug, M., Arnold, E. A., Lewis, L. A., Swofford, D. L., Hibbett, D. et al., 'Contemporaneous radiations of fungi and plants linked to symbiosis', *Nature Communications*, 9 (2018), 5451.

Lutzoni, F., Pagel, M., Reeb, V., 'Major fungal lineages are derived from lichen symbiotic ancestors', *Nature*, 411 (2001), pp. 937–40.

Ly, C., Greb, A. C., Cameron, L. P., Wong, J. M., Barragan, E. V., Wilson, P. C., Burbach, K. F., Zarandi, S., Sood, A., Paddy, M. R. et al., 'Psychedelics promote structural and functional neural plasticity', *Cell Reports*, 23 (2018), pp. 3170–82.

Lyons, T., Carhart-Harris, R. L., 'Increased nature relatedness and decreased authoritarian political views after psilocybin for treatment-resistant depression', *Journal of Psychopharmacology*, 32 (2018), pp. 811–19.

Ma, Z., Guo, D., Xu, X., Lu, M., Bardgett, R. D., Eissenstat, D. M., McCormack, L. M., Hedin, L. O., 'Evolutionary history resolves global organization of root functional traits', *Nature*, 555 (2018), pp. 94–7.

MacLean, K. A., Johnson, M. W., Griffiths, R. R., 'Mystical experiences occasioned by the hallucinogen psilocybin lead to increases in the personality domain of openness', *Journal of Psychopharmacology*, 25 (2011), pp. 1453–61.

Mangold, C. A., Ishler, M. J., Loreto, R. G., Hazen, M. L., Hughes, D. P., 'Zombie ant death grip due to hypercontracted mandibular muscles', *Journal of Experimental Biology*, 222 (2019), jeb200683.

Manicka, S., Levin, M., 'The Cognitive Lens: a primer on conceptual tools for analysing information processing in developmental and regenerative morphogenesis', *Philosophical Transactions of the Royal Society B*, 374 (2019), 20180369.

Manoharan, L., Rosenstock, N. P., Williams, A., Hedlund, K., 'Agricultural management practices influence AMF diversity and community composition with cascading effects on plant productivity', *Applied Soil Ecology*, 115 (2017), pp. 53–9.

Mardhiah, U., Caruso, T., Gurnell, A., Rillig, M. C., 'Arbuscular mycorrhizal fungal hyphae reduce soil erosion by surface water flow in a greenhouse experiment', *Applied Soil Ecology*, 99 (2016), pp. 137–40.

Margonelli, L., *Underbug: An Obsessive Tale of Termites and Technology* (New York, Farrar, Strauss and Giroux, 2018).

Margulis, L., *Symbiosis in Cell Evolution: Life and Its Environment on the Early Earth* (San Francisco, CA, W. H. Freeman, 1981).

Margulis, L., 'Gaia Is a Tough Bitch', in ed. John Brockman, *The Third Culture: Beyond the Scientific Revolution* (New York, Touchstone, 1996).

Margulis, L., *The Symbiotic Planet: A New Look at Evolution* (London, Phoenix, 1999).

Markram, H., Muller, E., Ramaswamy, S., Reimann, M. W., Abdellah, M., Sanchez, C., Ailamaki, A., Alonso-Nanclares, L., Antille, N., Arsever, S. et al., 'Reconstruction and simulation of neocortical microcircuitry', *Cell*, 163 (2015), pp. 456–92.

Marley, G., *Chanterelle Dreams, Amanita Nightmares: The Love, Lore, and Mystique of Mushrooms* (White River Junction, VT, Chelsea Green, 2010).

Márquez, L. M., Redman, R. S., Rodriguez, R. J., Roossinck, M. J., 'A virus in a fungus in a plant: three-way symbiosis required for thermal tolerance', *Science* 315 (2007), pp. 513–15.

Martin, F. M., Uroz, S., Barker, D. G., 'Ancestral alliances: plant mutualistic symbioses with fungi and bacteria', *Science* 356 (2017), eaad4501.

Martinez-Corral, R., Liu., Prindle, A., Süel, G. M., Garcia-Ojalvo, J., 'Metabolic basis of brain-like electrical signalling in bacterial communities', *Philosophical Transactions of the Royal Society B*, 374 (2019), 20180382.

Martínez-García, L. B., De Deyn, G. B., Pugnaire, F. I., Kothamasi, D., van der Heijden, M. G., 'Symbiotic soil fungi enhance ecosystem resilience to climate change', *Global Change Biology*, 23 (2017), pp. 5228–36.

Masiulionis, V. E., Weber, R. W., Pagnocca, F. C., 'Foraging of *Psilocybe* basidiocarps by the leaf-cutting ant *Acromyrmex lobicornis* in Santa Fé, Argentina', *SpringerPlus*, 2 (2013), 254.

Mateus, I. D., Masclaux, F. G., Aletti, C., Rojas, E. C., Savary, R., Dupuis, C., Sanders, I. R., 'Dual RNA-seq reveals large-scale non-conserved genotype × genotype-specific genetic reprograming and molecular crosstalk in the mycorrhizal symbiosis', *ISME Journal*, 13 (2019), pp. 1226–38.

Matossian, M. K., 'Ergot and the Salem Witchcraft Affair: An outbreak of a type of food poisoning known as convulsive ergotism may have led to the 1692 accusations of witchcraft', *American Scientist*, 70 (1982), pp. 355–7.

Matsuura, K., Yashiro, T., Shimizu, K., Tatsumi, S., Tamura, T., 'Cuckoo fungus mimics termite eggs by producing the cellulose-digesting enzyme β-Glucosidase', *Current Biology*, 19 (2009), pp. 30–6.

Matsuura, Y., Moriyama, M., Łukasik, P., Vanderpool, D., Tanahashi, M., Meng, X.-Y., McCutcheon, J. P., Fukatsu, T., 'Recurrent symbiont recruitment from fungal parasites in cicadas', *Proceedings of the National Academy of Sciences*, 115 (2018), E5970–E5979.

Maugh, T. H., 'The scent makes sense', *Science*, 215 (1982), p. 1224.

Maxman, A., 'CRISPR might be the banana's only hope against a deadly fungus', *Nature* (2019), www.nature.com/articles/d41586-019-02770-7 [accessed 29 October 2019].

Mazur, S., 'Lynne Margulis: "Intimacy of Strangers & Natural Selection"', *Scoop* (2009), www.scoop.co.nz/stories/HL0903/S00194/lynn-margulis-intimacy-of-strangers-natural-selection.htm [accessed 29 October 2019].

Mazzucato, L., Camera, L. G., Fontanini, A., 'Expectation-induced modulation of metastable activity underlies faster coding of sensory stimuli', *Nature Neuroscience*, 22 (2019), pp. 787–96.

McCoy, P., *Radical Mycology: A Treatise on Working and Seeing with Fungi* (Portland, OR, Chthaeus Press, 2016).

McFall-Ngai, M., 'Adaptive immunity: care for the community', *Nature*, 445 (2007), p. 153.

McGann, J. P., 'Poor human olfaction is a 19th-century myth', *Science*, 356 (2017), eaam7263.

McGuire, K. L., 'Common ectomycorrhizal networks may maintain monodominance in a tropical rain forest', *Ecology*, 88 (2007), pp. 567–74.

McKenna, D., *Brotherhood of the Screaming Abyss* (Clearwater, MN, North Star Press of St. Cloud Inc., 2012).

McKenna, T., *Food of the Gods: The Search for the Original Tree of Knowledge* (New York, Bantam, 1992).

McKenna, T., McKenna, D. (Oss OT and Oeric ON), *Psilocybin: Magic Mushroom Grower's Guide* (Berkeley, CA, AND/OR Press, 1976).

McKenzie, R. N., Horton, B. K., Loomis, S. E., Stockli, D. F., Planavsky, N. J., Lee, C.-T. A., 'Continental arc volcanism as the principal driver of icehouse–greenhouse variability', *Science*, 352 (2016), pp. 444–7.

McKerracher, L., Heath, I., 'Fungal nuclear behavior analysed by ultraviolet microbeam irradiation', *Cell Motility and the Cytoskeleton*, 6 (1986a), pp. 35–47.

McKerracher, L., Heath, I., 'Polarized cytoplasmic movement and inhibition of saltations induced by calcium-mediated effects of microbeams in fungal hyphae', *Cell Motility and the Cytoskeleton*, 6 (1986b), pp. 136–45.

Meeßen, J., Backhaus, T., Brandt, A., Raguse, M., Böttger, U., de Vera, J. P., de la Torre, R., 'The effect of high-dose ionizing radiation on the isolated photobiont of the astrobiological model lichen *Circinaria gyrosa*', *Astrobiology*, 17 (2017), pp. 154–62.

Mejía, L. C., Herre, E. A., Sparks, J. P., Winter, K., García, M. N., Bael, S. A., Stitt, J., Shi, Z., Zhang, Y., Guiltinan, M. J. et al., 'Pervasive effects of a dominant foliar endophytic fungus on host genetic and phenotypic expression in a tropical tree', *Frontiers in Microbiology*, 5 (2014), 479.

Merckx, V., 'Mycoheterotrophy: An Introduction', in ed. V. Merckx, *Mycoheterotrophy – The Biology of Plants Living on Fungi* (Springer International Publishing, 2013), pp. 1–18.

Merleau-Ponty, M., *Phenomenology of Perception* (London, Routledge Classics, 2002).

Meskkauskas, A., McNulty, L. J., Moore, D., 'Concerted regulation of all hyphal tips generates fungal fruit body structures: experiments with computer visualizations produced by a new mathematical model of hyphal growth', *Mycological Research*, 108 (2004), pp. 341–53.

Metzner, R., 'Introduction: Visionary Mushrooms of the Americas', in ed. R. Metzner, *Sacred Mushroom of Visions: Teonanacatl* (Rochester, VT, Park Street Press, 2005), pp. 1–48.

Miller, M. J., Albarracin-Jordan, J., Moore, C., Capriles, J. M., 'Chemical evidence for the use of multiple psychotropic plants in a 1,000-year-old ritual bundle from South America', *Proceedings of the National Academy of Sciences*, 116 (2019), pp. 11207–12.

Mills, B. J., Batterman, S. A., Field, K. J., 'Nutrient acquisition by symbiotic fungi governs Palaeozoic climate transition', *Philosophical Transactions of the Royal Society B*, 373 (2017), 20160503.

Milner, D. S., Attah, V., Cook, E., Maguire, F., Savory, F. R., Morrison, M., Müller, C. A., Foster, P. G., Talbot, N. J., Leonard, G. et al., 'Environment-dependent fitness gains can be driven by horizontal gene transfer of transporter-encoding genes', *Proceedings of the National Academy of Sciences*, 116 (2019), 201815994.

Moeller, H. V., Neubert, M. G., 'Multiple friends with benefits: an optimal mutualist management strategy?' *American Naturalist*, 187 (2016), E1–12.

Mohajeri, H. M., Brummer, R. J., Rastall, R. A., Weersma, R. K., Harmsen, H. J., Faas, M., Eggersdorfer, M., 'The role of the microbiome for human health: from basic science to clinical applications', *European Journal of Nutrition*, 57 (2018), pp. 1–14.

Mohan, J. E., Cowden, C. C., Baas, P., Dawadi, A., Frankson, P. T., Helmick, K., Hughes, E., Khan, S., Lang, A., Machmuller, M. et al., 'Mycorrhizal fungi mediation of terrestrial ecosystem responses to global change: mini-review', *Fungal Ecology*, 10 (2014), pp. 3–19.

Moisan, K., Cordovez, V., van de Zande, E. M., Raaijmakers, J. M., Dicke, M., Lucas-Barbosa, D., 'Volatiles of pathogenic and non-pathogenic soil-borne fungi affect plant development and resistance to insects', *Oecologia*, 190 (2019), pp. 589–604.

Monaco, E., 'The Secret History of Paris's Catacomb Mushrooms', *Atlas Obscura* (2017), www.atlasobscura.com/articles/paris-catacomb-mushrooms [accessed 29 October 2019].

Mondo, S. J., Lastovetsky, O. A., Gaspar, M. L., Schwardt, N. H., Barber, C. C., Riley, R., Sun, H., Grigoriev, I. V., Pawlowska, T. E., 'Bacterial endosymbionts influence host sexuality and reveal reproductive genes of early divergent fungi', *Nature Communications*, 8 (2017), 1843.

Money, N. P., 'More g's than the Space Shuttle: ballistospore discharge', *Mycologia*, 90 (1998), p. 547.

Money, N. P., 'Fungus punches its way in', *Nature*, 401 (1999), pp. 332–3.

Money, N. P., 'The fungal dining habit: a biomechanical perspective', *Mycologist*, 18 (2004a), pp. 71–6.

Money, N. P., 'Theoretical biology: mushrooms in cyberspace', *Nature*, 431 (2004b), p. 32.

Money, N. P., *Triumph of the Fungi: A Rotten History* (Oxford, Oxford University Press, 2007).

Money, N. P., 'Against the naming of fungi', *Fungal Biology*, 117 (2013), pp. 463–5.

Money, N. P., *Fungi: A Very Short Introduction* (Oxford, Oxford University Press, 2016).

Money, N. P., *The Rise of Yeast* (Oxford, Oxford University Press, 2018).

Montañez, I., 'A Late Paleozoic climate window of opportunity', *Proceedings of the National Academy of Sciences*, 113 (2016), pp. 2334–6.

Montiel-Castro, A. J., González-Cervantes, R. M., Bravo-Ruiseco, G., Pacheco-López, G., 'The microbiota-gut-brain axis: neurobehavioral correlates, health and sociality', *Frontiers in Integrative Neuroscience*, 7 (2013), 70.

Moore, D., Hock, B., Greening, J. P., Kern, V. D., Frazer, L., Monzer, J., 'Gravimorphogenesis in agarics', *Mycological Research*, 100 (1996), pp. 257–73.

Moore D., 'Graviresponses in fungi', *Advances in Space Research*, 17 (1996), pp. 73–82.

Moore, D., 'Principles of mushroom developmental biology', *International Journal of Medicinal Mushrooms*, 7 (2005), pp. 79–101.

Moore, D., *Fungal Biology in the Origin and Emergence of Life* (Cambridge, Cambridge University Press, 2013a).

Moore, D., *Slayers, Saviors, Servants, and Sex: An Exposé of Kingdom Fungi* (Springer International Publishing, 2013b).

Moore, D., Robson, G. D., Trinci, A. P. J., *21st-Century Guidebook to Fungi* (Cambridge, Cambridge University Press, 2011).

Mousavi, S. A., Chauvin, A., Pascaud, F., Kellenberger, S., Farmer, E. E., 'Glutamate receptor-like genes mediate leaf-to-leaf wound signalling', *Nature*, 500 (2013), pp. 422–6.

Muday, G. K., Brown-Harding, H., 'Nervous system-like signaling in plant defense', *Science*, 361 (2018), pp. 1068–9.

Mueller, R. C., Scudder, C. M., Whitham, T. G., Gehring, C. A., 'Legacy effects of tree mortality mediated by ectomycorrhizal fungal communities', *New Phytologist*, 224 (2019), pp. 155–65.

Muir, J., *The Yosemite* (New York, The Century Company, 1912), vault.sierraclub.org/john_muir_exhibit/writings/the_yosemite/ [accessed 29 October 2019].

Myers, N., 'Conversations on plant sensing: notes from the field', *NatureCulture*, 3 (2014), pp. 35–66.

Naef, R., 'The volatile and semi-volatile constituents of agarwood, the infected heartwood of *Aquilaria* species: a review', *Flavour and Fragrance Journal*, 26 (2011), pp. 73–87.

Nakagaki, T., Yamada, H., Tóth, A., 'Maze-solving by an amoeboid organism', *Nature*, 407 (2000), p. 470.

Nelsen, M. P., DiMichele, W. A., Peters, S. E., Boyce, K. C., 'Delayed fungal evolution did not cause the Paleozoic peak in coal production', *Proceedings of the National Academy of Sciences*, 113 (2016), pp. 2442–7.

Nelson, M. L., Dinardo, A., Hochberg, J., Armelagos, G. J., 'Mass spectroscopic characterization of tetracycline in the skeletal remains of an ancient population from Sudanese Nubia 350–550 CE', *American Journal of Physical Anthropology*, 143 (2010), pp. 151–4.

Newman, E. I., 'Mycorrhizal links between plants: their functioning and ecological significance', *Advances in Ecological Research*, 18 (1988), pp. 243–70.

Nikolova, I., Johanson, K. J., Dahlberg, A., 'Radiocaesium in fruitbodies and mycorrhizae in ectomycorrhizal fungi', *Journal of Environmental Radioactivity*, 37 (1997), pp. 115–25.

Niksic, M., Hadzic, I., Glisic, M., 'Is *Phallus impudicus* a mycological giant?' *Mycologist*, 18 (2004), pp. 21–2.

Noë, R., Hammerstein, P., 'Biological markets', *Trends in Ecology & Evolution*, 10 (1995), pp. 336–9.

Noë, R., Kiers, T. E., 'Mycorrhizal markets, firms, and co-ops', *Trends in Ecology & Evolution*, 33 (2018), pp. 777–89.

Nordbring-Hertz, B., 'Morphogenesis in the nematode-trapping fungus *Arthrobotrys oligospora* – an extensive plasticity of infection structures', *Mycologist*, 18 (2004), pp. 125–33.

Nordbring-Hertz, B., Jansson, H., Tunlid, A., 'Nematophagous Fungi' in *Encyclopedia of Life Sciences* (Chichester, John Wiley, 2011).

Novikova, N., Boever, P., Poddubko, S., Deshevaya, E., Polikarpov, N., Rakova, N., Coninx, I., Mergeay, M., 'Survey of environmental biocontamination on board the International Space Station', *Research in Microbiology*, 157 (2006), pp. 5–12.

O'Malley, M. A., 'Endosymbiosis and its implications for evolutionary theory', *Proceedings of the National Academy of Sciences*, 112 (2015), pp. 10270–7.

O'Regan, H. J., Lamb, A. L., Wilkinson, D. M., 'The missing mushrooms: searching for fungi in ancient human dietary analysis', *Journal of Archaeological Science*, 75 (2016), pp. 139–43.

Oettmeier, C., Brix, K., Döbereiner, H.-G., '*Physarum polycephalum* – a new take on a classic model system', *Journal of Physics D: Applied Physics*, 50 (2017), p. 41.

Oliveira, A. G., Stevani, C. V., Waldenmaier, H. E., Viviani, V., Emerson, J. M., Loros, J. J., Dunlap, J. C., 'Circadian control sheds light on fungal bioluminescence', *Current Biology*, 25 (2015), pp. 964–8.

Olsson, S., 'Nutrient Translocation and Electrical Signalling in Mycelia', in eds. N. A. R. Gow, G. D. Robson and G. M. Gadd, *The Fungal Colony* (Cambridge, Cambridge University Press, 2009), pp. 25–48.

Olsson, S., Hansson, B., 'Action potential-like activity found in fungal mycelia is sensitive to stimulation', *Naturwissenschaften*, 82 (1995), pp. 30–1.

Oolbekkink, G. T., Kuyper, T. W., 'Radioactive caesium from Chernobyl in fungi', *Mycologist*, 3 (1989), pp. 3–6.

Orrell, P., *Linking Above and Below-Ground Interactions in Agro-Ecosystems: An Ecological Network Approach*, PhD thesis, University of Newcastle, Newcastle (2018), theses.ncl.ac.uk/jspui/handle/10443/4102 [accessed 29 October 2019].

Osborne, O. G., De-Kayne, R., Bidartondo, M. I., Hutton, I., Baker, W. J., Turnbull, C. G., Savolainen, V., 'Arbuscular mycorrhizal fungi promote coexistence and niche divergence of sympatric palm species on a remote oceanic island', *New Phytologist*, 217 (2018), pp. 1254–66.

Ott, J., 'Pharmaka, philtres, and pheromones. Getting high and getting off', *MAPS*, 12 (2002), pp. 26–32.

Otto, S., Bruni, E. P., Harms, H., Wick, L. Y., 'Catch me if you can: dispersal and foraging of *Bdellovibrio bacteriovorus* 109J along mycelia', *ISME Journal*, 11 (2017), pp. 386–93.

Ouellette, N. T., 'Flowing crowds', *Science*, 363 (2019), pp. 27–8.

Oukarroum, A., Gharous, M., Strasser, R. J., 'Does *Parmelina tiliacea* lichen photosystem II survive at liquid nitrogen temperatures?' *Cryobiology*, 74 (2017), pp. 160–2.

Ovid, *Ovid: The Metamorphoses*, Gregory, H., trans. (New York, Viking Press, 1958).

Pagán, O. R., 'The brain: a concept in flux', *Philosophical Transactions of the Royal Society B*, 374 (2019), 20180383.

Paglia, C., *Sexual Personae: Art and Decadence from Nefertiti to Emily Dickinson* (New Haven, CT, Yale University Press, 2001).

Pan, X., Pike, A., Joshi, D., Bian, G., McFadden, M. J., Lu, P., Liang, X., Zhang, F., Raikhel, A. S., Xi, Z., 'The bacterium *Wolbachia* exploits host innate immunity to establish a symbiotic relationship with the dengue vector mosquito *Aedes aegypti*', *ISME Journal*, 12 (2017), pp. 277–88.

Patra, S., Banerjee, S., Terejanu, G., Chanda, A., 'Subsurface pressure profiling: a novel mathematical paradigm for computing colony pressures on substrate during fungal infections', *Scientific Reports*, 5 (2015), 12928.

Peay, K. G., 'The mutualistic niche: mycorrhizal symbiosis and community dynamics', *Annual Review of Ecology, Evolution, and Systematics*, 47 (2016), pp. 1–22.

Peay, K. G., Kennedy, P. G., Talbot, J. M., 'Dimensions of biodiversity in the Earth mycobiome', *Nature Reviews Microbiology*, 14 (2016), pp. 434–47.

Peintner, U., Poder, R., Pumpel, T., 'The iceman's fungi', *Mycological Research*, 102 (1998), pp. 1153–62.

Pennazza, G., Fanali, C., Santonico, M., Dugo, L., Cucchiarini, L., Dachà, M., D'Amico, A., Costa, R., Dugo, P., Mondello, L., 'Electronic nose and GC–MS analysis of volatile compounds in *Tuber magnatum* Pico: evaluation of different storage conditions', *Food Chemistry*, 136 (2013), pp. 668–74.

Pennisi, E., 'Chemicals released by bacteria may help gut control the brain, mouse study suggests', *Science* (2019a), www.sciencemag.org/news/2019/10/chemicals-released-bacteria-may-help-gut-control-brain-mouse-study-suggests [accessed 29 October 2019].

Pennisi, E., 'Algae suggest eukaryotes get many gifts of bacteria DNA', *Science* 363 (2019b), pp. 439–40.

Peris, J. E., Rodríguez, A., Peña, L., Fedriani, J., 'Fungal infestation boosts fruit aroma and fruit removal by mammals and birds', *Scientific Reports*, 7 (2017), 5646.

Perrottet, T., 'Mt. Rushmore', *Smithsonian Magazine* (2006), www.smithsonianmag.com/travel/mt-rushmore-116396890/ [accessed 29 October 2019].

Petri, G., Expert, P., Turkheimer, F., Carhart-Harris, R., Nutt, D., Hellyer, P., Vaccarino, F., 'Homological scaffolds of brain functional networks', *Journal of the Royal Society Interface*, 11 (2014), 20140873.

Pfeffer, C., Larsen, S., Song, J., Dong, M., Besenbacher, F., Meyer, R., Kjeldsen, K., Schreiber, L., Gorby, Y. A., El-Naggar, M. Y. et al., 'Filamentous bacteria transport electrons over centimetre distances', *Nature*, 491 (2012), pp. 218–21.

Phillips, R. P., Brzostek, E., Midgley, M. G., 'The mycorrhizal-associated nutrient economy: a new framework for predicting carbon–nutrient couplings in temperate forests', *New Phytologist*, 199 (2013), pp. 41–51.

Pickles, B., Egger, K., Massicotte, H., Green, D., 'Ectomycorrhizas and climate change', *Fungal Ecology*, 5 (2012), pp. 73–84.

Pickles, B. J., Wilhelm, R., Asay, A. K., Hahn, A. S., Simard, S. W., Mohn, W. W., 'Transfer of 13C between paired Douglas fir seedlings reveals plant kinship effects and uptake of exudates by ectomycorrhizas', *New Phytologist*, 214 (2017), pp. 400–11.

Pion, M., Spangenberg, J., Simon, A., Bindschedler, S., Flury, C., Chatelain, A., Bshary, R., Job, D., Junier, P., 'Bacterial farming by the fungus *Morchella crassipes*', *Proceedings of the Royal Society B*, 280 (2013), 20132242.

Pirozynski, K. A., Malloch, D. W., 'The origin of land plants: a matter of mycotrophism', *Biosystems*, 6 (1975), pp. 153–64.

Pither, J., Pickles, B. J., Simard, S. W., Ordonez, A., Williams, J. W., 'Below-ground biotic interactions moderated the postglacial range dynamics of trees', *New Phytologist*, 220 (2018), pp. 1148–60.

Policha, T., Davis, A., Barnadas, M., Dentinger, B. T., Raguso, R. A., Roy, B. A., 'Disentangling visual and olfactory signals in mushroom-mimicking *Dracula* orchids using realistic three-dimensional printed flowers', *New Phytologist*, 210 (2016), pp. 1058–71.

Pollan, M., 'The Intelligent Plant', *New Yorker* (2013), michaelpollan.com/articles-archive/the-intelligent-plant/ [accessed 29 October 2019].

Pollan, M., *How to Change Your Mind: The New Science of Psychedelics* (London, Penguin, 2018).

Popkin, G., 'Bacteria Use Brainlike Bursts of Electricity to Communicate', *Quanta* (2017), www.quantamagazine.org/bacteria-use-brainlike-bursts-of-electricity-to-communicate-20170905/ [accessed 29 October 2019].

Porada, P., Weber, B., Elbert, W., Pöschl, U., Kleidon, A., 'Estimating impacts of lichens and bryophytes on global biogeochemical cycles', *Global Biogeochemical Cycles*, 28 (2014), pp. 71–85.

Potts, S. G., Biesmeijer, J. C., Kremen, C., Neumann, P., Schweiger, O., Kunin, W. E., 'Global pollinator declines: trends, impacts and drivers', *Trends in Ecology & Evolution*, 25 (2010), pp. 345–53.

Poulsen, M., Hu, H., Li, C., Chen, Z., Xu, L., Otani, S., Nygaard, S., Nobre, T., Klaubauf, S., Schindler, P. M. et al., 'Complementary symbiont contributions to plant decomposition in a fungus-farming termite', *Proceedings of the National Academy of Sciences*, 111 (2014), pp. 14500–5.

Powell, J. R., Rillig, M. C., 'Biodiversity of arbuscular mycorrhizal fungi and ecosystem function', *New Phytologist*, 220 (2018), pp. 1059–75.

Powell, M., *Medicinal Mushrooms: A Clinical Guide* (Bath, Mycology Press, 2014).

Pozo, M. J., López-Ráez, J. A., Azcón-Aguilar, C., García-Garrido, J. M., 'Phytohormones as integrators of environmental signals in the regulation of mycorrhizal symbioses', *New Phytologist*, 205 (2015), pp. 1431–6.

Prasad, S., 'An ingenious way to combat India's suffocating pollution', *Washington Post* (2018), www.washingtonpost.com/news/theworldpost/wp/2018/08/01/india-pollution/ [accessed 29 October 2019].

Pressel, S., Bidartondo, M. I., Ligrone, R., Duckett, J. G., 'Fungal symbioses in bryophytes: new insights in the twenty-first century', *Phytotaxa*, 9 (2010), pp. 238–53.

Prigogine, I., Stengers, I., *Order Out of Chaos: Man's New Dialogue with Nature* (New York, Bantam, 1984).

Prindle, A., Liu, J., Asally, M., Ly, S., Garcia-Ojalvo, J., Süel, G. M., 'Ion channels enable electrical communication in bacterial communities', *Nature*, 527 (2015), pp. 59–63.

Purschwitz, J., Müller, S., Kastner, C., Fischer, R., 'Seeing the rainbow: light sensing in fungi', *Current Opinion in Microbiology*, 9 (2006), pp. 566–71.

Quéré, C., Andrew, R. M., Friedlingstein, P., Sitch, S., Hauck, J., Pongratz, J., Pickers, P., Korsbakken, J., Peters, G. P., Canadell, J. G. et al., 'Global Carbon Budget 2018', *Earth System Science Data Discussions* (2018), doi.org/10.5194/essd-2018-120.

Quintana-Rodriguez, E., Rivera-Macias, L. E., Adame-Alvarez, R. M., Torres, J., Heil, M., 'Shared weapons in fungus–fungus and fungus–plant interactions? Volatile organic compounds of plant or fungal origin exert direct antifungal activity *in vitro*', *Fungal Ecology*, 33 (2018), pp. 115–21.

Quirk. J., Andrews, M., Leake, J., Banwart, S., Beerling, D., 'Ectomycorrhizal fungi and past high CO_2 atmospheres enhance mineral weathering through increased below-ground carbon-energy fluxes', *Biology Letters*, 10 (2014), 20140375.

Rabbow, E., Horneck, G., Rettberg, P., Schott, J.-U., Panitz, C., L'Afflitto, A., von Heise-Rotenburg, R., Willnecker, R., Baglioni, P., Hatton, J. et al., 'EXPOSE, an astrobiological exposure facility on the International Space Station – from proposal to flight', *Origins of Life and Evolution of Biospheres*, 39 (2009), pp. 581–98.

Raes, J., 'Crowdsourcing Earth's microbes', *Nature*, 551 (2017), pp. 446–7.

Rambold, G., Stadler, M., Begerow, D., 'Mycology should be recognised as a field in biology at eye level with other major disciplines – a memorandum', *Mycological Progress*, 12 (2013), pp. 455–63.

Ramsbottom, J., *Mushrooms and Toadstools* (London, Collins, 1953).

Raverat, G., *Period Piece: A Cambridge Childhood* (London, Faber, 1952).

Rayner, A., *Degrees of Freedom* (London, World Scientific, 1997).

Rayner, A., Griffiths, G. S., Ainsworth, A. M., 'Mycelial Interconnectedness', in eds. N. A. R. Gow and G. M. Gadd, *The Growing Fungus* (London, Chapman and Hall, 1995), pp. 21–40.

Rayner, M., *Trees and Toadstools* (London, Faber, 1945).

Read, D., 'Mycorrhizal fungi: the ties that bind', *Nature*, 388 (1997), pp. 517–18.

Read, N., 'Fungal Cell Structure and Organization', in eds. C. C. Kibbler, R. Barton, N. A. R. Gow, S. Howell, S., MacCallum, D. M. Manuel, and R. J. *Oxford Textbook of Medical Mycology* (Oxford, Oxford University Press, 2018), pp. 23–34.

Read, N. D., Lichius, A., Shoji, J., Goryachev, A. B., 'Self-signalling and self-fusion in filamentous fungi', *Current Opinion in Microbiology*, 12 (2009), pp. 608–15.

Redman, R. S., Rodriguez, R. J., 'The Symbiotic Tango: Achieving Climate-Resilient Crops via Mutualistic Plant – Fungus Relationships', in ed. S. Doty, *Functional Importance of the Plant Microbiome, Implications for Agriculture, Forestry and Bioenergy* (Springer International Publishing, 2017), pp. 71–87.

Rees, B., Shepherd, V. A., Ashford, A. E., 'Presence of a motile tubular vacuole system in different phyla of fungi', *Mycological Research*, 98 (1994), pp. 985–92.

Reid, C. R., Latty, T., Dussutour, A., Beekman, M., 'Slime mold uses an externalized spatial "memory" to navigate in complex environments', *Proceedings of the National Academy of Sciences*, 109 (2012), pp. 17490–4.

Relman, D. A., '"Til death do us part": coming to terms with symbiotic relationships', *Nature Reviews Microbiology*, 6 (2008), pp. 721–4.

Reynaga-Peña, C. G., Bartnicki-García, S., 'Cytoplasmic contractions in growing fungal hyphae and their morphogenetic consequences', *Archives of Microbiology*, 183 (2005), pp. 292–300.

Reynolds, H. T., Vijayakumar, V., Gluck-Thaler, E., Korotkin, H., Matheny, P., Slot, J. C., 'Horizontal gene cluster transfer increased hallucinogenic mushroom diversity', *Evolution Letters*, 2 (2018), pp. 88–101.

Rich, A., 'Notes toward a Politics of Location', in *Blood, Bread, and Poetry: Selected Prose, 1979–1985* (New York, W. W. Norton, 1994).

Richards, T. A., Leonard, G., Soanes, D. M., Talbot, N. J., 'Gene transfer into the fungi', *Fungal Biology Reviews*, 25 (2011), pp. 98–110.

Rillig, M. C., Aguilar-Trigueros, C. A., Camenzind, T., Cavagnaro, T. R., Degrune, F., Hohmann, P., Lammel, D. R., Mansour, I., Roy, J., van der Heijden, M. G. et al., 'Why farmers should manage the arbuscular mycorrhizal symbiosis: a response to Ryan & Graham (2018), "Little evidence that farmers should consider abundance or diversity of arbuscular mycorrhizal fungi when managing crops"', *New Phytologist*, 222 (2019), pp. 1171–5.

Rillig, M. C., Lehmann, A., Lehmann, J., Camenzind, T., Rauh, C., 'Soil biodiversity effects from field to fork', *Trends in Plant Science*, 23 (2018), pp. 17–24.

Riquelme, M., 'Tip growth in filamentous fungi: a road trip to the apex', *Microbiology*, 67 (2012), pp. 587–609.

Ritz, K., Young, I., 'Interactions between soil structure and fungi', *Mycologist*, 18 (2004), pp. 52–9.

Robinson, J. M., 'Lignin, land plants, and fungi: biological evolution affecting Phanerozoic oxygen balance', *Geology*, 18 (1990), pp. 607–10.

Rodriguez, R., White, J. F., Arnold, A., Redman, R., 'Fungal endophytes: diversity and functional roles', *New Phytologist*, 182 (2009), pp. 314–30.

Rodriguez-Romero, J., Hedtke, M., Kastner, C., Müller, S., Fischer, R., 'Fungi, hidden in soil or up in the air: light makes a difference', *Microbiology*, 64 (2010), pp. 585–610.

Rogers, R., *The Fungal Pharmacy* (Berkeley, CA, North Atlantic Books, 2012).

Roper, M., Dressaire, E., 'Fungal biology: bidirectional communication across fungal networks', *Current Biology*, 29 (2019), R130–2.

Roper, M., Lee, C., Hickey, P. C., Gladfelter, A. S., 'Life as a moving fluid: fate of cytoplasmic macromolecules in dynamic fungal syncytia', *Current Opinion in Microbiology*, 26 (2015), pp. 116–22.

Roper, M., Seminara, A., 'Mycofluidics: the fluid mechanics of fungal adaptation', *Annual Review of Fluid Mechanics*, 51 (2017), pp. 1–28.

Roper, M., Seminara, A., Bandi, M., Cobb, A., Dillard, H. R., Pringle, A., 'Dispersal of fungal spores on a cooperatively generated wind', *Proceedings of the National Academy of Sciences*, 107 (2010), pp. 17474–9.

Roper, M., Simonin, A., Hickey, P. C., Leeder, A., Glass, L. N., 'Nuclear dynamics in a fungal chimera', *Proceedings of the National Academy of Sciences*, 110 (2013), pp. 12875–80.

Ross, A. A., Müller, K. M., Weese, J. S., Neufeld, J. D., 'Comprehensive skin microbiome analysis reveals the uniqueness of human skin and evidence for phylosymbiosis within the class Mammalia', *Proceedings of the National Academy of Sciences*, 115 (2018), E5786–95.

Ross, S., Bossis, A., Guss, J., Agin-Liebes, G., Malone, T., Cohen, B., Mennenga, S., Belser, A., Kalliontzi, K., Babb, J. et al., 'Rapid and sustained symptom reduction following psilocybin treatment for anxiety and depression in patients with life-threatening cancer: a randomized controlled trial', *Journal of Psychopharmacology*, 30 (2016), pp. 1165–80.

Roughgarden, J., *Evolution's Rainbow* (Berkeley, University of California Press, 2013).

Rouphael, Y., Franken, P., Schneider, C., Schwarz, D., Giovannetti, M., Agnolucci, M., Pascale, S., Bonini, P., Colla, G., 'Arbuscular mycorrhizal fungi act as biostimulants in horticultural crops', *Scientia Horticulturae*, 196 (2015), pp. 91–108.

Rubini, A., Riccioni, C., Arcioni, S., Paolocci, F., 'Troubles with truffles: unveiling more of their biology', *New Phytologist*, 174 (2007), pp. 256–9.

Russell, B., *Portraits from Memory and Other Essays* (New York, Simon and Schuster, 1956).

Ryan, M. H., Graham, J. H., 'Little evidence that farmers should consider abundance or diversity of arbuscular mycorrhizal fungi when managing crops', *New Phytologist*, 220 (2018), pp. 1092–1107.

Sagan, L., 'On the origin of mitosing cells', *Journal of Theoretical Biology*, 14 (1967), pp. 225–74.

Salvador-Recatalà, V., Tjallingii, F. W., Farmer, E. E., 'Real-time, *in vivo* intracellular recordings of caterpillar-induced depolarization waves in sieve elements using aphid electrodes', *New Phytologist*, 203 (2014), pp. 674–84.

Sample, I., 'Magma shift may have caused mysterious seismic wave event', *Guardian* (2018), www.theguardian.com/science/2018/nov/30/magma-shift-mysterious-seismic-wave-event-mayotte [accessed 29 October 2019].

Samorini, G., *Animals and Psychedelics: The Natural World and the Instinct to Alter Consciousness* (Rochester, VT, Park Street Press, 2002).

Sancho, L. G., de la Torre, R., Pintado, A., 'Lichens, new and promising material from experiments in astrobiology', *Fungal Biology Reviews*, 22 (2008), pp. 103–9.

Sapp, J., *Evolution by Association* (Oxford, Oxford University Press, 1994).

Sapp, J., 'The dynamics of symbiosis: an historical overview', *Canadian Journal of Botany*, 82 (2004), pp. 1046–56.

Sapp, J., *The New Foundations of Evolution* (Oxford, Oxford University Press, 2009).

Sapp, J., 'The symbiotic self', *Evolutionary Biology*, 43 (2016), pp. 596–603.

Sapsford, S. J., Paap, T., Hardy, G. E., Burgess, T. I., 'The "chicken or the egg": which comes first, forest tree decline or loss of mycorrhizae?' *Plant Ecology*, 218 (2017), pp. 1093–1106.

Sarrafchi, A., Odhammer, A. M., Salazar, L., Laska, M., 'Olfactory sensitivity for six predator odorants in cd-1 mice, human subjects, and spider monkeys', *PLOS ONE*, 8 (2013), e80621.

Saupe, S., 'Molecular genetics of heterokaryon incompatibility in filamentous ascomycetes', *Microbiology and Molecular Biology Reviews*, 64 (2000), pp. 489–502.

Scharf, C., 'How the Cold War Created Astrobiology', *Nautilus* (2016), nautil.us/issue/32/space/how-the-cold-war-created-astrobiology-rp [accessed 29 October 2019].

Scharlemann, J. P., Tanner, E. V., Hiederer, R., Kapos, V., 'Global soil carbon: understanding and managing the largest terrestrial carbon pool', *Carbon Management*, 5 (2014), pp. 81–91.

Schenkel, D., Maciá-Vicente, J. G., Bissell, A., Splivallo, R., 'Fungi indirectly affect plant root architecture by modulating soil volatile organic compounds', *Frontiers in Microbiology*, 9 (2018), 1847.

Schmieder, S. S., Stanley, C. E., Rzepiela, A., van Swaay, D., Sabotič, J., Nørrelykke, S. F., deMello, A. J., Aebi, M., Künzler, M., 'Bidirectional propagation of signals and nutrients in fungal networks via specialized hyphae', *Current Biology*, 29 (2019), pp. 217–28.

Schmull, M., Dal-Forno, M., Lücking, R., Cao, S., Clardy, J., Lawrey, J. D., '*Dictyonema huaorani* (Agaricales: Hygrophoraceae), a new lichenized basidiomycete from Amazonian Ecuador with presumed hallucinogenic properties', *Bryologist*, 117 (2014), pp. 386–94.

Schultes, R., Hofmann, A., Rätsch, C., *Plants of the Gods: Their Sacred, Healing, and Hallucinogenic Powers*, 2nd edition (Rochester, VT, Healing Arts Press, 2001).

Schultes, R. E., 'Teonanacatl: the narcotic mushroom of the Aztecs', *American Anthropologist*, 42 (1940), pp. 429–43.

Seaward, M., 'Environmental Role of Lichens', in ed. T. H. Nash, *Lichen Biology* (Cambridge, Cambridge University Press, 2008), pp. 274–98.

Selosse, M.-A., 'Prototaxites: a 400-Myr-old giant fossil, a saprophytic holobasidiomycete, or a lichen?' *Mycological Research*, 106 (2002), pp. 641–4.

Selosse, M.-A., Schneider-Maunoury, L., Martos, F., 'Time to re-think fungal ecology? Fungal ecological niches are often prejudged', *New Phytologist*, 217 (2018), pp. 968–72.

Selosse, M.-A., Schneider-Maunoury, L., Taschen E., Rousset, F., Richard, F., 'Black truffle, a hermaphrodite with forced unisexual behaviour', *Trends in Microbiology*, 25 (2017), pp. 784–7.

Selosse, M.-A., Strullu-Derrien, C., Martin, F. M., Kamoun, S., Kenrick, P., 'Plants, fungi and oomycetes: a 400-million-year affair that shapes the biosphere', *New Phytologist*, 206 (2015), pp. 501–6.

Selosse, M.-A., Tacon, L. F., 'The land flora: a phototroph–fungus partnership?' *Trends in Ecology & Evolution*, 13 (1998), pp. 15–20.

Sergeeva, N. G., Kopytina, N. I., 'The first marine filamentous fungi discovered in the bottom sediments of the oxic/anoxic interface and in the bathyal zone of the Black Sea', *Turkish Journal of Fisheries and Aquatic Sciences*, 14 (2014), pp. 497–505.

Sheldrake, M., Rosenstock, N. P., Revillini, D., Olsson, P. A., Wright, S. J., Turner, B. L, 'A phosphorus threshold for mycoheterotrophic plants in tropical forests', *Proceedings of the Royal Society B*, 284 (2017), 20162093.

Shepherd, V., Orlovich, D., Ashford, A., 'Cell-to-cell transport via motile tubules in growing hyphae of a fungus', *Journal of Cell Science*, 105 (1993), pp. 1173–8.

Shomrat, T., Levin, M., 'An automated training paradigm reveals long-term memory in planarians and its persistence through head regeneration', *Journal of Experimental Biology*, 216 (2013), pp. 3799–3810.

Shukla, V., Joshi, G. P., Rawat, M. S. M., 'Lichens as a potential natural source of bioactive compounds: a review', *Phytochemical Reviews*, 9 (2010), pp. 303–14.

Siegel, R. K., *Intoxication: The Universal Drive for Mind-Altering Substances* (Rochester, VT, Park Street Press, 2005).

Silvertown, J., 'A new dawn for citizen science', *Trends in Ecology & Evolution*, 24 (2009), pp. 467–71.

Simard, S., 'Mycorrhizal Networks Facilitate Tree Communication, Learning, and Memory', in eds. F. Baluska, M. Gagliano and G. Witzany, *Memory and Learning in Plants* (Springer International Publishing, 2018), pp. 191–213.

Simard, S., Asay, A., Beiler, K., Bingham, M., Deslippe, J., He, X., Phillip, L., Song, Y., Teste, F., 'Resource Transfer between Plants through Ectomycorrhizal Fungal Networks', in ed. T. Horton, *Mycorrhizal Networks* (Springer International Publishing, 2015), pp. 133–76.

Simard, S. W., Beiler, K. J., Bingham, M. A., Deslippe, J. R., Philip, L. J., Teste, F. P., 'Mycorrhizal networks: mechanisms, ecology and modelling', *Fungal Biology Reviews*, 26 (2012), pp. 39–60.

Simard, S., Perry, D. A., Jones, M. D., Myrold, D. D., Durall, D. M., Molina, R., 'Net transfer of carbon between ectomycorrhizal tree species in the field', *Nature*, 388 (1997), pp. 579–82.

Singh, H., *Mycoremediation* (New York, John Wiley, 2006).

Slayman, C., Long, W., Gradmann, D., '"Action potentials" in *Neurospora crassa*, a mycelial fungus', *Biochimica et Biophysica Acta*, 426 (1976), pp. 732–44.

Smith, S. E., Read, D. J., *Mycorrhizal Symbiosis* (London, Academic Press, 2008).

Solé, R., Moses, M., Forrest, S., 'Liquid brains, solid brains', *Philosophical Transactions of the Royal Society B*, 374 (2019), 20190040.

Soliman, S., Greenwood, J. S, Bombarely, A., Muelle, L. A., Tsao, R. Mosser, D. D., Raizada, M. N., 'An endophyte constructs fungicide-containing extracellular barriers for its host plant', *Current Biology*, 25 (2015), pp. 2570–6.

Song, Y., Zeng, R., 'Interplant communication of tomato plants through underground common mycorrhizal networks', *PLOS ONE*, 5 (2010), e11324.

Song, Y., Simard, S. W., Carroll, A., Mohn, W. W., Zeng, R., 'Defoliation of interior Douglas fir elicits carbon transfer and stress signalling to ponderosa pine neighbors through ectomycorrhizal networks', *Scientific Reports*, 5 (2015a), 8495.

Song, Y., Ye, M., Li, C., He, X., Zhu-Salzman, K., Wang, R., Su, Y., Luo, S., Zeng, R., 'Hijacking common mycorrhizal networks for herbivore-induced defence signal transfer between tomato plants', *Scientific Reports*, 4 (2015b), 3915.

Southworth, D., He, X.-H., Swenson, W., Bledsoe, C., Horwath, W., 'Application of network theory to potential mycorrhizal networks', *Mycorrhiza*, 15 (2005), pp. 589–95.

Spanos, N. P., Gottleib, J., 'Ergotism and the Salem village witch trials', *Science*, 194 (1976), pp. 1390–4.

Splivallo, R., Novero, M., Bertea, C. M., Bossi, S., Bonfante, P., 'Truffle volatiles inhibit growth and induce an oxidative burst in *Arabidopsis thaliana*', *New Phytologist*, 175 (2007), pp. 417–24.

Splivallo, R., Fischer, U., Göbel, C., Feussner, I., Karlovsky, P., 'Truffles regulate plant root morphogenesis via the production of auxin and ethylene', *Plant Physiology*, 150 (2009), pp. 2018–29.

Splivallo, R., Ottonello, S., Mello, A., Karlovsky, P., 'Truffle volatiles: from chemical ecology to aroma biosynthesis', *New Phytologist*, 189 (2011), pp. 688–99.

Spribille, T., Tuovinen, V., Resl, P., Vanderpool, D., Wolinski, H., Aime, C. M., Schneider, K., Stabentheiner, E., Toome-Heller, M., Thor, G. et al., 'Basidiomycete yeasts in the cortex of ascomycete macrolichens', *Science*, 353 (2016), pp. 488–92.

Spribille, T., 'Relative symbiont input and the lichen symbiotic outcome', *Current Opinion in Plant Biology*, 44 (2018), pp. 57–63.

Stamets, P., *Psilocybin Mushrooms of the World* (Berkeley, CA, Ten Speed Press, 1996).

Stamets, P., 'Global Ecologies, World Distribution and Relative Potency of Psilocybin Mushrooms', in ed. R. Metzner, *Sacred Mushroom of Visions: Teonanacatl* (Rochester, VT, Park Street Press, 2005), pp. 69–75.

Stamets, P., *Mycelium Running* (Berkeley, CA, Ten Speed Press, 2011).

Stamets, P. E., Naeger, N. L., Evans, J. D., Han, J. O., Hopkins, B. K., Lopez, D., Moershel, H. M., Nally, R., Sumerlin, D., Taylor, A. W. et al., 'Extracts of polypore mushroom mycelia reduce viruses in honey bees', *Scientific Reports*, 8 (2018), 3936.

State of the World's Fungi (Kew, Royal Botanic Gardens, 2018), stateoftheworldsfungi. org [accessed 29 October 2019].

Steele, E. J., Al-Mufti, S., Augustyn, K. A., Chandrajith, R., Coghlan, J. P., Coulson, S. G., Ghosh, S., Gillman, M., Gorczynski, R. M., Klyce, B. et al., 'Cause of Cambrian explosion – terrestrial or cosmic?' *Progress in Biophysics and Molecular Biology*, 136 (2018) pp. 3–23.

Steidinger, B., Crowther, T., Liang, J., Nuland, V. M., Werner, G., Reich, P., Nabuurs, G., de-Miguel, S., Zhou, M., Picard, N. et al., 'Climatic controls of decomposition drive the global biogeography of forest–tree symbioses', *Nature*, 569 (2019), pp. 404–8.

Steinberg, G., 'Hyphal growth: a tale of motors, lipids and the *Spitzenkörper*', *Eukaryotic Cell*, 6 (2007), pp. 351–60.

Steinhardt, J. B., *Mycelium is the Message: Open Science, Ecological Values and Alternative Futures with Do-It-Yourself Mycologists* (PhD thesis, University of California, Santa Barbara, 2018).

Stierle, A., Strobel, G., Stierle, D., 'Taxol and taxane production by *Taxomyces andreanae*, an endophytic fungus of Pacific yew', *Science*, 260 (1993), pp. 214–16.

Stough, J. M., Yutin, N., Chaban, Y. V., Moniruzzaman, M., Gann, E. R., Pound, H. L., Steffen, M. M., Black, J. N., Koonin, E. V., Wilhelm, S. W. et al., 'Genome and environmental activity of a chrysochromulina parva virus and its virophages', *Frontiers in Microbiology*, 10 (2019), 703.

Strullu-Derrien, C., Selosse, M.-A., Kenrick, P., Martin, F. M., 'The origin and evolution of mycorrhizal symbioses: from palaeomycology to phylogenomics', *New Phytologist*, 220 (2018), pp. 1012–30.

Studerus, E., Kometer, M., Hasler, F., Vollenweider, F. X., 'Acute, subacute and long-term subjective effects of psilocybin in healthy humans: a pooled analysis of experimental studies', *Journal of Psychopharmacology*, 25 (2011), pp. 1434–52.

Stukeley, W., *Memories of Sir Isaac Newton's Life* (Unpublished, 1752; available from website of the Royal Society: ttp.royalsociety.org/ttp/ttp.html?id=1807da00- [accessed 29 October 2019].

Suarato, G., Bertorelli, R., Athanassiou, A., 'Borrowing from nature: biopolymers and biocomposites as smart wound care materials', *Frontiers in Bioengineering and Biotechnology*, 6 (2018), 137.

Sudbery, P., Gow, N., Berman, J., 'The distinct morphogenic states of *Candida albicans*', *Trends in Microbiology*, 12 (2004), pp. 317–24.

Swift, R. S., 'Sequestration of carbon by soil', *Soil Science*, 166 (2001), pp. 858–71.

Taiz, L., Alkon, D., Draguhn, A., Murphy, A., Blatt, M., Hawes, C., Thiel, G., Robinson, D. G., 'Plants neither possess nor require consciousness', *Trends in Plant Science*, 24 (2019), pp. 677–87.

Takaki, K., Yoshida, K., Saito, T., Kusaka, T., Yamaguchi, R., Takahashi, K., Sakamoto, Y., 'Effect of electrical stimulation on fruit body formation in cultivating mushrooms', *Microorganisms*, 2 (2014), pp. 58–72.

Talou, T., Gaset, A., Delmas, M., Kulifaj, M., Montant, C., 'Dimethyl sulphide: the secret for black truffle hunting by animals?' *Mycological Research*, 94 (1990), pp. 277–8.

Tanney, J. B., Visagie, C. M., Yilmaz, N., Seifert, K. A., 'Aspergillus subgenus *Polypaecilum* from the built environment', *Studies in Mycology*, 88 (2017), pp. 237–67.

Taschen, E., Rousset, F., Sauve, M., Benoit, L., Dubois, M.-P., Richard, F., Selosse, M.-A., 'How the truffle got its mate: insights from genetic structure in spontaneous and planted Mediterranean populations of *Tuber melanosporum*', *Molecular Ecology*, 25 (2016), pp. 5611–27.

Taylor, A., Flatt, A., Beutel, M., Wolff, M., Brownson, K., Stamets, P., 'Removal of *Escherichia coli* from synthetic stormwater using mycofiltration', *Ecological Engineering*, 78 (2015), pp. 79–86.

Taylor, L., Leake, J., Quirk, J., Hardy, K., Banwart, S., Beerling, D., 'Biological weathering and the long-term carbon cycle: integrating mycorrhizal evolution and function into the current paradigm', *Geobiology*, 7 (2009), pp. 171–91.

Taylor, T., Klavins, S., Krings, M., Taylor, E., Kerp, H., Hass, H., 'Fungi from the Rhynie chert: a view from the dark side', *Transactions of the Royal Society of Edinburgh: Earth Sciences*, 94 (2007), pp. 457–73.

Temple, R., 'The prehistory of panspermia: astrophysical or metaphysical?' *International Journal of Astrobiology*, 6 (2007), pp. 169–80.

Tero, A., Takagi, S., Saigusa, T., Ito, K., Bebber, D. P., Fricker, M. D., Yumiki, K., Kobayashi, R., Nakagaki, T., 'Rules for biologically inspired adaptive network design', *Science*, 327 (2010), pp. 439–42.

Terrer, C., Vicca, S., Hungate, B. A., Phillips, R. P., Prentice, I. C., 'Mycorrhizal association as a primary control of the CO_2 fertilization effect', *Science*, 353 (2016), pp. 72–4.

Thierry, G., 'Lab-grown mini brains: we can't dismiss the possibility that they could one day outsmart us', *Conversation* (2019), theconversation.com/lab-grown-mini-brains-we-cant-dismiss-the-possibility-that-they-could-one-day-outsmart-us-125842 [accessed 29 October 2019].

Thirkell, T. J., Charters, M. D., Elliott, A. J., Sait, S. M., Field, K. J., 'Are mycorrhizal fungi our sustainable saviours? Considerations for achieving food security', *Journal of Ecology*, 105 (2017), pp. 921–9.

Thirkell, T. J., Pastok, D., Field, K. J., 'Carbon for nutrient exchange between arbuscular mycorrhizal fungi and wheat varies according to cultivar and changes in atmospheric carbon dioxide concentration', *Global Change Biology* (2019), DOI: 10.1111/gcb.14851.

Thomas, P., Büntgen, U., 'First harvest of Périgord black truffle in the UK as a result of climate change', *Climate Research*, 74 (2017), pp. 67–70.

Tilman, D., Balzer, C., Hill, J., Befort, B. L., 'Global food demand and the sustainable intensification of agriculture', *Proceedings of the National Academy of Sciences*, 108 (2011), pp. 20260–4.

Tilman, D., Cassman, K. G., Matson, P. A., Naylor, R., Polasky, S., 'Agricultural sustainability and intensive production practices', *Nature*, 418 (2002), pp. 671–7.

Tkavc, R., Matrosova, V. Y., Grichenko, O. E., Gostinčar, C., Volpe, R. P., Klimenkova, P., Gaidamakova, E. K., Zhou, C. E., Stewart, B. J., Lyman, M. G. et al., 'Prospects for fungal bioremediation of acidic radioactive waste sites: characterization and genome sequence of *Rhodotorula taiwanensis* MD1149', *Frontiers in Microbiology*, 8 (2018), 2528.

Tlalka, M., Hensman, D., Darrah, P., Watkinson, S., Fricker, M. D., 'Noncircadian oscillations in amino acid transport have complementary profiles in assimilatory and foraging hyphae of *Phanerochaete velutina*', *New Phytologist*, 158 (2003), pp. 325–35.

Tlalka, M., Bebber, D. P., Darrah, P. R., Watkinson, S. C., Fricker, M. D., 'Emergence of self-organised oscillatory domains in fungal mycelia', *Fungal Genetics and Biology*, 44 (2007), pp. 1085–95.

Toju, H., Guimarães, P. R., Olesen, J. M., Thompson, J. N., 'Assembly of complex plant–fungus networks', *Nature Communications*, 5 (2014), 5273.

Toju, H., Yamamoto, S., Tanabe, A. S., Hayakawa, T., Ishii, H. S., 'Network modules and hubs in plant-root fungal biomes', *Journal of the Royal Society Interface*, 13 (2016), 20151097.

Toju, H., Peay, K. G., Yamamichi, M., Narisawa, K., Hiruma, K., Naito, K., Fukuda, S., Ushio, M., Nakaoka, S., Onoda, Y. et al., 'Core microbiomes for sustainable agroecosystems', *Nature Plants*, 4 (2018), pp. 247–57.

Toju, H., Sato, H., 'Root-associated fungi shared between arbuscular mycorrhizal and ectomycorrhizal conifers in a temperate forest', *Frontiers in Microbiology*, 9 (2018), 433.

Tolkien, J. R. R., *The Lord of the Rings* (London, Harper Collins, 2014).

Tornberg, K., Olsson, S., 'Detection of hydroxyl radicals produced by wood-decomposing fungi', *FEMS Microbiology Ecology*, 40 (2002), pp. 13–20.

Torri, L., Migliorini, P., Masoero, G., 'Sensory test vs. electronic nose and/or image analysis of whole bread produced with old and modern wheat varieties adjuvanted by means of the mycorrhizal factor', *Food Research International*, 54 (2013), pp. 1400–8.

Toyota, M., Spencer, D., Sawai-Toyota, S., Jiaqi, W., Zhang, T., Koo, A. J., Howe, G. A., Gilroy, S., 'Glutamate triggers long-distance, calcium-based plant defense signaling', *Science*, 361 (2018), pp. 1112–5.

Trappe, J., 'Foreword', in ed. T. Horton, *Mycorrhizal Networks* (Springer International Publishing, 2015).

Trappe, J. M., 'A. B. Frank and mycorrhizae: the challenge to evolutionary and ecologic theory', *Mycorrhiza*, 15 (2005), pp. 277–81.

Trewavas, A., 'Response to Alpi et al.: Plant neurobiology – all metaphors have value', *Trends in Plant Science*, 12 (2007), pp. 231–33.

Trewavas, A., *Plant Behaviour and Intelligence* (Oxford, Oxford University Press, 2014).

Trewavas, A., 'Intelligence, cognition, and language of green plants', *Frontiers in Psychology*, 7 (2016), 588.

Trivedi, D. K., Sinclair, E., Xu, Y., Sarkar, D., Walton-Doyle, C., Liscio, C., Banks, P., Milne, J., Silverdale, M., Kunath, T. et al., 'Discovery of volatile biomarkers of Parkinson's disease from sebum', *ACS Central Science*, 5 (2019), pp. 599–606.

Tsing, A. L., *The Mushroom at the End of the World* (Princeton, NJ, Princeton University Press, 2015).

Tuovinen, V., Ekman, S., Thor, G., Vanderpool, D., Spribille, T., Johannesson, H., 'Two basidiomycete fungi in the cortex of wolf lichens', *Current Biology*, 29 (2019), pp. 476–83.

Tyne, D., Manson, A. L., Huycke, M. M., Karanicolas, J., Earl, A. M., Gilmore, M. S., 'Impact of antibiotic treatment and host innate immune pressure on enterococcal adaptation in the human bloodstream', *Science Translational Medicine*, 11 (2019), eaat8418.

Umehata, H., Fumagalli, M., Smail, I., Matsuda, Y., Swinbank, A. M., Cantalupo, S., Sykes, C., Ivison, R. J., Steidel, C. C., Shapley, A. E. et al., 'Gas filaments of the cosmic web located around active galaxies in a protocluster', *Science*, 366 (2019), pp. 97–100.

Vadder, F., Grasset, E., Holm, L., Karsenty, G., Macpherson, A. J., Olofsson, L. E., Bäckhed, F., 'Gut microbiota regulates maturation of the adult enteric nervous system via enteric serotonin networks', *Proceedings of the National Academy of Sciences*, 115 (2018), pp. 6458–63.

Vahdatzadeh, M., Deveau, A., Splivallo, R., 'The role of the microbiome of truffles in aroma formation: a meta-analysis approach', *Applied and Environmental Microbiology*, 81 (2015), pp. 6946–52.

Vajda, V., McLoughlin, S., 'Fungal proliferation at the cretaceous – tertiary boundary', *Science*, 303 (2004), p. 1489.

Valles-Colomer, M., Falony, G., Darzi, Y., Tigchelaar, E., Wang, J., Tito, R. Y., Schiweck, C., Kurilshikov, A., Joossens, M., Wijmenga, C. et al., 'The neuroactive potential of the human gut microbiota in quality of life and depression', *Nature Microbiology* (2019), pp. 623–32.

van Delft, F. C., Ipolitti, G., Nicolau, D. V., Perumal, A., Kašpar, O., Kheireddine, S., Wachsmann-Hogiu, S., Nicolau, D. V., 'Something has to give: scaling combinatorial computing by biological agents exploring physical networks encoding NP-complete problems', *Journal of the Royal Society Interface Focus*, 8 (2018), 20180034.

van der Heijden, M. G., Bardgett, R. D., Straalen, N. M., 'The unseen majority: soil microbes as drivers of plant diversity and productivity in terrestrial ecosystems', *Ecology Letters*, 11 (2008), pp. 296–310.

van der Heijden, M. G., Horton, T. R., 'Socialism in soil? The importance of mycorrhizal fungal networks for facilitation in natural ecosystems', *Journal of Ecology*, 97 (2009), pp. 1139–50.

van der Heijden, M. G., Walder, F., 'Reply to "Misconceptions on the application of biological market theory to the mycorrhizal symbiosis"', *Nature Plants*, 2 (2016), 16062.

van der Heijden, M. G., 'Underground networking', *Science*, 352 (2016), pp. 290–1.

van der Heijden, M. G., Dombrowski, N., Schlaeppi, K., 'Continuum of root–fungal symbioses for plant nutrition', *Proceedings of the National Academy of Sciences*, 114 (2017), pp. 11574–6.

van der Linde, S., Suz, L. M., Orme, D. C., Cox, F., Andreae, H., Asi, E., Atkinson, B., Benham, S., Carroll, C., Cools, N. et al., 'Environment and host as large-scale controls of ectomycorrhizal fungi', *Nature*, 558 (2018), pp. 243–8.

Van Tyne, D., Manson, A. L., Huycke, M. M., Karanicolas, J., Earl, A. M., Gilmore, M. S., 'Impact of antibiotic treatment and host innate immune pressure on enterococcal adaptation in the human bloodstream', *Science Translational Medicine*, 487 (2019), eaat8418.

Vannini, C., Carpentieri, A., Salvioli, A., Novero, M., Marsoni, M., Testa, L., Pinto, M., Amoresano, A., Ortolani, F., Bracale, M. et al., 'An interdomain network: the endobacterium of a mycorrhizal fungus promotes antioxidative responses in both fungal and plant hosts', *New Phytologist*, 211 (2016), pp. 265–75.

Venner, S., Feschotte, C., Biémont, C., 'Dynamics of transposable elements: towards a community ecology of the genome', *Trends in Genetics*, 25 (2009), pp. 317–23.

Verbruggen, E., Röling, W. F., Gamper, H. A., Kowalchuk, G. A., Verhoef, H. A., van der Heijden, M. G., 'Positive effects of organic farming on below-ground mutualists: large-scale comparison of mycorrhizal fungal communities in agricultural soils', *New Phytologist*, 186 (2010), pp. 968–79.

Vetter, W., Roberts, D., 'Revisiting the organohalogens associated with 1979-samples of Brazilian bees (*Eufriesea purpurata*)', *Science of the Total Environment*, 377 (2007), pp. 371–7.

Vita, F., Taiti, C., Pompeiano, A., Bazihizina, N., Lucarotti, V., Mancuso, S., Alpi, A., 'Volatile organic compounds in truffle (*Tuber magnatum* Pico): comparison of samples from different regions of Italy and from different seasons', *Scientific Reports*, 5 (2015), 12629.

Viveiros de Castro, E., 'Exchanging perspectives: the transformation of objects into subjects in amerindian ontologies', *Common Knowledge* (2004), pp. 463–84.

von Bertalanffy, L., *Modern Theories of Development: An Introduction to Theoretical Biology* (London, Humphrey Milford, 1933).

von Humboldt, A., *Kosmos: Entwurf einer physischen Weltbeschreibung* (Stuttgart and Tübingen, J. G. Cotta'schen Buchhandlungen, 1845) archive.org/details/b29329693_0001 [accessed 29 October 2019].

von Humboldt, A., *Cosmos: A Sketch of Physical Description of the Universe* (London, Henry G. Bohn, 1849).

Wadley, G., Hayden, B., 'Pharmacological influences on the Neolithic Transition', *Journal of Ethnobiology*, 35 (2015), pp. 566–84.

Wagg, C., Bender, F. S., Widmer, F., van der Heijden, M. G., 'Soil biodiversity and soil community composition determine ecosystem multifunctionality', *Proceedings of the National Academy of Sciences*, 111 (2014), pp. 5266–70.

Wainwright, M., 'Moulds in Folk Medicine', *Folklore*, 100 (1989a), pp. 162–6.

Wainwright, M., 'Moulds in ancient and more recent medicine', *Mycologist*, 3 (1989b), pp. 21–3.

Wainwright, M., Rally, L., Ali, T., 'The scientific basis of mould therapy', *Mycologist*, 6 (1992), pp. 108–10.

Walder, F., Niemann, H., Natarajan, M., Lehmann, M. F., Boller, T., Wiemken, A., 'Mycorrhizal networks: common goods of plants shared under unequal terms of trade', *Plant Physiology*, 159 (2012), pp. 789–97.

Walder, F., van der Heijden, M. G., 'Regulation of resource exchange in the arbuscular mycorrhizal symbiosis', *Nature Plants*, 1 (2015), 15159.

Waller, L. P., Felten, J., Hiiesalu, I., Vogt-Schilb, H., 'Sharing resources for mutual benefit: crosstalk between disciplines deepens the understanding of mycorrhizal symbioses across scales', *New Phytologist*, 217 (2018), pp. 29–32.

Wang, B., Yeun, L., Xue, J., Liu, Y., Ané, J., Qiu, Y., 'Presence of three mycorrhizal genes in the common ancestor of land plants suggests a key role of mycorrhizas in the colonisation of land by plants', *New Phytologist*, 186 (2010), pp. 514–25.

Wasson, G., Kramrisch, S., Ott, J., Ruck, C., *Persephone's Quest: Entheogens and the Origins of Religion* (New Haven, CT, Yale University Press, 1986).

Wasson, G., Hofmann, A., Ruck, C., *The Road to Eleusis: Unveiling the Secret of the Mysteries* (Berkeley, CA, North Atlantic Books, 2009).

Wasson, V. P., Wasson, G., *Mushrooms, Russia and History* (New York, Pantheon, 1957).

Watanabe, S., Tero, A., Takamatsu, A., Nakagaki, T., 'Traffic optimisation in railroad networks using an algorithm mimicking an amoeba-like organism, *Physarum plasmodium*', *Biosystems*, 105 (2011), pp. 225–32.

Watkinson, S. C., Boddy, L., Money, N., *The Fungi* (London, Academic Press, 2015).

Watts, J., 'Scientists identify vast underground ecosystem containing billions of micro-organisms', *Guardian* (2018), www.theguardian.com/science/2018/dec/10/tread-softly-because-you-tread-on-23bn-tonnes-of-micro-organisms [accessed 29 October 2019].

Watts-Williams, S. J., Cavagnaro, T. R., 'Nutrient interactions and arbuscular mycorrhizas: a meta-analysis of a mycorrhiza-defective mutant and wild-type tomato genotype pair', *Plant and Soil*, 384 (2014), pp. 79–92.

Wellman, C. H., Strother, P. K., 'The terrestrial biota prior to the origin of land plants (embryophytes): a review of the evidence', *Palaeontology*, 58 (2015), pp. 601–27.

Weremijewicz, Janos, D. P., 'Common mycorrhizal networks amplify competition by preferential mineral nutrient allocation to large host plants', *New Phytologist*, 212 (2016), pp. 461–71.

Werner, G. D., Kiers, T. E., 'Partner selection in the mycorrhizal mutualism', *New Phytologist*, 205 (2015), pp. 1437–42.

Werner, G. D., Strassmann, J. E., Ivens, A. B., Engelmoer, D. J., Verbruggen, E., Queller, D. C., Noë, R., Johnson, N., Hammerstein, P., Kiers, T. E., 'Evolution of microbial markets', *Proceedings of the National Academy of Sciences*, 11 (2014), pp. 1237–44.

Werrett, S., *Thrifty Science: Making the Most of Materials in the History of Experiment* (Chicago, University of Chicago Press, 2019).

West, M., 'Putting the "I" in science', *Nature* (2019), www.nature.com/articles/d41586-019-03051-z [accessed 29 October 2019].

Westerhoff, H. V., Brooks, A. N., Simeonidis, E., García-Contreras, R., He, F., Boogerd, F. C., Jackson, V. J., Goncharuk, V., Kolodkin, A., 'Macromolecular networks and intelligence in microorganisms', *Frontiers in Microbiology*, 5 (2014), 379.

Weyrich, L. S., Duchene, S., Soubrier, J., Arriola, L., Llamas, B., Breen, J., Morris, A. G., Alt, K. W., Caramelli, D., Dresely, V. et al., 'Neanderthal behaviour, diet and disease inferred from ancient DNA in dental calculus', *Nature*, 544 (2017), pp. 357–361.

Whiteside, M. D., Werner, G. D. A., Caldas, V. E. A., Van't Padje, A., Dupin, S. E., Elbers, B., Bakker, M., Wyatt, G. A. K., Klein, M., Hink, M. A. et al., 'Mycorrhizal fungi respond to resource inequality by moving phosphorus from rich to poor patches across networks', *Current Biology*, 29 (2019), pp. 2043–50.

Whittaker, R., 'New concepts of kingdoms of organisms', *Science*, 163 (1969), pp. 150–60.

Wiens, F., Zitzmann, A., Lachance, M.-A., Yegles, M., Pragst, F., Wurst, F. M., von Holst, D., Guan, S., Spanagel, R., 'Chronic intake of fermented floral nectar by wild treeshrews', *Proceedings of the National Academy of Sciences*, 105 (2008), pp. 10426–31.

Wilkinson, D. M., 'The evolutionary ecology of mycorrhizal networks', *Oikos*, 82 (1998), pp. 407–10.

Willerslev, R., *Soul Hunters: Hunting, Animism, and Personhood among the Siberian Yukaghirs* (Berkeley, University of California Press, 2007).

Wilson, G. W., Rice, C. W., Rillig, M. C., Springer, A., Hartnett, D. C., 'Soil aggregation and carbon sequestration are tightly correlated with the abundance of arbuscular mycorrhizal fungi: results from long-term field experiments', *Ecology Letters*, 12 (2009), pp. 452–61.

Winkelman, M. J., 'The mechanisms of psychedelic visionary experiences: hypotheses from evolutionary psychology', *Frontiers in Neuroscience*, 11 (2017), 539.

Wipf, D., Krajinski, F., Tuinen, D., Recorbet, G., Courty, P., 'Trading on the arbuscular mycorrhiza market: from arbuscules to common mycorrhizal networks', *New Phytologist*, 223 (2019), pp. 1127–42.

Wisecaver, J. H., Slot, J. C., Rokas, A., 'The evolution of fungal metabolic pathways', *PLOS Genetics*, 10 (2014), e1004816.

Witt, P., 'Drugs alter web-building of spiders: a review and evaluation', *Behavioral Science*, 16 (1971), pp. 98–113.

Wolfe, B. E., Husband, B. C., Klironomos, J. N., 'Effects of a belowground mutualism on an aboveground mutualism', *Ecology Letters*, 8 (2005), pp. 218–23.

Wright, C. K., Wimberly, M. C., 'Recent land use change in the Western Corn Belt threatens grasslands and wetlands', *Proceedings of the National Academy of Sciences*, 110 (2013), pp. 4134–9.

Wulf, A., *The Invention of Nature* (New York, Alfred A. Knopf, 2015).

Wyatt, G. A., Kiers, T. E., Gardner, A., West, S. A., 'A biological market analysis of the plant–mycorrhizal symbiosis', *Evolution*, 68 (2014), pp. 2603–18.

Yano, J. M., Yu, K., Donaldson, G. P., Shastri, G. G., Ann, P., Ma, L., Nagler, C. R., Ismagilov, R. F., Masmanian, S. K., Hsiao, E. Y., 'Indigenous bacteria from the gut microbiota regulate host serotonin biosynthesis', *Cell*, 161 (2015), pp. 264–76.

Yon, D., 'Now You See It', *Quanta* (2019), aeon.co/essays/how-our-brain-sculpts-experience-in-line-with-our-expectations? [accessed 29 October 2019].

Yong, E., 'The Guts That Scrape the Skies', *National Geographic* (2014), www.nationalgeographic.com/science/phenomena/2014/09/23/the-guts-that-scrape-the-skies/ [accessed 29 October 2019].

Yong, E., *I Contain Multitudes: The Microbes Within Us and a Grander View of Life* (New York, Ecco Press, 2016).

Yong, E., 'How the Zombie Fungus Takes Over Ants' Bodies to Control Their Minds', *Atlantic* (2017), www.theatlantic.com/science/archive/2017/11/how-the-zombie-fungus-takes-over-ants-bodies-to-control-their-minds/545864/ [accessed 29 October 2019].

Yong, E., 'This Parasite Drugs Its Hosts with the Psychedelic Chemical in Shrooms', *Atlantic* (2018), www.theatlantic.com/science/archive/2018/07/massospora-parasite-drugs-its-hosts/566324/ [accessed 29 October 2019].

Yong, E., 'The Worst Disease Ever Recorded', *Atlantic* (2019), www.theatlantic.com/science/archive/2019/03/bd-frogs-apocalypse-disease/585862/ [accessed 29 October 2019].

Young, R. M., *Darwin's Metaphor* (Cambridge, Cambridge University Press, 1985).

Yuan, X., Xiao, S., Taylor, T. N., 'Lichen-like symbiosis 600 million years ago', *Science*, 308 (2005), pp. 1017–20.

Yun-Chang, W., 'Mycology in Ancient China', *Mycologist*, 1 (1985), pp. 59–61.

Zabinski, C. A., Bunn, R. A., 'Function of Mycorrhizae in Extreme Environments', in eds. Z. Solaiman, L. Abbott and A. Varma, *Mycorrhizal Fungi: Use in Sustainable Agriculture and Land Restoration* (Springer International Publishing, 2014), pp. 201–14.

Zhang, M. M., Poulsen, M., Currie, C. R., 'Symbiont recognition of mutualistic bacteria by *Acromyrmex* leaf-cutting ants', *ISME Journal*, 1 (2007), pp. 313–20.

Zhang, S., Lehmann, A., Zheng, W., You, Z., Rillig, M. C., 'Arbuscular mycorrhizal fungi increase grain yields: a meta-analysis', *New Phytologist*, 222 (2019), pp. 543–55.

Zhang, Y., Kastman, E. K., Guasto, J. S., Wolfe, B. E., 'Fungal networks shape dynamics of bacterial dispersal and community assembly in cheese rind microbiomes', *Nature Communications*, 9 (2018), 336.

Zheng, C., Ji, B., Zhang, J., Zhang, F., Bever, J. D., 'Shading decreases plant carbon preferential allocation towards the most beneficial mycorrhizal mutualist', *New Phytologist*, 205 (2015), pp. 361–8.

Zheng, P., Zeng, B., Zhou, C., Liu, M., Fang, Z., Xu, X., Zeng, L., Chen, J., Fan, S., Du, X. et al., 'Gut microbiome remodeling induces depressive-like behaviors through a pathway mediated by the host's metabolism', *Molecular Psychiatry*, 21 (2016), pp. 786–96.

Zhu, K., McCormack, L. M., Lankau, R. A., Egan, F. J., Wurzburger, N., 'Association of ectomycorrhizal trees with high carbon-to-nitrogen ratio soils across temperate forests is driven by smaller nitrogen not larger carbon stocks', *Journal of Ecology*, 106 (2018), pp. 524–35.

Zhu, L., Aono, M., Kim, S.-J., Hara, M., 'Amoeba-based computing for traveling salesman problem: Long-term correlations between spatially separated individual cells of *Physarum polycephalum*', *Biosystems*, 12 (2013), pp. 1–10.

Zobel, M., 'Eltonian niche width determines range expansion success in ectomycorrhizal conifers', *New Phytologist*, 220 (2018), pp. 947–9.

Index

Page references in *italics* indicate images.